AF411371

Power System Dynamic and Stability Issues in Modern Power Systems Facing Energy Transition

Power System Dynamic and Stability Issues in Modern Power Systems Facing Energy Transition

Editors

Cosimo Pisani
Giorgio Maria Giannuzzi

MDPI • Basel • Beijing • Wuhan • Barcelona • Belgrade • Manchester • Tokyo • Cluj • Tianjin

Editors
Cosimo Pisani
Terna Italian Transmission
System Operator
Italy

Giorgio Maria Giannuzzi
Terna Italian Transmission
System Operator
Italy

Editorial Office
MDPI
St. Alban-Anlage 66
4052 Basel, Switzerland

This is a reprint of articles from the Special Issue published online in the open access journal *Energies* (ISSN 1996-1073) (available at: https://www.mdpi.com/journal/energies/special_issues/stability_issues).

For citation purposes, cite each article independently as indicated on the article page online and as indicated below:

LastName, A.A.; LastName, B.B.; LastName, C.C. Article Title. *Journal Name* **Year**, *Volume Number*, Page Range.

ISBN 978-3-0365-6035-9 (Hbk)
ISBN 978-3-0365-6036-6 (PDF)

Contents

About the Editors

Cosimo Pisani

Cosimo Pisani was born in Benevento, Italy, in 1985. He received an M.Sc. degree (Hons.) in energy engineering from the University of Sannio, Benevento, in 2010, and a Ph.D. degree in electrical engineering from the University of Naples Federico II, Naples, Italy, in 2014. During his Ph.D. in collaboration with TERNA, he investigated some dynamic stability issues of large interconnected power system, such as the European power system. From May 2014 to March 2016, he was a Research Fellow with the University of Sannio. Since March 2016, he has been with TERNA. He is currently the Head of Stability and Network Logics in the Dispatching and Switching Department. He is the author or co-author of over 80 scientific articles in the IEEE and CIGRE communities. His research interests include applications of the dynamic stability of power systems, wide-area monitoring and protection systems, high-voltage direct current systems, and power system restoration. He is a Leader of the WAMS task force within the ENTSO-E System Protection and Dynamics and was recently appointed convener of the same subgroup. He was a member of the CIGRE C2.17 working group "Wide Area Monitoring Systems–Support for Control Room Applications". He is currently a member of the CIGRE C2.18 working group "Wide Area Monitoring Protection and Control Systems–Decision Support for System Operators" and the Italian representative of the CIGRE Study Committee C4 System Technical Performance.

Giorgio Maria Giannuzzi

Giorgio M. Giannuzzi received a Ph.D. degree in electrical engineering from the University of Rome, Rome, Italy. He worked at ABB until 2000 in Rome and Milan (Italy), Helsinki (Finland), Ladenburg (Germany), and Raleigh (USA). Since 2001, he has been working at TERNA, Rome, as an expert in dynamic studies, protection, telecontrol, substation automation, wide area defense systems, and wide-area monitoring systems. Under his guidance, the TERNA EMS (optimal power flow security and market constrained, optimal reactive power flow, dynamic security assessment tool, dynamic and static security verification software, and operator training simulator) was designed. He supervised the Italian Grid Code technical enclosures (primary and secondary frequency regulation, load shedding, protection and automation, and defense plans). He has been a member of the UCTE Expert Group on power system stability and, in 2010, joined the ENTSO-E System Protection and Dynamics Group. Since 2014, he has been Convenor, coordinating the European evaluation overdispersed generation impact on system security and load shedding. He is currently responsible for the Engineering Department of the National Dispatching Centre. His research interests include defense systems, WAMS, WADC, WAPS, protection systems, modal analysis, simulations and tools (RMS and EMT), and dynamics of generators (AVR, Prime Movers, PSS, dynamic models).

Review

Enabling Technologies for Enhancing Power System Stability in the Presence of Converter-Interfaced Generators

Giorgio M. Giannuzzi [1,†], Viktoriya Mostova [2,†], Cosimo Pisani [1,*,†], Salvatore Tessitore [1,†] and Alfredo Vaccaro [2,†]

[1] Terna Rete Italia S.p.A., V.le Egidio Galbani, 70, 00156 Rome, Italy
[2] Department of Engineering, University of Sannio, Piazza Roma 21, 82100 Benevento, Italy
* Correspondence: cosimo.pisani@terna.it
† These authors contributed equally to this work.

Abstract: The growing attention to environmental issues is leading to an increasing integration of renewable energy sources into electrical grids. This integration process could contribute to power system decarbonization, supporting the diversification of primary energy sources and enhancing the security of energy supply, which is threatened by the uncertain costs of conventional energy sources. Despite these environmental and economical benefits, many technological and regulatory problems should be fixed in order to significantly increase the level of penetration of renewable power generators, which are connected to power transmission and distribution systems via power electronic interfaces. Indeed, these converter-interfaced generators (CIGs) perturb grid operation, especially those fueled by non-programmable energy sources (e.g., wind and solar generators), affecting the system stability and making power systems more vulnerable to dynamic perturbations. To face these issues, the conventional operating procedures based on pre-defined system conditions, which are currently adopted in power system operation tools, should be enhanced in order to allow the "online" solution of complex decision-making problems, providing power system operators with the necessary measures and alerts to promptly adjust the system. A comprehensive analysis of the most promising research directions and the main enabling technologies for addressing this complex issue is presented in this paper.

Keywords: power system stability; frequency stability; rotor angle stability; voltage stability; power system inertia; converter-interfaced generation; renewable power generators

Citation: Giannuzzi, G.M.; Mostova, V.; Pisani, C.; Tessitore, S.; Vaccaro, A. Enabling Technologies for Enhancing Power System Stability in the Presence of Converter-Interfaced Generators. *Energies* **2022**, *15*, 8064. https://doi.org/10.3390/en15218064

Academic Editor: Abu-Siada Ahmed

Received: 29 September 2022
Accepted: 24 October 2022
Published: 30 October 2022

Publisher's Note: MDPI stays neutral with regard to jurisdictional claims in published maps and institutional affiliations.

1. Introduction

The need for mitigating the effects of climate change, the progressive depletion of fossil resources, and the growing energy demand of developing countries has led to an exponential increase of renewable power generators connected to transmission and distribution grids by power electronic converters.

In particular, it is expected that the globally installed capacity of electricity generation from wind and photovoltaics generators will increase to about 3.8 TW by 2030 [1]. It is well known that the large scale penetration of these converter-interfaced generators (CIGs) in existing power grids requires major revisions of the conventional policies currently adopted for power system planning, operation, and regulation [2]. The importance of this revision process has been confirmed in many countries, where the rapid growth of grid-connected CIGs without defining and implementing proper corrective actions aimed at enhancing the grid flexibility, sensibly affected the power system stability and security.

Moreover, the modern technological advances pushed the pervasive integration of power electronic converters in both transmission and distribution power systems for enabling a wide range of applications, which are not only restricted to the grid connection of distributed generators. For example, in transmission systems, power converters are the technological backbone supporting flexible AC transmission systems (FACTS) and high voltage DC transmission systems (HVDC), while, in distribution networks, they are extensively deployed in microgrids and DC distribution networks. Finally, they are widely used in many end-user applications such as electric vehicles and variable frequency drives [3].

The technical impacts induced by the large-scale penetration of power electronic converters in modern power systems should be carefully analyzed, especially those related to CIGs, which in the medium/long term are expected to replace all the conventional synchronous generators (SG). In particular, the main features differentiating CIGs from SGs are the nominal power, which is extremely lower, the generator inertia, which is zero for static generators, the active power regulation, which is limited for renewable power generators, and the dynamic response, which is extremely faster [4]. These intrinsic differences reduce the grid flexibility, make power systems more vulnerable to dynamic perturbations, and sensibly increase the complexity of power system operation.

To address these critical issues, wide-area monitoring protection and control systems (WAMPACs) are being increasingly applied by transmission system operators worldwide for real-time power system monitoring and early detection of dynamic perturbations. WAMPACs involve the use of system-wide information to avoid large disturbances and reduce the probability of catastrophic events by supporting the application of adaptive protection and control strategies aimed at increasing network capacity and minimizing wide-area disturbances [5]. Moreover, based on measured data processing, online decision support systems have also been applied in the task of promptly identify mitigation actions aimed at protecting the power system, avoiding large blackouts. These solutions allow processing large sensor data-streaming, providing system operators with the necessary measures and alerts to adjust the system, i.e., reduce impacts of load and renewable power generation fluctuations, for example, due to intermittency. In addition, several indices have been proposed for real-time monitoring of the grid stability, whose estimation can greatly enhance the situational awareness of TSOs [6]. In this context, this paper, starting from a comprehensive classification of the main phenomena that could affect power system stability, analyzes the relevant research directions and the most promising technologies and methodologies aimed at enhancing the dynamic grid performance in the presence of a massive penetration of power converter-based distributed generators. Moreover, the concept of electrical system stability and its classification are properly revised in order to consider the potential effects deriving by the large-scale replacement of conventional synchronous generators with distributed renewable generation systems. In this context, particular emphasis is given to the analysis of the critical issues characterizing modern power systems, such as the inertia decrease and the vulnerability to dynamic perturbations over multiple time scales, and their potential impacts on power system stability, flexibility and resilience. Finally, the most promising enabling technologies for improving system stability are analyzed in details.

The rest of the paper is organized as follows: Section 2 defines and classifies the power system stability according to the modern approaches proposed in the literature. The main technical impacts and the most challenging issues induced by a large-scale pervasion of CIGs in power systems are discussed in Section 3. Section 4 is focused on the analysis of the impacts of CIGs on power system stability, and the review of the most promising mitigation technologies used by TSOs. The concluding remarks are summarized in Section 5.

2. Definition and Classification of Power System Stability

Power system stability is defined as "the ability of the system, for a given initial operating condition, to return to a state of operating equilibrium after a physical disturbance, with most of the variables of the system constrained so that the entire system remains intact" [7]. This definition highlights the ability of the system to withstand both minor disturbances, such as load variations and, more importantly, major disturbances such as the failure of a generator or a short circuit on a line, since these events entail significant structural changes with obvious effects on the stability of the system. Studies and constant analysis of the network are essential in order to implement appropriate planning measures. However, designing a power system that can withstand any single or multiple contingency, resulting as robust against any disturbance event, is technically and economically unfeasible. Hence, power system planning is often based on a proper trade-off between reliability and economy, considering only the events with the highest probability of occurrence and/or those inducing severe grid perturbations.

Hence, in order to analyze the dynamic events that could compromise the system stability, and develop effective strategies aimed at mitigating their effects, it is extremely important to rigorously classify the forms of stability considering various features such as the disturbance magnitude, the time dynamics, the involved devices, and, most importantly, the physical nature of the instability [7].

Recently, the stability classification has been revised and expanded in order to take into account the grid effects of CIGs. In particular, it has been introduced two additional categories to the three existing categories of rotor angle stability, frequency stability, and voltage stability, namely resonance stability and converter-driven stability. This enhancement was necessary because, in contrast to conventional generators, which are characterized by rather slow electromechanical phenomena, CIGs are characterized by much faster dynamics, ranging from a few microseconds to several milliseconds, inducing complex electromagnetic phenomena [8].

More details about this classification, which is schematically depicted in Figure 1, are discussed in the follow sub-sections.

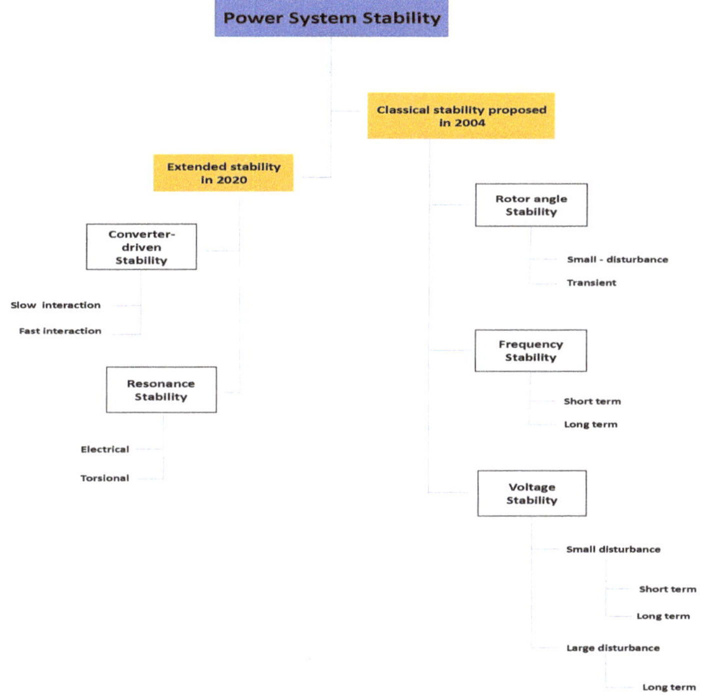

Figure 1. Power system stability classification (2004) and subsequent extension (2020) performed by the Task Force by the IEEE Power System Dynamic Performance Committee and the CIGRE Study Committee 38.

Rotor Angle Stability: The ability of the power system to maintain synchronism after a small disturbance (small-signal stability) or a large disturbance (transient stability). After a disturbance, there is an imbalance between mechanical and electromagnetic torque that causes the acceleration or deceleration of the generator rotors. This produces an angular difference that causes part of the load to be transferred from the slower machines to the faster machines, and if the system is unable to synchronize by absorbing kinetic energy, rotor angular instability occurs.

The change in electromagnetic torque can be divided into two components: a synchronisation torque in phase with the rotor angle deviation, whose absence causes non-oscillatory instability, and a damping torque in phase with the speed deviation, whose absence causes oscillatory instability. For the existence of a stability condition, the presence of both components is required for each synchronous machine.

Frequency Stability: The ability of the system to keep the frequency close to the nominal value after a large imbalance between generation and load. It is ensured by proper coordinating the control and protection devices, and can be classified into short-term and long-term frequency stability, depending on the timing of the event. Frequency stability is closely related to rotor angle stability, since rotor speed is related to grid frequency. It should also be taken into account that during frequency excursions, the voltage also fluctuates significantly, which can lead to untimely or poorly coordinated operation of the protection systems [9].

Voltage Stability: the ability of the power system to maintain voltage magnitudes at all buses as close as possible to the nominal value after a disturbance. Voltage instability induces a drop or rise in voltage at some buses, which can lead to cascading outages or load losses, and depends on the ability to maintain or restore the balance between load demand and supply [8]. Voltage stability can be divided into two subcategories: one considers the magnitude of the disturbance (voltage stability for small disturbances and voltage stability for large disturbances), while the other distinguishes the duration of instability (short-term voltage stability and long-term voltage stability). Voltage stability is highly load dependent, since after a disturbance the load on the high-voltage grid is increased with the aim of restoring power consumption beyond the capacity of the transmission and generation system. This leads to an increase in reactive power consumption, and consequently to an increase in voltage drop [10]. Although the most common form of voltage instability is related to voltage reduction, overvoltage instability could also be possible, and it is mainly caused by the network capacitance, which hinders the absorption of reactive power by synchronous compensators [11].

Resonance Stability: resonance occurs when energy exchange occurs in an oscillatory pattern; when oscillations exceed certain predetermined thresholds, there is a risk that resonance instability will occur. Depending on whether the resonance phenomenon manifests itself mechanically (torsion of the drive shaft) or electrically, a further subdivision is made [8].

(a) Torsional resonance: The resonance is due to torsional interactions between the series-compensated lines and the mechanical shaft of the turbine generator. The vibrations generated by the interaction between the fast-acting control devices and the system stabilizers can be poorly damped, undamped, or even negatively damped and increasing, threatening the mechanical integrity of the shaft.

(b) Electrical resonance: This resonance occurs in certain types of generators that have been found to be particularly susceptible to the phenomenon of self-excitation, in which the series-connected capacitor forms a resonant circuit with the actual inductance of the generator at sub-synchronous frequencies. The resonance results in both large currents and large voltage fluctuations that can be fatal to the electrical equipment of the generator, as well as to the transmission system.

Converter-Driven Stability: Unstable oscillations can occur in the system due to cross-couplings with the electromagnetic dynamics of the machines and electromagnetic transients in the network. The couplings are mainly due to the very fast dynamics of the converter controls on which the CIG is based. Depending on whether the interactions that occur are slow (less than 10 Hz) or fast (10 to hundreds of Hz), the phenomena can be further classified according to frequency levels [8] as follows:

(a)   Stability of converters with fast interactions: interactions can cause oscillations at high/very high frequencies (hundreds of Hz to several kHz), a phenomenon called harmonic instability. An example of such instability is the controllers of synthetic inertial systems.

(b)   Stability due to slow-response inverters: dynamic interactions involve control systems of power electronics devices with slow-response system components.

3. Grid Impacts of Large-Scale CIGs Penetration

The main feature characterizing power systems with high CIGs penetration is certainly the uncertainties induced by the intermittence of the renewable power generators, whose injected power profiles are not programmable and depend on the dynamics of the particular energy source (e.g., wind speed, solar radiation). Although these profiles can be predicted over different time horizons, these predictions are often affected by large forecasting errors, which raises the power system operation complexities [12].

Another remarkable feature of these power systems is the different time scales characterizing the system dynamics. While the dynamics of the conventional power systems settles in a range of 10^{-2} to 10^3 s, the dynamics of the CIGs-based system goes down to 10^{-4} or 10^{-6} s by switching from electromechanical to electromagnetic dynamics [13]. This feature leads to undesirable dynamic coupling, which is much more likely in MV and LV networks characterized by a high R/X ratio.

Systems with high CIGs penetration also have a low tolerance to frequency and voltage deviations. For example, the limits of a wind generator are far lower than those of a conventional generator (50.2 Hz versus 51.5 Hz for frequency and 1.1 p.u. versus 1.3 p.u. for voltage), which could trigger frequent grid disconnections, and dynamic perturbation phenomena [14]. Moreover, these generators are characterized by low inertia and low short-circuit current. In particular, the low inertia is due to the presence of the converter, which decouples the frequency from the grid and no longer has the characteristics of traditional frequency response based on rotational kinetic energy. Low system inertia, especially at the regional level, causes very rapid and high amplitude frequency deviations with the resulting risk of cascading events [15]. The low short-circuit current, on the other hand, poses considerable challenges to power system protection. In a network with high CIGs penetration, unlike conventional grids, no sub-transient period occurs during a short-circuit event, hence requiring the modification of the protection and control devices. Finally, the non-linear effects induced by the discrete switching operation or saturation of the power converters should be carefully analyzed, since they can trigger cascading events of the protective systems, which require specific functions, e.g., low and high voltage ride-through (LVRT-HVRT) [16].

4. Enabling Technologies for Power System Stability Enhancement

4.1. Rotor Angle Stability

High CIGs penetration sensibly affects the active and reactive power flows, hence influencing the electromechanical vibration characteristics. To assess whether this influence is positive or negative, factors such as penetration level, type, location, force, operation, and the control strategy/parameters (PTLSOC) [4] should be analyzed. As far as transient stability is concerned, both positive and negative effects have been reported in the literature, depending mainly on the size and location of the disturbance and the type of the implemented control [17,18]. In general, there are many case studies reporting positive effects of CIGs on transient stability due to greater flexibility in active and reactive power control [19].

Additionally, the presence of HVDC converters strongly affects power system stability. In this context, [20] proposes the installation of an additional POD (Power Oscillations Damping) controller since the introduction of HVDC lines into the system induced inter-area oscillation modes around 1 Hz, which are different from the typical modes of traditional transmission lines around 0.2 Hz. Therefore, to mitigate these higher frequency oscillations, it is necessary to improve the design of POD controllers so as to avoid unwanted interactions with modes of a different nature that are close to those of the inter-area target modes. Reference [21] evaluates the impact of control structure parameters on the system's intrinsic mode damping and frequency. Indeed, the control system of HVDC converters can be likened to a first-order dynamic system and can be described through two parameters: the control system gain (k_{DC}) and the system time constant (T_{DC}). The study sought to identify parameter pairs that constitute a compromise between network stability and reactivity. Reference [22] proposes a control strategy that is based on a combined approach, which identifies the desired value of the synchronization power by a model-based approach. This control strategy allows improving both the inertial and the frequency response, as demonstrated by detailed simulation studies developed by considering realistic operation scenarios.

Since unstable oscillations can even lead to a system collapse, real-time power system monitoring represents an effective tool for reliably assessing the system stability. To this aim, the spread of phasor measurement units (PMUs) in power transmission systems enables the development of advanced monitoring functions for early detection of dynamic perturbations. In this context, an interesting application is based on the deployment of modal analysis for damping estimation, which is currently used by a Finnish power system operator [23]. Additionally, ref. [24] analyzes real-time monitoring using PMUs, but proposes a decentralized architecture to estimate modal properties highlighting the advantages of such an approach, especially in the detection of local oscillations. PMUs can also be used to detect dangerous oscillations in advance and take appropriate measures. In this regard, in [25], through an optimization of the Hilbert solution method, the measurements provided by PMUs are processed in order to estimate and classify the oscillation components, and the characteristic parameters such as frequency and damping. The proposed method estimated the frequency and damping values with a very good accuracy and also had the advantage of being able to classify oscillations characterized by very close frequency values, so that distinct peaks in the signal spectrum could not be displayed.

The recent wide area measurement system (WAMS) technology can also be used to monitor inter-area oscillations. Reference [26] highlights the advantages of using WAMS technology over conventional power system stabilizers (PSSs). In particular, it analyzes a novel wide-area control technique aimed at modulating the active power injections in the task of damping the critical frequency oscillations, which include the inter-area oscillations and the transient frequency swing. The advantage of using WAMS technology is that multiple remote signals can be used for designing effective control strategies, whereas the PSSs process only local signals. This results in greater robustness of the control actions, since even if the centralized controller link fails, the local controllers continue to operate. Reference [27] evaluates the benefits deriving by the installation of a wide-area power oscillation damper for simultaneously damping both forced oscillations and inter-area modes. The proposed technique is based on an event-triggered strategy, which activates the adaptive control scheme, modulating the voltage set-points of the power oscillation damper. Time synchronized measurements can also enable the development of wide-area damping controllers (WADC), which can be used to damp inter-area oscillations, enhancing the grid observability and controllability [28].

An effective method for analyzing the grid resilience and assessing the correct operation of the main grid components is based on restoration tests, which are based on measurement campaigns aimed at analyzing low-frequency oscillations (LFO) in active power, reactive power and voltage magnitude profiles. Among the possible causes of this phenomenon, Reference [29] focuses on the mutual coupling between the restoration path and the transmission grid in normal operation, identifying, through electromechanical simulations, the main parameters influencing the amplitude and phase of LFOs.

Reference [30] proposes a tool to detect synchronization torque deficiency as a method of prevention against aperiodic rotor angle instability due to small disturbances (ASD). In a first step, the potential buses that can influence the stability margin of the ASD generator are identified by using a Thevenin impedance analysis with respect to a change in nodal admittance; subsequently, the buses are reduced by implementing a sensitivity analysis based on self-propagating graphs, and, finally, the optimal mitigation measures are identified.

4.2. Frequency Stability

The replacement of large synchronous generators with CIGs leads to a dramatic reduction of power system inertia, which could sensibly threaten the frequency stability, especially when the penetration of wind and photovoltaic systems exceed some critical levels [9]. Low system inertia causes a higher rate of change of frequency (ROCOF), and lower/higher nadir/zenith frequency, hence requesting more complex and faster protection systems. In this context, many technical solutions are currently under investigation, such as ultrafast control, virtual inertia, and grid-forming converter control [14]. These techniques aim at allowing generators to support the grid in the presence of transient perturbations by improving the frequency stability [31]. Other approaches propose intentional islanding as a preventive measure aimed at improving the system resilience [32], which has been shown to be an effective method for preventing cascading phenomena induced by HILP (High Impact Low Probability) events, facilitating the identification of pre-contingency mitigation actions.

Dynamic Security Assessment (DSA) is another relevant tool that has been attracted a great amount of attention from both researchers and industry in recent years, mainly due to its effectiveness in accurately describing the grid dynamics in the presence of both external and internal perturbations. However, its application requires the analysis of a large number of operational and failure events, making its online deployment extremely challenging. Hence, DSA is traditionally performed offline, by deploying Monte Carlo simulations for analyzing the dynamic grid evolution in the presence of "credible" contingencies, and estimating the corresponding stability limits. The study [33] draws attention to the possibility of using DSA as an "online" decision support tool by deploying high performance computing (HPC) paradigms in the task of promptly solving multiple contingency analyses under different operation scenarios, according to the strictly time-constraints requested by real-time power system operation functions [34,35]. The adoption of HPC-based DSA has been reported in many case studies: (i) the PJM control centre analyses a 13,500-bus system and can process up to 3000 contingencies every 15 min, (ii) EPRI simulates 1000 contingencies of a 20,000-bus system in about 27 min, (iii) iTesla, which combines offline and online simulations for reducing the total amount of scenarios to be analyzed [36].

To improve power system transient stability, it is possible to adjust in real-time the control strategies of CIGs, by dynamically modulating the generated active and reactive power profiles. In particular, the study [37] analyses the factors mainly influencing power system transient stability, and suggests performing a sensitivity analysis based on a multivariate optimization procedure in order to identify a proper mapping between the active and reactive power profiles that should be generated during a grid contingency.

The frequency stability is strongly related to the inertia of the power system. Reference [38] presents a method for "online" inertia estimation, using the obtained estimation for assessing the grid vulnerability to dynamic perturbations. The proposed estimation method, which is based on the integration of the Wold's decomposition method and then Goertzel's algorithm, uses a second-order infinite impulse response filter to describe the grid dynamics, assuming logistic distributions, and deploying first-order autoregressive models. The proposed method exhibited good performances in a realistic operation scenario, by reliably identifying the actual system inertia, and also detecting critical operation states in the presence of sensible penetration of renewable power generators. Another interesting approach aimed at managing low inertia power systems is proposed in [39], which analyzes the role of synchronous compensators (SC) in the task of increasing the short-circuit power, enhancing the voltage magnitude dynamic after severe grid contingencies. Detailed experimental results obtained on real case studies are also presented in this paper, demonstrating the effectiveness of SCs in enhancing the dynamic performance of low-inertia transmission systems.

Active Power Gradient (APG) control represents another promising technique for enhancing the frequency stability, as demonstrated in [40], which proposed two interesting mathematical methods for adjusting the parameters of APG controllers in order to reduce the frequency deviations after severe power imbalances. Both methods are formulated by a constrained single-objective optimization problem, but the first method aimed at minimizing the instantaneous variation of the kinetic energy of the synchronous areas, while the second one aimed at minimizing the spatial shifts between the dynamic trajectories of the temporal responses in different synchronous areas.

4.3. Voltage Stability

The massive penetration of CIGs into the transmission and distribution grid makes the voltage regulation process much more complex. In this context, short-term voltage stability could be compromised due to the fast dynamic response of the converters, which could induce severe transient voltage disturbances (e.g., over/undervoltages) [41].

Among the possible mitigation techniques that could be deployed for enhancing the voltage stability, those based on synchrophasors-based data processing are the most promising. These techniques allow detecting voltage instability and implementing load reduction techniques by processing time-synchronized measurements acquired by a network of PMUs. However, the application of these techniques is still embryonic, and several open problems should be fixed in order to obtain a complete and reliable system observability. In this context, the uncertainties induced by measurement errors, the accuracy and integrity of the time-synchronization sources, and the vulnerability of PMUs to cyberattacks are some of the relevant issue to address [42,43].

A useful analysis of the main effects induced by DFIG-based wind generators on voltage stability is presented in [44], which identifies the effects of these generators on maximum loadability limits and the maximum scalable active power demand, clearly emphasizing that an increasing penetration of wind power generators can overload critical grid equipment, hence limiting their hosting capacity. Similarly, in [45], the impact on the long-term voltage stability (LTVS) of photovoltaic generation is assessed. In particular, this study showed that solar generation can also have both negative and beneficial effects on voltage stability. Furthermore, the parameters that most affect LTVS were identified, including solar irradiation and ambient temperature, inverter power, reactive power gain and current-limiting strategies.

A further cause of voltage instability is the decommissioning of conventional synchronous generators, which lowers the available reactive power reserve. In this context, Reference [46] performs a comprehensive voltage stability analysis computing the VQ stability margins, emphasizing the fundamental role played by the reactive power control strategy and reactive power capacity of converter-interfaced technologies in ensuring voltage stability. According to this important result, several studies suggest the deployment of additional devices aimed at enhancing voltage stability by improving reactive power reserves, such as synchronous capacitors, synchronous static compensators (STATCOM), switchable and non-switchable shunt capacitors and reactors, and on-load tap changing transformers (OLTCs) [47].

4.4. Converter-Driven Stability

Many experimental studies developed worldwide demonstrated converter-driven stability issues that could induce unstable power system operations due to a number of complex dynamic phenomena, which include sub-synchronous oscillations between wind generators and series compensated lines [48], and harmonic instability induced by solar generators [49]. These phenomena can be generated by a combination of multiple effects, such as interactions between the CIGs control units, power system weakness, converter-interfaced load dynamics, and congested power lines [50]. In particular, the fast dynamic of the CIGs power/voltage control unit could induce rapid variations of the grid frequency and transient distortions of the voltage waveform, which could trigger the grid protections [51]. In this context, it is extremely important to accurately analyze and classify all the potential sources of converter-driven instabilities, in order to design and implement robust corrective actions aimed at mitigating the dynamic perturbation effects of these phenomena on correct power system operation.

4.5. Resonance Stability

Resonance stability issues are mainly related to the torsional impacts of flexible alternating current transmission systems and/or high-voltage direct current transmission systems, and the electrical resonances originating from the grid interactions of the CIGs controllers. The analysis of these oscillatory phenomena can be performed according to different computing techniques, which include time-domain electromagnetic power system simulation, state-space analysis, and frequency-domain impedance-based methods. In particular, electromagnetic power system simulation requires the numerical solution of the differential equations describing the dynamics of the power electronic components of the CIGs. Although these techniques allow accurately modeling the grid oscillations, they could not allow analyzing the intrinsic phenomena ruling the oscillations [52]. State-space analysis-based methods allow overcoming this limitation, by analyzing the eigenvalues of the state-matrix in the task of detecting unstable oscillation modes, determining the main parameters influencing their evolution [53]. Despite these benefits, developing a state-space model for power systems in the presence of a large number of CIGs is a challenging issue. In this context, the complexities in describing the reactive effects of the power transmission system, and the dynamic of the power electronic converters are some of the most relevant issues to address. The deployment of frequency-domain impedance-based methods represent a promising alternative approach for resonance stability analysis. The mathematical backbone supporting these approaches is based on the Nyquist stability criterion of the single port-impedance, which can be easily developed by using different modeling approaches, such as small-disturbance models, harmonic linearization, and the measurement methods [53]. Recently, more effective modeling approaches based on the s-domain nodal admittance matrix analysis have been proposed in the task of accurately analyzing the resonance stability of power systems in the presence of a large-scale penetration of CIGs [54]. The application of these modeling approaches allows identifying the main phenomena affecting resonance stability, and analyzing the potential technical solutions aimed at enhancing the grid robustness against these phenomena. In this context, the most promising mitigation techniques include the deployment of static var compensators for damping the torsional resonance, and the smart coordination of the CIGs controllers in the task of damping the electrical resonance [50].

5. Open Problems and Future Research Directions

The large scale integration of CIGs in power transmission and distribution networks, and the consequent decommitment of conventional synchronous generators, is reducing the system inertia inducing new and complex stability issues. In particular, an increasing number of both conventional perturbation phenomena (i.e., characterized by a time-frame of seconds), and new fast stability phenomena (i.e., characterized by shorter time scales, ranging from micro to milliseconds) have been observed in recent years. These phenomena could significantly affect power system stability, generating major challenges to system operators, which should revise their network codes in order to ensure that the power electronic interfaces of both generators and loads should provide proper grid stabilizing capabilities [55], defining new resilience requirements aimed at enhancing system stability in daily operation and during severe grid perturbations. To this aim, it is also important to define reliable markets schemes and effective regulations in order to stimulate the availability of grid flexibility sources at reasonable costs. Moreover, it is extremely important to support the large scale deployment of the enabling technologies for grid stability enhancement, in order to properly mitigate the impacts of the increasing number of CIGs. Finally, in the authors' opinion, the most promising research directions include the development of:

- Measurement-based methods for the online estimation of the power system inertia, and for assessing its (minimum) critical value.
- Condition monitoring systems aimed at predicting and monitoring the power system stability.
- Decision support systems for online control of the grid flexibility sources and the distributed energy resources in the task of enhancing the system stability, security and resiliency.
- Harmonization and interoperability standards defining the technical capabilities, the connection specifications, and the grid support services (e.g., grid forming, fast frequency reserves) of inverter-based resources.
- Advanced modeling platforms for real-time simulation of CIGs-dominated power systems.
- Market schemes aimed at ensuring the availability of grid flexibility resources aimed at maintaining the power system stability, security and resiliency.

6. Conclusions

This paper analyzed the classification of power system stability according to the recent revisions proposed in the literature in the task of characterizing the dynamic perturbation phenomena induced by a large penetration of converter-interfaced generators in transmission and distribution grids. Starting from this analysis, the potential grid impacts deriving by the replacement of conventional synchronous generators with distributed renewable power generators are presented and discussed, outlining the new challenging issues, the most promising research directions and the main mitigation technologies. The presented analyses demonstrated that the conventional strategies currently adopted in power system operation and planning should be revised in order to enhance the grid robustness to dynamic perturbations, which could induce severe consequences, especially in low-inertia systems. In this context, it is expected that the grid control and protection techniques will undergo profound changes to adapt to a new way of managing the grid, as will operation and planning criteria. In this scenario, it is pivotal to conduct new studies on the stability of the system and to investigate the effects that dynamics over several time scales and any undesirable coupling have on the grid.

Author Contributions: Conceptualization, A.V. and C.P.; investigation, G.M.G., C.P., S.T. and A.V.; writing—original draft preparation, V.M. and A.V.; writing—review and editing, V.M., S.T.; supervision, G.M.G., C.P. and A.V. All authors have read and agreed to the published version of the manuscript.

Funding: This research received no external funding.

Data Availability Statement: Not applicable.

Conflicts of Interest: The authors declare no conflict of interest.

Abbreviations

The following abbreviations are used in this manuscript:

APG	Active Power Gradient
ASD	Aperiodic Small Disturbance
CIG	Converter Interfaced Generation
DFIG	Doubly-Fed Induction Generator
DSA	Dynamic Security Assessment
FACTS	Flexible Alternating Current Transmission System
HIL	Hardware In Loop
HILP	High Impact Low Prequency
HPC	High Performance Computing
HVDC	High Voltage Transmission System

HVRT	High Voltage Ride-Through
LFO	Low Frequency Oscillation
LTVS	Long-Term Voltage Stability
LQR	Linear Quadratic Regulator
LVRT	Low Voltage Ride-Through
OLTC	On-Load Tap Changing
PMU	Phasor Measurement Unit
POD	Power Oscillations Damping
PSS	Power System Stabiliser
ROCOF	Rate Of Change Of Frequency
SC	Synchronous Compensator
SCADA	Supervisory Control And Data Acquisition
SG	Synchronous Generators
STATCOM	Static Synchronous Compensator
SVC	Static VAr Compensator
TSO	Transmissin System Operator
WADC	Wide Area Damping Controller
WAMS	Wide Area Measurement System
WAPOD	Wide Area Power Oscillation Damper

References

1. Shair, J.; Li, H.; Hu, J.; Xie, X. Power system stability issues, classifications and research prospects in the context of high-penetration of renewables and power electronics. *Renew. Sustain. Energy Rev.* **2021**, *145*, 111111. [CrossRef]
2. Guerrero, J.M.; Blaabjerg, F.; Zhelev, T.; Hemmes, K.; Monmasson, E.; Jemei, S.; Comech, M.P.; Granadino, R.; Frau, J.I. Distributed generation: Toward a new energy paradigm. *IEEE Ind. Electron. Mag.* **2010**, *4*, 52–64. [CrossRef]
3. Matallana, A.; Ibarra, E.; López, I.; Andreu, J.; Garate, J.; Jordà, X.; Rebollo, J. Power module electronics in HEV/EV applications: New trends in wide-bandgap semiconductor technologies and design aspects. *Renew. Sustain. Energy Rev.* **2019**, *113*, 109264. [CrossRef]
4. Yi, W.; Hill, D.J.; Song, Y. Impact of high penetration of renewable resources on power system transient stability. In Proceedings of the 2019 IEEE Power & Energy Society General Meeting (PESGM), Atlanta, GA, USA, 4–8 August 2019; IEEE: Piscataway, NJ, USA, 2019; pp. 1–5.
5. Annakage, U.; Rajapakse, A.; Bhargava, B.; Chaudhuri, N.; Mehrizi-sani, A.; Hauser, C.; Wadduwage, D.; Ribeiro Campos Andrade, S.; Pathirana, V.; Katsaros, K.; et al. *Application of Phasor Measurement Units for Monitoring Power System Dynamic Performance*; Technical Report; Cigré: Paris, France, 2017.
6. O'Sullivan, J.; Rogers, A.; Flynn, D.; Smith, P.; Mullane, A.; O'Malley, M. Studying the maximum instantaneous non-synchronous generation in an island system—Frequency stability challenges in Ireland. *IEEE Trans. Power Syst.* **2014**, *29*, 2943–2951. [CrossRef]
7. Kundur, P.; Paserba, J.; Ajjarapu, V.; Andersson, G.; Bose, A.; Canizares, C.; Hatziargyriou, N.; Hill, D.; Stankovic, A.; Taylor, C.; et al. Definition and classification of power system stability IEEE/CIGRE joint task force on stability terms and definitions. *IEEE Trans. Power Syst.* **2004**, *19*, 1387–1401.
8. Hatziargyriou, N.; Milanovic, J.; Rahmann, C.; Ajjarapu, V.; Canizares, C.; Erlich, I.; Hill, D.; Hiskens, I.; Kamwa, I.; Pal, B.; et al. Definition and classification of power system stability–revisited & extended. *IEEE Trans. Power Syst.* **2020**, *36*, 3271–3281.
9. Bryant, J.S.; Sokolowski, P.; Jennings, R.; Meegahapola, L. Synchronous Generator Governor Response: Performance Implications Under High Share of Inverter-Based Renewable Energy Sources. *IEEE Trans. Power Syst.* **2021**, *36*, 2721–2724. [CrossRef]
10. Nourollah, S.; Gharehpetian, G.B. Coordinated Load Shedding Strategy to Restore Voltage and Frequency of Microgrid to Secure Region. *IEEE Trans. Smart Grid* **2019**, *10*, 4360–4368. [CrossRef]
11. Vournas, C.; Tagkoulis, N. Investigation of synchronous generator underexcited operation in isolated systems. In Proceedings of the 2018 XIII International Conference on Electrical Machines (ICEM), Alexandroupoli, Greece, 3–6 September 2018; IEEE: Piscataway, NJ, USA, 2018; pp. 270–276.
12. Shair, J.; Xie, X.; Liu, W.; Li, X.; Li, H. Modeling and stability analysis methods for investigating subsynchronous control interaction in large-scale wind power systems. *Renew. Sustain. Energy Rev.* **2021**, *135*, 110420. [CrossRef]
13. Meegahapola, L.; Sguarezi, A.; Bryant, J.S.; Gu, M.; Conde, D.E.R.; Cunha, R.B.A. Power system stability with power-electronic converter interfaced renewable power generation: Present issues and future trends. *Energies* **2020**, *13*, 3441. [CrossRef]
14. Fernández-Guillamón, A.; Gómez-Lázaro, E.; Muljadi, E.; Molina-García, Á. Power systems with high renewable energy sources: A review of inertia and frequency control strategies over time. *Renew. Sustain. Energy Rev.* **2019**, *115*, 109369. [CrossRef]
15. Mancarella, P.; Billimoria, F. The fragile grid: The physics and economics of security services in low-carbon power systems. *IEEE Power Energy Mag.* **2021**, *19*, 79–88. [CrossRef]
16. Murad, M.A.A.; Ortega, A.; Milano, F. Impact on power system dynamics of PI control limiters of VSC-based devices. In Proceedings of the 2018 Power Systems Computation Conference (PSCC), Dublin, Ireland, 11–15 June 2018; IEEE: Piscataway, NJ, USA, 2018; pp. 1–7.

17. Wang, L.; Xie, X.; Jiang, Q.; Liu, H.; Li, Y.; Liu, H. Investigation of SSR in practical DFIG-based wind farms connected to a series-compensated power system. *IEEE Trans. Power Syst.* **2014**, *30*, 2772–2779. [CrossRef]
18. Liu, H.; Xie, X.; He, J.; Xu, T.; Yu, Z.; Wang, C.; Zhang, C. Subsynchronous interaction between direct-drive PMSG based wind farms and weak AC networks. *IEEE Trans. Power Syst.* **2017**, *32*, 4708–4720. [CrossRef]
19. Munkhchuluun, E.; Meegahapola, L.; Vahidnia, A. The large disturbance rotor angle stability with DFIG wind farms. In Proceedings of the 2019 9th International Conference on Power and Energy Systems (ICPES), Perth, Australia, 10–12 December 2019; IEEE: Piscataway, NJ, USA, 2019; pp. 1–6.
20. Xing, Y.; Marinescu, B.; Belhocine, M.; Xavier, F. Power oscillations damping controller for hvdc inserted in meshed ac grids. In Proceedings of the 2018 IEEE PES Innovative Smart Grid Technologies Conference Europe (ISGT-Europe), Sarajevo, Bosnia and Herzegovina, 21–25 October 2018; IEEE: Piscataway, NJ, USA, 2018; pp. 1–6.
21. Michi, L.; Carlini, E.; Scirocco, T.B.; Bruno, G.; Gnudi, R.; Pecoraro, G.; Pisani, C. AC Transmission Emulation Control Strategy in VSC-HVDC systems: General criteria for optimal tuning of control system. In Proceedings of the 2019 AEIT HVDC International Conference (AEIT HVDC), Florence, Italy, 9–10 May 2019; IEEE: Piscataway, NJ, USA, 2019; pp. 1–6.
22. Lauria, D.; Mottola, F.; Pisani, C.; Del Pizzo, F.; Carlini, E.M.; Giannuzzi, G.M. HVDC tie links in smart power grids: A versatile control strategy of VSC converters. In Proceedings of the 2021 AEIT HVDC International Conference (AEIT HVDC), Genoa, Italy, 27–28 May 2021; IEEE: Piscataway, NJ, USA, 2021; pp. 1–6.
23. Seppänen, J.; Nikkilä, A.J.; Turunen, J.; Haarla, L. Operational experiences of wams-based damping estimation methods. In Proceedings of the 2017 IEEE Manchester PowerTech, Manchester, UK, 18–22 June 2017; IEEE: Piscataway, NJ, USA, 2017; pp. 1–6.
24. Singh, R.S.; Baudette, M.; Hooshyar, H.; Vanfretti, L.; Almas, M.S.; Løvlund, S. 'In Silico' testing of a decentralized PMU data-based power systems mode estimator. In Proceedings of the 2016 IEEE Power and Energy Society General Meeting (PESGM), Boston, MA, USA, 17–21 July 2016; IEEE: Piscataway, NJ, USA, 2016; pp. 1–5.
25. Carlini, E.M.; Giannuzzi, G.M.; Zaottini, R.; Pisani, C.; Tessitore, S.; Liccardo, A.; Angrisani, L. Parameter Identification of Interarea Oscillations in Electrical Power Systems via an Improved Hilbert Transform Method. In Proceedings of the 2020 55th International Universities Power Engineering Conference (UPEC), Torino, Italy, 1–4 September 2020; IEEE: Piscataway, NJ, USA, 2020; pp. 1–6.
26. Xie, R.; Kamwa, I.; Chung, C.Y. A Novel Wide-Area Control Strategy for Damping of Critical Frequency Oscillations via Modulation of Active Power Injections. *IEEE Trans. Power Syst.* **2021**, *36*, 485–494. [CrossRef]
27. Surinkaew, T.; Shah, R.; Muyeen, S.M.; Mithulananthan, N.; Emami, K.; Ngamroo, I. Novel Control Design for Simultaneous Damping of Inter-Area and Forced Oscillation. *IEEE Trans. Power Syst.* **2021**, *36*, 451–463. [CrossRef]
28. Zhang, C.; Zhao, Y.; Zhu, L.; Liu, Y.; Farantatos, E.; Patel, M.; Hooshyar, H.; Pisani, C.; Zaottini, R.; Giannuzzi, G. Implementation and Hardware-In-the-Loop Testing of A Wide-Area Damping Controller Based on Measurement-Driven Models. In Proceedings of the 2021 IEEE Power & Energy Society General Meeting (PESGM), Washington, DC, USA, 25–29 July 2021; IEEE: Piscataway, NJ, USA, 2021; pp. 1–5.
29. Benato, R.; Giannuzzi, G.M.; Masiero, S.; Pisani, C.; Poli, M.; Sanniti, F.; Talomo, S.; Zaottini, R. Analysis of Low Frequency Oscillations Observed During a Power System Restoration Test. In Proceedings of the 2021 AEIT International Annual Conference (AEIT), Milan, Italy, 4–8 October 2021; IEEE: Piscataway, NJ, USA, 2021; pp. 1–6.
30. Dmitrova, E.; Wittrock, M.L.; Jóhannsson, H.; Nielsen, A.H. Early prevention method for power system instability. *IEEE Trans. Power Syst.* **2014**, *30*, 1784–1792. [CrossRef]
31. Gao, H.; Xin, H.; Huang, L.; Li, Z.; Huang, W.; Ju, P.; Wu, C. Common-Mode Frequency in Inverter-Penetrated Power Systems: Definition, Analysis, and Quantitative Evaluation. *IEEE Trans. Power Syst.* **2022**, *37*, 4846–4860. [CrossRef]
32. Biswas, S.; Bernabeu, E.; Picarelli, D. Proactive islanding of the power grid to mitigate high-impact low-frequency events. In Proceedings of the 2020 IEEE Power & Energy Society Innovative Smart Grid Technologies Conference (ISGT), Washington, DC, USA, 17–20 February 2020; IEEE: Piscataway, NJ, USA, 2020; pp. 1–5.
33. Konstantelos, I.; Jamgotchian, G.; Tindemans, S.H.; Duchesne, P.; Cole, S.; Merckx, C.; Strbac, G.; Panciatici, P. Implementation of a massively parallel dynamic security assessment platform for large-scale grids. *IEEE Trans. Smart Grid* **2016**, *8*, 1417–1426. [CrossRef]
34. Li, H.; Diao, R.; Zhang, X.; Lin, X.; Lu, X.; Shi, D.; Wang, Z.; Wang, L. An Integrated Online Dynamic Security Assessment System for Improved Situational Awareness and Economic Operation. *IEEE Access* **2019**, *7*, 162571–162582. [CrossRef]
35. Fereidouni, A.; Susanto, J.; Mancarella, P.; Hong, N.; Smit, T.; Sharafi, D. Online Security Assessment of Low-Inertia Power Systems: A Real-Time Frequency Stability Tool for the Australian South-West Interconnected System. In Proceedings of the 2021 31st Australasian Universities Power Engineering Conference (AUPEC), Virtual, 26–30 September 2021; pp. 1–9. [CrossRef]
36. Josz, C.; Fliscounakis, S.; Maeght, J.; Panciatici, P. AC power flow data in MATPOWER and QCQP format: iTesla, RTE snapshots, and PEGASE. *arXiv* **2016**, arXiv:1603.01533.
37. Farrokhseresht, N.; van der Meer, A.A.; Torres, J.R.; van der Meijden, M.A.; Palensky, P. Increasing the share of wind power by sensitivity analysis based transient stability assessment. In Proceedings of the 2019 2nd International Conference on Smart Grid and Renewable Energy (SGRE), Doha, Qatar, 19–21 November 2019; IEEE: Piscataway, NJ, USA, 2019; pp. 1–6.

38. Donnini, G.; Carlini, E.; Giannuzzi, G.; Zaottini, R.; Pisani, C.; Chiodo, E.; Lauria, D.; Mottola, F. On the Estimation of Power System Inertia accounting for Renewable Generation Penetration. In Proceedings of the 2020 AEIT International Annual Conference (AEIT), Catania, Italy, 23–25 September 2020; IEEE: Piscataway, NJ, USA, 2020; pp. 1–6.

39. Carlini, E.; Neri, S.; Marietti, A.; Borriello, A.; Barison, L.; Campisano, L. Integration of synchronous compensator in Terna's remote control and remote operation system. In Proceedings of the 2019 AEIT International Annual Conference (AEIT), Florence, Italy, 18–20 September 2019; IEEE: Piscataway, NJ, USA, 2019; pp. 1–4.

40. Perilla, A.; Gusain, D.; Torres, J.R.; Palensky, P.; van der Meijden, M.; Gonzalez-Longatt, F. Optimal Tuning of Active Power Gradient Control for Frequency Support in Multi-Energy Systems. In Proceedings of the 2020 IEEE PES Innovative Smart Grid Technologies Europe (ISGT-Europe), Delft, The Netherlands, 26–28 October 2020; IEEE: Piscataway, NJ, USA, 2020; pp. 889–893.

41. Sajadi, A.; Kolacinski, R.M.; Clark, K.; Loparo, K.A. Transient stability analysis for offshore wind power plant integration planning studies—Part II: Long-term faults. *IEEE Trans. Ind. Appl.* **2018**, *55*, 193–202. [CrossRef]

42. Vournas, C.D.; Lambrou, C.; Mandoulidis, P. Voltage Stability Monitoring from a Transmission Bus PMU. *IEEE Trans. Power Syst.* **2017**, *32*, 3266–3274. [CrossRef]

43. Ramapuram Matavalam, A.R.; Singhal, A.; Ajjarapu, V. Monitoring Long Term Voltage Instability Due to Distribution and Transmission Interaction Using Unbalanced μ PMU and PMU Measurements. *IEEE Trans. Smart Grid* **2020**, *11*, 873–883. [CrossRef]

44. Adetokun, B.B.; Muriithi, C.M.; Ojo, J.O. Voltage stability assessment and enhancement of power grid with increasing wind energy penetration. *Int. J. Electr. Power Energy Syst.* **2020**, *120*, 105988. [CrossRef]

45. Munkhchuluun, E.; Meegahapola, L.; Vahidnia, A. Long-term voltage stability with large-scale solar-photovoltaic (PV) generation. *Int. J. Electr. Power Energy Syst.* **2020**, *117*, 105663. [CrossRef]

46. Qi, B.; Hasan, K.N.; Milanović, J.V. Identification of Critical Parameters Affecting Voltage and Angular Stability Considering Load-Renewable Generation Correlations. *IEEE Trans. Power Syst.* **2019**, *34*, 2859–2869. [CrossRef]

47. Huang, L.; Xin, H.; Dörfler, F. H∞-Control of Grid-Connected Converters: Design, Objectives and Decentralized Stability Certificates. *IEEE Trans. Smart Grid* **2020**, *11*, 3805–3816. [CrossRef]

48. Adams, J.; Pappu, V.A.; Dixit, A. Ercot experience screening for Sub-Synchronous Control Interaction in the vicinity of series capacitor banks. In Proceedings of the 2012 IEEE Power and Energy Society General Meeting, San Diego, CA, USA, 22–26 July 2012; pp. 1–5. [CrossRef]

49. Li, C. Unstable Operation of Photovoltaic Inverter from Field Experiences. *IEEE Trans. Power Deliv.* **2018**, *33*, 1013–1015. [CrossRef]

50. Kong, L.; Xue, Y.; Qiao, L.; Wang, F. Review of Small-Signal Converter-Driven Stability Issues in Power Systems. *IEEE Open Access J. Power Energy* **2022**, *9*, 29–41. [CrossRef]

51. Paolone, M.; Gaunt, T.; Guillaud, X.; Liserre, M.; Meliopoulos, S.; Monti, A.; Van Cutsem, T.; Vittal, V.; Vournas, C. Fundamentals of power systems modelling in the presence of converter-interfaced generation. *Electr. Power Syst. Res.* **2020**, *189*, 106811. [CrossRef]

52. Xing, F.; Xu, Z.; Zhang, Z.; Dan, Y.; Zhu, Y. Resonance Stability Analysis of Large-Scale Wind Power Bases with Type-IV Wind Generators. *Energies* **2020**, *13*, 15. [CrossRef]

53. Prajapat, G.P.; Senroy, N.; Kar, I.N. Modal analysis of grid connected DFIG with detailed structural modeling of wind turbine. In Proceedings of the 2017 IEEE Innovative Smart Grid Technologies—Asia (ISGT-Asia), Auckland, New Zealand, 4–7 December 2017; pp. 1–6. [CrossRef]

54. Xu, Z.; Wang, S.; Xing, F.; Xiao, H. Study on the method for analyzing electric network resonance stability. *Energies* **2018**, *11*, 13. [CrossRef]

55. ENTSO-E. *ENTSO-E Position Paper Stability Management in Power Electronics Dominated Systems: A Prerequisite to the Success of the Energy Transition*; ENTSO-E: Brussels, Belgium, 2022.

energies

MDPI

Article

Impact of Passive-Components' Models on the Stability Assessment of Inverter-Dominated Power Grids

Davide del Giudice [1], Federico Bizzarri [1,2], Samuele Grillo [1,*], Daniele Linaro [1] and Angelo Maurizio Brambilla [1]

[1] DEIB, Politecnico di Milano, 20133 Milan, Italy
[2] Advanced Research Center on Electronic Systems for Information and Communication Technologies E. De Castro (ARCES), University of Bologna, 40126 Bologna, Italy

Abstract: Power systems are experiencing some profound changes, which are posing new challenges in many different ways. One of the most significant of such challenges is the increasing presence of inverter-based resources (IBRs), both as loads and generators. This calls for new approaches and a wide reconsideration of the most commonly established practices in almost all the levels of power systems' analysis, operation, and planning. This paper focuses specifically on the impacts on stability analyses of the numerical models of power system passive components (e.g., lines, transformers, along with their on-load tap changers). Traditionally, loads have been modelled as constant power loads, being this both a conservative option for what concerns stability results and a computationally convenient simplification. However, compared to their counterparts above, in some operating conditions IBRs can effectively be considered real constant power loads, whose behaviour is much more complex in terms of the equivalent impedance seen by the network. This has an impact on the way passive network components should be modelled to attain results and conclusions consistent with the real power system behaviour. In this paper, we investigate these issues on the IEEE14 bus test network. To begin with, we assess the effects of constant-power and constant-impedance load models. Then, we replace a transmission line with a DC line connected to the network through two modular multilevel converters (MMCs), which account for the presence of IBRs in modern grids. Lastly, we analyse how and to which extent inaccurate modelling of MMCs and other passive components can lead to wrong stability analyses and transient simulations.

Keywords: load modelling; line modelling; power system analysis; transient stability; small-signal stability; inverter-based resources; modular multilevel converters

Citation: del Giudice, D.; Bizzarri, F.; Grillo, S.; Linaro, D.; Brambilla, A.M. Impact of Passive-Components' Models on the Stability Assessment of Inverter-Dominated Power Grids. *Energies* **2022**, *15*, 6348. https://doi.org/10.3390/en15176348

Academic Editors: Cosimo Pisani and Giorgio Maria Giannuzzi

Received: 29 July 2022
Accepted: 24 August 2022
Published: 31 August 2022

Publisher's Note: MDPI stays neutral with regard to jurisdictional claims in published maps and institutional affiliations.

1. Introduction

During this last decade, power electronics converters have been integrated at an ever-rising pace in generation, transmission, and distribution power systems. This technology lends itself to a plethora of applications: among others, it could be used as a grid interface for specific kinds of loads (i.e., converter-connected loads (CCLs)) or as a means for energy conversion in generation systems fed by fuel cells or renewable energy sources, such as wind or solar photovoltaic (i.e., converter-interfaced generation (CIG)). Both usages can be grouped into the broad class of inverter-based resources (IBRs).

The increasing presence of IBRs has numerous ramifications in different facets of modern electric power systems, such as planning and operation. According to [1], IBRs are significantly changing power system dynamic behaviour (and, most notably, oscillatory behaviour during disturbances [2]) to the extent that the basic stability terms developed in the literature have been recently revised to consider the fast response of converters. Moreover, due to the progressive phase-out of conventional synchronous generators, CIG is expected to have a prominent role in supporting the stability of future grids through the provision of specific services, namely frequency and voltage regulation. In the light of the above, the interactions of IBRs with the power grid must be accurately studied to ensure a correct operation at all times.

Before the arrival of IBRs, conventional generation and transmission systems were typically simulated through single-phase equivalent representations and static load and transformer models. This modelling approach was originally imposed during the development of the first power system simulators to overcome the deficiency in the capability of computational systems to perform calculations using detailed, dynamic three-phase models. Based on this simulation setting, generator, transformer, line, and load models have been developed over the years to accurately simulate and predict the complex behaviour of electricity grids. In particular, some models have become customarily based on particular assumptions that were true in the past. For instance, a widely adopted approach is to describe lines with algebraic (i.e., static) representations and loads with constant power models to perform simulations of worst-case scenarios. As shown by the results of a survey collected in [3], about 70% of utilities and system operators in the world adopt only static load models for steady-state power system studies. The only exception is represented by utilities and system operators in the USA, which use a combination of static and dynamic models. In addition to this, static load models are typically set as constant power loads, and distributed generation is modelled as negative loads. This modelling approach is also dominant when it comes to dynamic power system analyses. Although the survey dates back to 2010, this practice does not seem to have changed up to at least 2018 and 2022, when [4,5], which also deal with the topic of load modelling, were respectively published. In [4] the typical values for the static models under different operating conditions are described. In [5] the impact of two static load models (i.e, exponential and polynomial model) on small-signal, transient, and frequency stability studies has been assessed. The conclusion of this work is that, in case no accurate data on loads are available, a constant power model should be adopted, as it provides the worst-case scenario and therefore constitutes a conservative approach. This same point is raised in [6,7]. Another practice established in power system analysis to typically accelerate simulations is to adopt static line models instead of dynamic ones [8,9].

In this article, we want to highlight how these approaches should be used with care in the case of IBR-dominated grids, as they might lead to a misguided power system stability assessment. Indeed, the growing presence of IBRs is significantly changing the behaviour of electric power systems. This begs the question as to whether previously developed load and passive elements models (and the assumptions in which they ground) are still adequate in the presence of IBR-dominated power grids or require revision—a question this paper attempts to answer. This is the same question the authors of [10] pose in the conclusion of their work. After describing the challenges IBRs bring to power systems stability studies, they suggest that the increasing penetration of IBRs may produce new ways through which instability occurs, calling for both more accurate models and new simulation techniques. In particular, the work presented here delves into the most commonly adopted models of loads and passive elements (i.e., lines, transformers, and shunts). After elaborating on their usage in conventional power systems, we assess how these models could be modified and adapted to reliably and accurately simulate and analyse the stability properties of IBR-dominated grids.

IBRs can be implemented through several converters characterised by different detailed three-phase models. In this paper, we choose the modular multilevel converter (MMC) as an example to drive the discussion on the impact of the models of passive components on the stability assessment of inverter-dominated power grids [11]. This converter, which was first introduced in [12], has become the technology of choice in high-voltage direct current (HVDC) and multi-terminal direct current (MTDC) transmission systems thanks to its scalability to high voltages and powers, lower switching activity of the sub-modules composing their legs, high voltage waveform quality, and efficiency [13,14]. It is worth noting that the choice above does not limit the validity of the proposed analysis only to MMCs. Indeed, the aspects we highlight, the results and observations we draw, can be almost invariantly applied to other converter topologies.

To guide the reader through this work, we analyze the well-known IEEE 14-bus power test system in several scenarios. In each of these, the test system has been modified according to the study's needs (e.g., by considering constant impedance loads behind on-load tap changers instead of constant power loads, or by replacing a line with an HVDC link with two MMCs).

What emerges from our work is that: (i) constant power load models and algebraic models of passive elements, despite giving a worst-case approximation of the stability boundary conditions, fail to adequately describe more complex dynamics, especially in IBR-dominated power systems; and (ii) the relationship between MMC impedance and operating frequency is not trivial and has a profound impact on the stability of hybrid power systems, when these are studied by means of a small-signal frequency scan, eigenvalue computation, and large-signal transient stability. In particular, the frequency bandwidth to be considered in the presence of MMCs (and IBRs in general) is much larger (up to kHz) than that of the conventional pure electro-mechanical models (few tens of Hz). If inappropriate or inaccurate models of passive elements such as transmission lines, transformers, and shunts are used, this extended bandwidth might lead to poorly damped or even unstable modes. On the contrary, these features might not appear when more detailed models of passive elements are adopted.

The remainder of this work is structured as follows. In Section 2 we introduce two conventional load models: the constant power and constant impedance load. After deriving their small-signal models and proving that they influence the stability of a simple power system in different ways, we elaborate on the usage of these models and passive elements in power system simulations over time. In Section 3, we analyse the impact of different models of load and passive elements (i.e., lines and transformers) on the stability of the IEEE14 benchmark system. Section 4 analyzes the effect on stability due to the connection of an MMC-based HVDC link to the IEEE14 benchmark. In particular, two MMC models of different levels of accuracy are used to perform several studies, including eigenvalue analysis and transient simulation. Lastly, Section 5 summarises the main results of this work and suggests possible future research avenues.

2. Constant Power and Constant Impedance Conventional Loads

Figure 1 depicts a single-phase equivalent model of a simple power system described in the DQ-frame [15,16] and consisting of an infinite bus (which fixes the DQ-frame components of its voltage (e_d, e_q) to a given value regardless of power exchange), a line, and a load of two possible kinds: constant power load or constant impedance load. These loads work at the same power level in nominal operating conditions, which implies that the power flow (PF) solution does not change for both load implementations if the voltage at the load bus is at its nominal value.

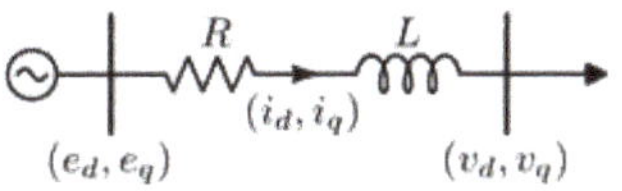

Figure 1. The single-phase schematic of a simple power system made up of an infinite bus, a line, and a load.

2.1. Small-Signal Models of the Constant Power and Constant Impedance Loads

The equations describing the constant power load are

$$v_d = \frac{i_d P - i_q Q}{i_d^2 + i_q^2} \, , \quad v_q = \frac{i_q P + i_d Q}{i_d^2 + i_q^2} \, ,$$

where P and Q are the loads' active and reactive power, (v_d, v_q) are the DQ-frame components of the bus voltage at which the load is connected, and (i_d, i_q) are the corresponding DQ-frame components of the load current.

The constant power load differential impedance at the $\left(\widehat{i}_d, \widehat{i}_q\right)$ PF solution is

$$
\begin{bmatrix}
\dfrac{\partial v_d}{\partial i_d} & \dfrac{\partial v_d}{\partial i_q} \\[2mm]
\dfrac{\partial v_q}{\partial i_d} & \dfrac{\partial v_q}{\partial i_q}
\end{bmatrix}
=
\begin{bmatrix}
R_p & X_p \\
X_p & -R_p
\end{bmatrix},
\tag{1}
$$

where

$$
R_p = \frac{P\left(\widehat{i}_q^2 - \widehat{i}_d^2\right) + 2Q\,\widehat{i}_d\widehat{i}_q}{\left(\widehat{i}_d^2 + \widehat{i}_q^2\right)^2}
\tag{2}
$$

$$
X_p = \frac{Q\left(\widehat{i}_q^2 - \widehat{i}_d^2\right) - 2P\,\widehat{i}_d\widehat{i}_q}{\left(\widehat{i}_d^2 + \widehat{i}_q^2\right)^2},
\tag{3}
$$

with the $\widehat{}$ symbol denoting values at the PF solution. By representing the $x + \jmath y$ generic complex number as the $[x, y]^{\mathrm{T}}$ two-element real vector, we can derive the $\widetilde{v}_d^{\,p} + \jmath\widetilde{v}_q^{\,p}$ small-signal voltage obtained by applying a small-signal current perturbation $\widetilde{i}_d^{\,p} + \jmath\widetilde{i}_q^{\,p}$ as

$$
\begin{bmatrix}
\widetilde{v}_d^{\,p} \\
\widetilde{v}_q^{\,p}
\end{bmatrix}
=
\begin{bmatrix}
R_p & X_p \\
X_p & -R_p
\end{bmatrix}
\begin{bmatrix}
\widetilde{i}_d^{\,p} \\
\widetilde{i}_q^{\,p}
\end{bmatrix},
\tag{4}
$$

where the ^{p} superscript refers to the voltages and currents of the constant power load, whereas the $\widetilde{}$ symbol denotes small-signal variables.

Consider now the constant impedance load $R_z + \jmath X_z$ with $R_z > 0$. Its small-signal voltage is

$$
\begin{bmatrix}
\widetilde{v}_d^{\,z} \\
\widetilde{v}_q^{\,z}
\end{bmatrix}
=
\begin{bmatrix}
R_z & -X_z \\
X_z & R_z
\end{bmatrix}
\begin{bmatrix}
\widetilde{i}_d^{\,z} \\
\widetilde{i}_q^{\,z}
\end{bmatrix}
\tag{5}
$$

where the ^{z} superscript refers to the voltages and currents of the constant impedance load. The signs of the matrix in (4) can be equated to those in (5) by imposing

$$
\begin{bmatrix}
\widetilde{v}_d^{\,p} \\
\widetilde{v}_q^{\,p}
\end{bmatrix}
=
\begin{bmatrix}
R_p & -X_p \\
X_p & R_p
\end{bmatrix}
\begin{bmatrix}
\widetilde{i}_d^{\,p} \\
-\widetilde{i}_q^{\,p}
\end{bmatrix}.
\tag{6}
$$

It is easy to see that this sign rearrangement means that the small-signal voltage of the constant power load can be expressed as $\widetilde{v}_d^{\,p} + \jmath\widetilde{v}_q^{\,p} = \left(R_p + \jmath X_p\right)\left(\widetilde{i}_d^{\,p} + \jmath\widetilde{i}_q^{\,p}\right)^{*}$, where the * symbol represents the complex conjugation operator. By comparing (5) with (6), one can immediately realize that the small-signal model of the constant power and constant impedance loads are deeply different even when their PF solutions are the same. As will be shown extensively in the following, this implies that the choice of load model may lead to different results in small-signal analyses and transient simulations.

2.2. Power System Small-Signal Stability with Constant Power and Constant Impedance

Let us now study the small-signal stability of the simple power system mentioned at the beginning of this Section. To do so, assume that the line in Figure 1 is described by a dynamic model. In a dynamic load model, the relationship between line voltage and current is described by a differential equation. On the contrary, in a static line model, said relationship is purely algebraic.

When the constant power load is used, the system small-signal behaviour can be modelled by the following differential equation:

$$
\begin{bmatrix}
L & 0 \\
0 & L
\end{bmatrix}
\frac{d}{dt}
\begin{bmatrix}
\widetilde{i}_d^{\,p} \\
\widetilde{i}_q^{\,p}
\end{bmatrix}
= -
\begin{bmatrix}
R_p & -X_p \\
X_p & R_p
\end{bmatrix}
\begin{bmatrix}
\widetilde{i}_d^{\,p} \\
-\widetilde{i}_q^{\,p}
\end{bmatrix}
-
\begin{bmatrix}
R & -\omega L \\
\omega L & R
\end{bmatrix}
\begin{bmatrix}
\widetilde{i}_d^{\,p} \\
\widetilde{i}_q^{\,p}
\end{bmatrix}
+
\begin{bmatrix}
\widetilde{e}_d \\
\widetilde{e}_q
\end{bmatrix},
\tag{7}
$$

where $R > 0$ and $L > 0$ are respectively the resistance and the inductance of the line, ω is the synchronous angular frequency, and $(\tilde{e}_d, \tilde{e}_q)$ are the small-signal components of the infinite bus voltage. The corresponding pair of eigenvalues is

$$\lambda_{\pm}^{p} = -\frac{R}{L} \pm \sqrt{\frac{R_p^2 + X_p^2}{L^2} - \omega^2}. \tag{8}$$

Through Equations (2) and (3) we obtain the following expression

$$R_p^2 + X_p^2 = \frac{P^2 + Q^2}{\left(\hat{i}_d^2 + \hat{i}_q^2\right)^2} = \frac{\left(\hat{v}_d^2 + \hat{v}_q^2\right)\left(\hat{i}_d^2 + \hat{i}_q^2\right)}{\left(\hat{i}_d^2 + \hat{i}_q^2\right)^2} = \frac{\hat{v}_d^2 + \hat{v}_q^2}{\hat{i}_d^2 + \hat{i}_q^2},$$

which can be used to recast Equation (8) and derive that, if $\dfrac{\hat{v}_d^2 + \hat{v}_q^2}{\hat{i}_d^2 + \hat{i}_q^2} > R^2 + L^2\omega^2$, we have

instability because one eigenvalue has a positive real part. Otherwise, we have complex conjugate eigenvalues with negative real parts. In both cases, the eigenvalues do not depend on the sign of P and Q.

On the contrary, if we adopt a constant impedance load, the small-signal model becomes

$$\begin{bmatrix} L & 0 \\ 0 & L \end{bmatrix} \frac{d}{dt} \begin{bmatrix} \tilde{i}_d^z \\ \tilde{i}_q^z \end{bmatrix} = -\begin{bmatrix} R_z & -X_z \\ X_z & R_z \end{bmatrix} \begin{bmatrix} \tilde{i}_d^z \\ \tilde{i}_q^z \end{bmatrix} - \begin{bmatrix} R & -\omega L \\ \omega L & R \end{bmatrix} \begin{bmatrix} \tilde{i}_d^z \\ \tilde{i}_q^z \end{bmatrix} + \begin{bmatrix} \tilde{e}_d \\ \tilde{e}_q \end{bmatrix}.$$

In this case, the $\lambda_{\pm}^z$ eigenvalues are

$$\lambda_{\pm}^z = -\frac{R_z + R}{L} \pm J\left(\omega + \frac{X_z}{L}\right). \tag{9}$$

Therefore, the system is stable because both eigenvalues *always* have *negative* real parts.

The above results reinforce the point raised in the previous subsection: the adoption of constant power and constant impedance load models leads to different behaviour in small-signal analyses.

2.3. On-Load Tap Changer

This subsection briefly outlines the operating principle of on-load tap changers (OLTCs) and reviews their most relevant representations developed in the literature. This element is adopted in some of the scenarios simulated in Section 3.

The target of an OLTC is to keep the bus voltage in the predefined dead-band by adjusting the transformer ratio in a given range over a discrete number of tap positions, thereby ensuring that loads connected to them operate at voltage levels close to their rated one. For example, in the OLTC model reported in [17], the dead-band can cover the $[0.99, 1.01]$ p.u. voltage interval, while the transformer ratio can vary in the $[0.88, 1.20]$ interval over 33 discrete positions (i.e., the ratio changes by 0.01 from one position to the next one). The decision process governing the tap control is the following. When the load voltage leaves a dead-band at time t_0, the first tap change takes place at time $t_0 + \tau_1$ and the subsequent ones at times $t_0 + \tau_1 + k\tau_2$ with $(k = 1, 2, \ldots)$. The delays τ_1 and τ_2 have values generally above 20 s and can differ from one OLTC to another. The delay is reset to τ_1 after the controlled voltage (i) has re-entered the dead-band or (ii) has jumped from one dead-band side to the other. The latter criterion prevents the tap from moving if the bus voltage oscillates too much or too frequently compared to the τ_1 and τ_2 delays.

The interested reader is referred to [18] and references therein for several OLTCs representations developed in the literature. The most accurate OLTC model, used in this paper, is implemented by a state machine, corresponding to a *hybrid dynamical system* with a decision block that generates *events* [19]. The main drawback of adopting this representation is that the PF and transient stability analyses must deal with a digital design that implements the state machine of the OLTC, which may lead to a complex simulation approach. In light of this, the majority of the remaining models in [18] were developed to attain a simplified, but still effective, version of OLTCs to achieve easy numerical simulations without resorting to a state machine implementation. For instance, the simplest OLTC representation (referred to in [18] as a continuous model) adopts a simple first-order differential equation to describe the tap variation process of the OLTCs to bring the voltage inside the dead band. So doing, the discrete tap switching process is transformed into a continuous one, thereby disregarding the previously mentioned delays.

It is important to point out that the connection of OLTC to loads is a common practice in electric power systems, whose effects are better detailed in the next subsection.

2.4. Conventional Load Modelling in Power System Simulations

The majority of the available power system test benches described in the literature and available on the web use (i) constant power loads and (ii) static models of passive components.

The first feature (i) contrasts with the fact that until the last decade the number of full-fledged constant power loads connected to distribution feeders was practically irrelevant. Thus, based just on the above, this approach seems rather conservative and one would be more inclined to adopt constant impedance load models instead. However, it is worth pointing out that the distribution feeders of electric power systems are often equipped with OLTCs (see the previous subsection) that restore voltage levels by bringing them inside a fixed voltage dead-band. Therefore, if only their steady-state behaviour is required (i.e., the PF solution), it seems reasonable to replace constant impedance loads behind OLTCs with constant power ones. Indeed, the *long-term* voltage restoration process of the OLTCs implies that at steady-state such loads operate at their nominal voltage and, thus, always withdraw their rated power. Constant power loads correspond somewhat to a synthesis of constant impedance loads behind OLTCs. However, it is worth pointing out that the number of loads whose active and reactive power is actually controlled became significant only with the introduction of IBRs (i.e., converter-connected loads (CCLs)). Indeed, the converters interfacing these resources to the grid can implement several controls, including power regulation. When IBRs employ this latter control, they constitute *full-fledged* constant power loads.

Although this modelling approach leads to correct PF results, it may lead to inaccurate transient simulations and small-signal analyses. Indeed, as better shown in the following, constant power loads and constant impedance loads behind OLTCs exhibit different dynamic behaviours. The adoption of the former model generally allows considering a worst-case scenario that real conventional power systems tend to. We believe this modelling choice was rooted in the 1970's when power system simulation tools were first developed [20]. A common practice to overcome the limited capabilities of computers in those years consisted in modelling constant impedance loads behind OLTCs as constant power loads to speed up simulations by incurring a minor loss in accuracy. This choice proved extremely effective until recent years when the penetration of IBRs in the power system rose drastically.

Concerning the second feature (ii), the dynamic models of passive components (e.g., transformers, lines, and shunts) were neglected during the development of conventional power system tools since the very fast, purely electrical dynamics of the grid were typically not of concern. This removed the stability problems due to the electrical dynamics of passive elements. Indeed, the electro-mechanical dynamic was the important aspect that had to be accurately simulated. As shown in the following, IBRs implement *full-fledged* constant power loads and act in a wide frequency bandwidth. This last trait imposes the adoption of dynamic models of passive components to analyse the stability of the system. Indeed, using algebraic models may lead to inaccurate results.

3. Simulation Results of the Conventional IEEE14 Power System

To show the role of constant power and constant impedance load models behind OLTCs on stability, we use as a benchmark the well-known IEEE14 power system, whose schematic is shown in Figure 2. All the parameters of the conventional version of IEEE14 power system can be found in [9].

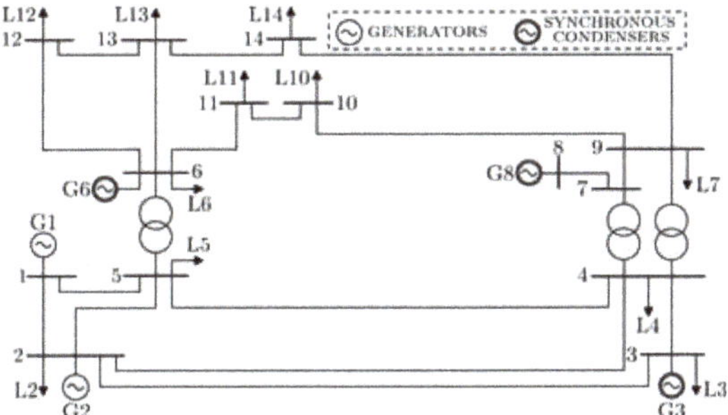

Figure 2. The schematic of the IEEE14 power system.

In the following, we modify this benchmark several times to simulate different case studies. In this section, we consider four scenarios. The first one, hereafter referred to as IEEE14-P, adopts the conventional version of the IEEE14 system with constant power loads. In the second scenario, defined as IEEE14-Z, we substitute the constant power loads with constant impedance ones behind OLTCs, whose nominal transformer ratio equals 1 (i.e., all bus voltages computed at PF are inside the respective dead-bands). The parameters of the constant impedance loads are tuned to absorb the same active and reactive power at the PF solution as the IEEE14-P scenario.

It is worth pointing out that the classic version of the IEEE14 test system adopts static models of passive elements. Compared to their IEEE14-P and IEEE14-Z counterparts, the third and fourth scenarios, hereafter respectively referred to as IEEE14-PD and IEEE14-ZD, employ dynamic models of inductances and capacitances of power lines, transformers, and shunts. These last two scenarios aim at showcasing the impact of alternative passive element modelling on power system stability.

3.1. The IEEE14-P and IEEE14-Z Scenarios

We performed a transient stability analysis of both IEEE14-P and IEEE14-Z scenarios. In both cases, we tripped the line between BUS2 and BUS5 after 200 s from the beginning of the simulations.

The results of these simulations are shown in Figure 3. By observing the waveforms related to the IEEE14-P scenario in the upper panel of the figure, one can notice that when the line is tripped, the rotor angular speed of the G2 synchronous generator suddenly decreases and stabilises after some time just below 0.999 [pu]. This behaviour stems from the fact that line tripping induces a load voltage drop and that all the loads in this scenario are of constant power type. Indeed, to guarantee constant power absorption despite the voltage drop, the currents flowing through the loads must increase. This leads to higher currents and power losses across the lines and transformers (and, thus, overall power demand) and, in turn, a decrease in system synchronous frequency. In the specific case of the IEEE14 benchmark, which does not implement turbine governors, the system frequency stabilises due to the contributions by damping in the swing equations of the synchronous generators and condensers.

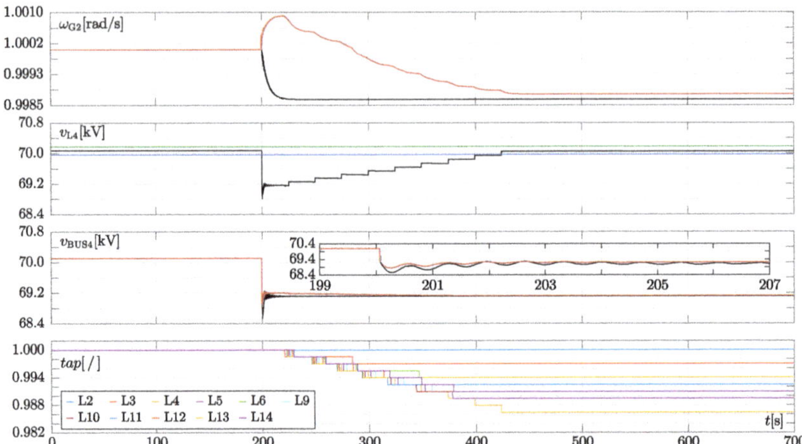

Figure 3. Simulation results of IEEE14-P and IEEE14-Z scenarios. Panels are described below from top to bottom. First panel: rotor angular speeds in [pu] of the G2 synchronous generator in the IEEE14-P and IEEE14-Z scenarios (black and red trace, respectively). Second panel: the black trace corresponds to the voltage magnitude of the constant impedance load L4 when connected to BUS4 through an OLTC in the IEEE14-Z scenario, whereas the green and blue traces respectively depict the upper and lower edge of the voltage dead-band inside which the corresponding OLTC restores the load voltage. Third panel: voltage magnitude at BUS4 in the IEEE14-P (black trace) and IEEE14-Z case studies (red trace). Fourth panel: the tap ratios of the OLTCs connected to each load in the IEEE14-Z scenario. X-axis: time [s].

It is important to note that this phenomenon leads to an iterative process in which the increase in line currents determines a decrease in load voltage, which, in turn, yields a further increase in line currents. Eventually, this process converges into a new stable working mode of the IEEE14-P scenario, as shown by the magnitude of the BUS4 voltage in Figure 3 (third panel from the top, black trace).

Consider now the IEEE14-Z scenario: contrary to the IEEE14-P case, the rotor speed of the G2 synchronous generator increases instead of decreasing right after the line tripping. Analogously to the previous scenario, this behaviour is coherent with the load model adopted and line tripping resulting in a load voltage drop. For instance, if we observe the magnitude of the voltage at BUS4 (third panel from the top), also in this scenario it decreases just after line tripping (although slightly less than in the IEEE14-P case). Since the power absorbed by constant impedance loads varies quadratically with voltage, the decrease in load voltage magnitudes leads to lower load powers. This implies that the currents flowing through the loads, lines, and transformers decrease. The same holds for the overall power demand, thereby leading to an increase in frequency, which stabilises once again due to the damping of the synchronous generators and condensers.

The traces in the second panel from the top in Figure 3 show that after some time delay the OLTC at BUS4 starts restoring the load voltage till bringing it inside the voltage dead-band (i.e., almost full voltage restoration to its value before the line outage is attained). As shown in the last panel of the figure, the other OLTCs do the same with different time schedules. The load voltage restoration process causes a progressive step increase in the power of the constant impedance loads and, thus, a corresponding decrease in grid frequency. At the end of this (long-term) process lasting more than 400 s (see traces of the other OLTC in the last panel of Figure 3), the steady-state behaviour of the power system is very close to that of the IEEE14-P case, but not identical, since load voltages are set to different values inside the OLTC dead-bands.

Similar results were already reported in [9], although with different and simpler OLTC models [18,21]. The previously described scenarios highlight that if the analysis focuses exclusively on power system steady-state behaviour (i.e., PF results), constant power loads can be adopted instead of constant impedance ones behind OLTCs, thus boosting simulation efficiency at the expense of a minor accuracy loss. However, if the dynamic behaviour is of interest (i.e., transient simulations or small-signal analyses are executed), these two models are not equivalent and may lead to quite different results. For instance, this difference is evident when applying a 20% overload to the IEEE14-P and IEEE14-Z scenarios and tripping the line between BUS2 and BUS5. In the first case, as reported in [9,22], the system PF solution is characterised by a pair of complex conjugated eigenvalues with positive real part equal to $0.006268 \pm J1.4357$ after line tripping, and thus it becomes unstable. On the contrary, in the second case, the system remains *stable* despite the overload (i.e., all eigenvalues have a negative real part before and after line tripping). In particular, the grid in this case can withstand a 58 % overload before becoming unstable.

To confirm these completely different results, we performed a transient stability analysis of the IEEE14-P and IEEE14-Z case studies by applying a 20% overload. Based on the above, we predict that the former scenario shall become unstable after applying a disturbance (i.e., line tripping), while the latter remains stable. A similar approach is used in the following section to demonstrate the actual stability or instability of the IEEE14 benchmark in other simulated scenarios after performing an eigenvalue analysis. Results are reported in Figure 4.

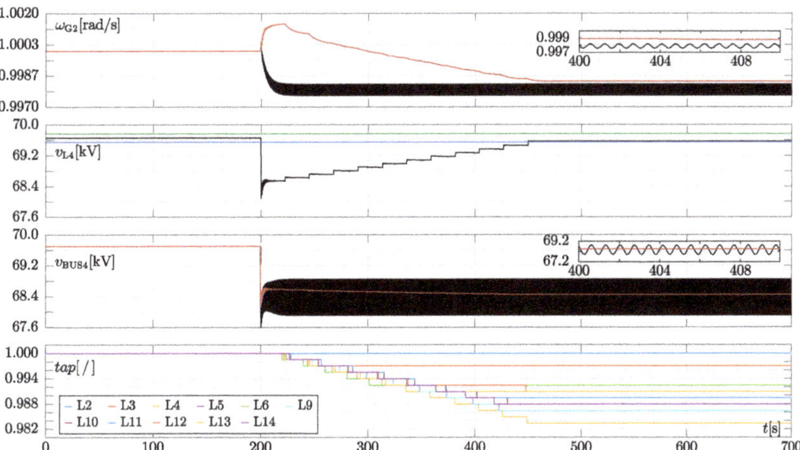

Figure 4. Simulation results of IEEE14-P and IEEE14-Z scenarios when a 20% overload is applied. The traces in each panel have the same meaning as those in Figure 3. Insets in the first and third panels are meant to show the oscillations that arise in the IEEE14-P scenario.

As expected, after line tripping, the rotor speed of G2 in the IEEE14-P scenario (black trace in the upper panel of Figure 4) starts oscillating, indicating that a periodic steady-state behaviour is observed since a stable stationary solution is not reached (i.e., the grid becomes "unstable"). As shown by the black trace in the third panel from the top, the voltage magnitude of BUS4 also oscillates. The same holds for the voltage at other buses, too. On the contrary, after line tripping, the grid in the IEEE14-Z scenario remains stable, without showing any oscillations. Note that OLTCs "fully" restore load powers and thus the absence of oscillations can not be ascribed to the fact that the power of constant impedance loads are lower after line tripping.

In summary, the results in the two case studies confirm that the use of constant impedance loads behind OLTCs instead of constant power ones leads to completely different stability results for what concerns both small-signal analyses and transient simulations. Based on this observation, one might wonder which load model actually correctly and accurately represents the real behaviour of the IEEE14 system and possibly other grids. To answer this question, we believe the following points are worth considering.

(i) The use of constant power load models leads to pessimistic estimations of the stability boundary conditions. From an engineering point of view, this could be a good feature since it gives safety margins. From this perspective, despite leading to inaccurate results, constant power load models can still be considered adequate and useful.

(ii) The use of constant power loads eliminates the need to implement the complex automata modelling the behaviour of OLTCs connected to constant impedance loads, which simplifies power flow and transient stability analyses and reduces the computational burden.

(iii) The tap switching process of the OLTCs occurs after relatively long time periods (i.e., tens of seconds). Thus, since the power restoration of the constant impedance loads connected to the OLTCs is a decidedly slow process, the electrical dynamics of passive elements can be neglected. Indeed, the only dynamics that should be considered are the electro-mechanical ones of the generators, which are retained when adopting constant power loads. Also, in this case, employing algebraic (i.e., static) models of passive elements facilitate and accelerate simulations while ensuring adequately accurate results.

The above considerations lead to the conclusion that the use of (i) constant power loads instead of constant impedance ones behind OLTC in distribution feeders and (ii) algebraic models of passive elements instead of dynamic ones have become customary in modelling and simulating power systems till today. However, this established practice may no longer be valid due to the increasing presence of IBRs. Indeed, due to the controls that they can implement, IBRs may represent *fast* constant power loads and generation; their AC impedances, which might be complex functions of frequency spanning in the range of several kHz, can deeply interact with the dynamics of passive elements. If algebraic models are used, these interactions may remain hidden; thus, simulations may not accurately reflect the true behaviour of the system. This aspect, which is dealt with in the following, suggests that the conventional model of loads and passive elements may require revision when used in IBR-dominated grids.

3.2. The IEEE14-PD and IEEE14-ZD Dynamic Test Systems

The case studies analyzed so far relied on static models of lines. In the following, we adopt dynamic line models by considering the IEEE14-PD and IEEE14-ZD scenarios and simulating the same outage described in the previous subsection (i.e., line tripping between BUS2 and BUS5 at 200 s).

The IEEE14-PD case gives extremely straightforward results: the adoption of dynamic models and constant power loads yields an *unstable* PF solution since several eigenvalues jump in the right-half portion of the complex plane. Note that this result anticipates that the introduction in the system of IBRs (e.g., MMCs), whose active and reactive power absorption is regulated to track a reference value (thereby mimicking constant power loads), will lead to similar instabilities. Since the attainment of an unstable PF solution prevents any further analysis of this case study, in the sequel we consider only the IEEE14-ZD scenario.

In this case, a completely different result is obtained because the PF solution is *stable*. The different behaviours observed in the IEEE14-PD and IEEE14-ZD scenarios are coherent with what we obtained by computing the eigenvalues of Equations (7) and (9). Indeed, as explained in Section 2.1, power systems with constant power loads are less robust than those with constant impedance ones from a stability point of view. To facilitate comparisons, the simulation results of the IEEE14-Z scenario are replicated in Figure 5 together with those of the IEEE14-ZD one.

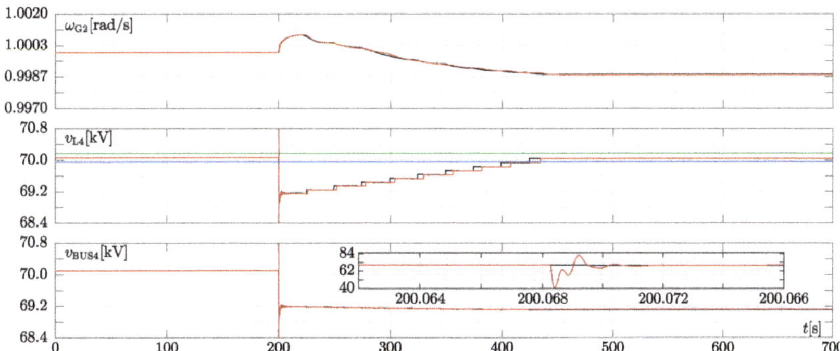

Figure 5. Simulation results of IEEE14-Z and IEEE14-ZD scenarios. The traces in each panel have an analogous meaning to those in Figure 3. In this case, the black and red traces in each panel respectively refer to the results of the IEEE14-Z and IEEE14-ZD scenarios. In the second and last panel, the voltage magnitude of BUS4 is upper and lower clipped to better show voltage steps at a tap position change.

The traces in the two cases almost overlap despite them adopting two different line models. It is worth pointing out that in Figure 5 the magnitude of the BUS4 voltage is clipped (upper and lower) to allow a better comparison of the waveform in the two cases. The inset in Figure 5 depicts the unclipped voltage magnitude just before and right after the line tripping.

4. Simulation Results of the Modified IEEE14 Power System with an MMC-Based HVDC Link

In this section, we show some simulation results of a modified version of the IEEE14 system. To take into account the effect of the growing presence of IBRs on power system stability, the line between BUS4 and BUS5 was replaced with an HVDC link comprising two MMCs, as shown in Figure 6.

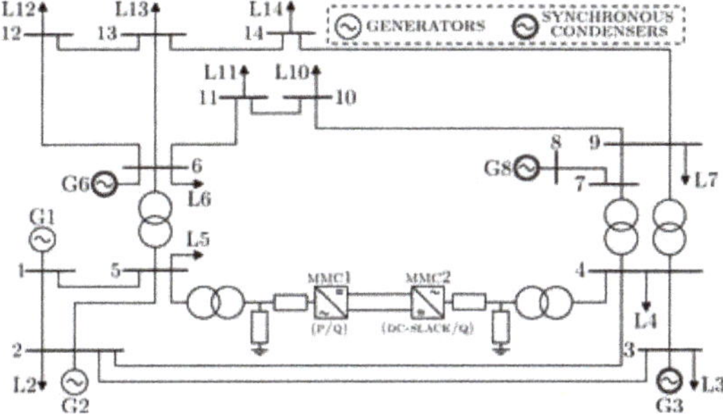

Figure 6. The schematic of the modified IEEE14 power system. Compared to its counterpart in Figure 2, the line between bus 4 and 5 has been substituted with an HVDC link made up of two MMCs.

This modification was proposed and described in [23–25]. In a nutshell, the MMC1 and MMC2 converters are respectively of P/Q and DC-SLACK/Q types. In other words, MMC1 is controlled to exchange a specific amount of active and reactive power. On the contrary, MMC2 regulates the DC-side voltage and reactive power exchange. In particular, the active and reactive power transmitted by the HVDC link is regulated to equal that of the replaced line. To do so, the active power reference value of MMC1 (i.e., sending end) is $P_o = 60.7\,\text{MW}$, while the reactive power reference value of both MMCs is $Q_o = \mp 13\,\text{MVAR}$. In addition, the DC-side voltage reference value of MMC2 (i.e., receiving end) is $V_{dc} = 400\,\text{kV}$. The interested reader is referred to Figure 1 of [26] and references therein for a possible MMC control scheme implementation.

When employed in HVDC systems, the sub-modules (SMs) in each arm of an MMC can be in the order of hundreds. In this case, adopting the most accurate MMC model (known as *full-physics* model) requires considering a large number of semiconductor devices, thereby leading to prohibitively high computational times [27]. To address this issue, several MMC models in the literature have been developed that implement different trade-offs between simulation speed and accuracy. Each of these models is better suited for analysing the efficiency of specific operating conditions and MMC controls (e.g., circulating current suppression strategies, protections, PLLs, and SM capacitor voltage balancing algorithms [28,29]). The interested reader can refer to [30–32] and references therein for a review of some of these models, including a description of their advantages and shortcomings when adopted in power system simulation.

In general, the analysis of converters can be divided into three stages [27]: component, system, and network level studies. Component level studies focus on the early design stage of the converter and the performances of its semiconductors (e.g., conduction and switching losses), with the time of investigation ranging from nano to milliseconds. On the contrary, system-level studies aim at evaluating the interactions between the converter and the power system connected to it, whose extension is usually limited to a few generators and loads. These analyses, which involve dynamics that last from milliseconds to several seconds, allow the validation of converter controls, filters, and protections. Finally, network-level studies analyse how the converter behaves in a large AC network and affects the electromechanical transients and the steady-state operation of the whole system. Thus, the time interval of interest ranges from a few seconds to several minutes. Such studies address, for instance, power flow and stability analyses.

The full-detailed and bi-value resistor MMC models allow reducing CPU times by resorting to simplified representations of the semiconductor devices in each SM, thus preserving the overall converter topology [30]. These models can be used as a benchmark to validate simpler ones, test MMC controls, protections, and behaviour during start-up, normal and abnormal operating conditions. However, due to the simplifications introduced, these representations cannot be used for component-level studies, which can only be performed straightforwardly with the full physics model.

The Thévenin equivalent [33] and switching function [34] models, which still rely on simplified SM representations, are good candidates for system-level studies because they further boost simulation speed by suitably grouping the SMs together, thus obtaining a more compact representation of the SM strings in each arm. Alternatively, in [32] a technique called isomorphism is proposed to boost MMC simulation efficiency that is compatible with any SM model used, thereby paving the way for detailed analyses if accurate SM representations are used. In a nutshell, this technique dynamically clusters while simulating the SMs in groups characterised by the same behaviour.

The average model decreases simulation time even more by neglecting the voltage and current ripples due to SMs commutations [35] and significantly simplifying the MMC topology by exploiting voltage and current controlled sources. This model is suitable for network-level studies which investigate the dynamic behaviour and stability of large grids comprising multiple MMCs.

All of the models described so far are represented in the ABC (i.e., three-phase) frame. On the contrary, MMC models formulated in the DQ0 frame, either based on an average representation [36] or dynamic phasors [37], grant a significant boost in simulation speed, making them suitable candidates for MMC simulation in large-scale power systems.

In general, the boost in simulation speed offered by the MMC models mentioned above may come at the cost of reduced simulation accuracy and/or the loss of internal MMC variables, which are no longer present due to the simplifications introduced in the SM model and the MMC topology. As previously stated, based on the study that needs to be performed (and, thus, the sought trade-off between simulation speed and accuracy), one MMC model might be more suitable than others.

In this paper, we modelled the MMCs of the HVDC system at two very different levels of accuracy: the former, hereafter referred to as *accurate*, is shown in Figure 1a–c of [26] and derives from [38]. This three-phase model substitutes the SMs in each arm of the MMC with a single equivalent circuit. It retains the main features of the converter and allows analyzing all of its upper-level controls, as well as the circulating current suppression strategy. The usage of an average model instead of one based on dynamic phasors is justified by the relatively limited size of the IEEE14 benchmark system. In addition, more complex MMC models, which allow us to analyse the individual behaviour of SMs, have not been adopted because their level of detail is excessive for the simulations described in the following sections. The second MMC model used in this work (hereafter referred to as *simplified*) is the one described in [39,40], which resorts to macro-models that retain only the main features at the points of connection of the converters and allow only the implementation of higher-level controllers, such as active and reactive power regulations and droop controls. This representation corresponds to an average model formulated in the DQ-frame. Some of its components are added to properly take into account the converter losses at the AC and DC sides. Despite leading to a lower computational burden, one of the drawbacks of this model is that it does not consider any dynamics. Indeed, the impedances connected to its AC side (e.g., the MMC transformer impedance) are referred to as the synchronous frequency and are *fixed* (i.e., they do not change with grid frequency). As shown in the following, this static model constitutes a deep simplification that could lead to largely different simulation results from those obtained with a more accurate MMC model. Indeed, the complex topology and control scheme of MMCs result in them having a multi-frequency response generating non-negligible harmonics.

Regardless of the representation employed, the key aspect of the analyses with these MMC models is that the MMC1 converter at BUS5 is always a *full-fledged* P/Q type. Even if a simplified version of the converter is used, its main features at the point of connection to the AC grid (i.e., absorbing a constant active power of $P_o = 60.7$ MW and contributing a reactive power of $Q_o = -13$ MVAR) are maintained. This represents the main difference with respect to the previously considered cases.

We want to stress that our goal here is not to investigate in detail all the possible controls that can be implemented in MMC1 and MMC2, but rather to show the impact that the two previously described MMC models and their controls might have on the overall power system stability in different scenarios.

4.1. Simplified MMC Model

To begin with, we used the simplified model of the MMCs, which, as previously stated, does not consider any dynamics. We instead considered the electrical dynamics of passive components, adopted constant impedance loads behind OLTCs, and swept the active power absorbed by the P/Q type MMC1 converter in the $[0, P_o]$ interval.

The reactive power of both MMCs was also swept in proportion accordingly. At each sample of the sweep, we performed an eigenvalue analysis. The top panel of Figure 7 reports the paths followed in the complex plane by the most relevant eigenvalues of the system (i.e., those closest to the right-half portion of the complex plane).

During the sweep, as soon as the active power goes above 14.9 MW, the power system becomes unstable since a complex-conjugate pair of eigenvalues with a natural frequency of about 1 kHz enters the right-half portion of the complex plane. It is worth pointing out that this power value is lower than the one the HVDC link is supposed to transmit (i.e., $P_o = 60.7$ MW). The onset of unstable behaviour at around 14.9 MW is visible in the top panel of Figure 7: eigenvalues are colour-coded based on the active power reference value. Moreover, when the active power sweep reaches 55.6 MW, another complex conjugate pair of eigenvalues enters the right-half plane.

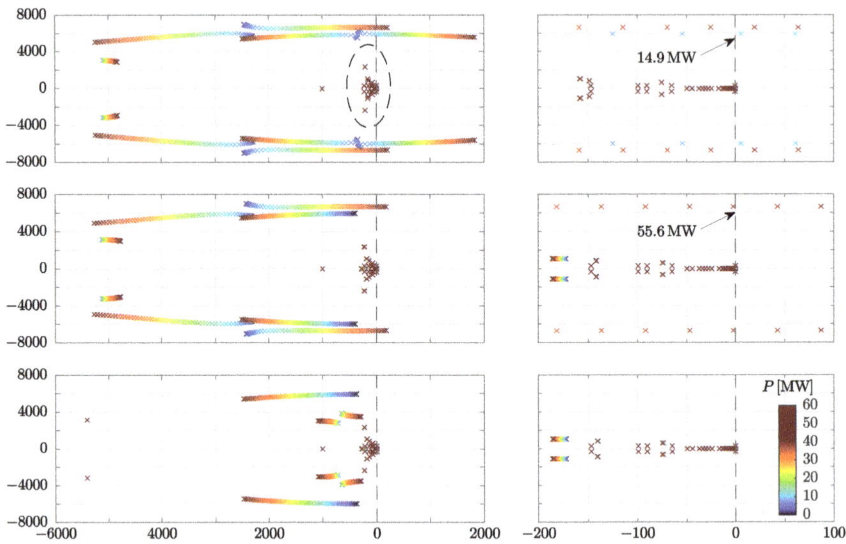

Figure 7. Evolution of the main eigenvalues of the IEEE14 power system shown in Figure 6 when the active power reference P of the P/Q type MMC1 converter at BUS5 is swept in the $[0, 60.7\,\text{MW}]$ interval. For each row, the panels on the right are an inset of those on the left. The vertical dotted line in each panel denotes the null real part of the eigenvalues. Eigenvalues are colour-coded based on P (see the colour bar in the bottom panel, right side) to better identify the power reference values that make the system unstable, which are highlighted by arrows in the panels on the right. Top row: base case. The real part of the eigenvalues inside the dashed ellipse is close to zero, but it is always negative regardless of P (the same holds for the other panels). Middle row: case in which MMC1 includes the droop control and low-pass filter described in Equation (11). Bottom row: compared to the previous case, a shunt capacitor is also added to the AC-side of MMC2. In all cases, dynamic models of lines and transformers are used, together with constant impedance load models behind OLTCs.

The unstable pairs of complex-conjugate eigenvalues that arise at 14.9 MW and 55.6 MW are due to different reasons, which have been identified through the analysis of the participation factors. In the former case, results indicate that instability is mainly due to BUS5 and the line that connects BUS2 and BUS5 [41–43] (i.e., where the P/Q type MMC1 is connected). On the contrary, in the latter case, the participation factors show that the second unstable mode is mainly due to BUS4 and the line that connects BUS3 and BUS4 (i.e., where the DC-SLACK/Q type MMC2 is connected). The relevant aspect that we underline is that the MMC1 and MMC2 converters are not unstable *per se* but they are when used in the IEEE14 benchmark due to their interactions with the grid.

For instance, a qualitative explanation of the onset of instability at BUS5 due to the P/Q type MMC1 is the following. From (1) and (2) it can be inferred that the MMC1 is characterised by a negative resistance, which is a common trait of converter-connected elements whose active and reactive power absorption is regulated. In the specific case of the IEEE14 benchmark, when the MMC1 active power reference exceeds 14.9 MW, its negative resistance becomes higher than the series resistance of the lines/transformers, thereby leading to undamped RLC oscillations [44].

A possible way to keep the eigenvalues of the MMC1 in the left half-plane is to regulate its actual active power exchange by adding an AC-voltage/active power droop. More specifically, with the addition of the droop the power absorbed by this converter becomes

$$P(t) = P_o + \gamma_P \left(\overbrace{\sqrt{v_d^2(t) + v_q^2(t)} - \sqrt{\widehat{v}_d^2 + \widehat{v}_q^2}}^{e(t)} \right), \tag{10}$$

where the electrical quantities refer to BUS5 and $\gamma_P > 0$. This latter parameter mirrors the constant of the proportional regulator used in the MMC to implement the droop. This control is such that the absorbed power P rises if the modulus of the BUS5 voltage increases with respect to its value at the PF solution (i.e., $\sqrt{\widehat{v}_d^2 + \widehat{v}_q^2}$) and vice versa otherwise. The introduction of the droop emulates a "constant impedance" load, which absorbs P_o at nominal bus voltage.

Since the droop modifies the MMC1 active power set-point (i.e., its transmitted power), we added a low-pass filter block to slowly restore power to its P_o nominal value. The overall model implementing the regulator is

$$\begin{aligned} \tau_L \frac{dx(t)}{dt} + x(t) - e(t) &= 0 \\ P(t) &= P_o + \gamma_P(e(t) - x(t)), \end{aligned} \tag{11}$$

whose transfer function is $H(s) = \dfrac{P(s)}{e(s)} = \gamma_P \dfrac{\tau_L s}{1 + \tau_L s}$. The addition of the low-pass filter emulates the same effects that an OLTC has on the restoration of the voltage, and, thus, power of a constant impedance load connected to it. In the light of the above, the combination of the droop and the low-pass filter transforms the P/Q type MMC1 into a constant impedance load behind an OLTC.

After this modification, we repeated the power sweep analysis that we did before. The loci described by the most relevant eigenvalues are reported in the middle panel of Figure 7. The beneficial effect due to the introduction of the droop and low-pass filter block is evident. Indeed, the unstable complex-conjugate pair of eigenvalues that previously arose at 14.9 MW is now absent.

However, the eigenvalue pair that enters the right-half plane when P_o goes above 55.6 MW is still present. The analysis of the participation factors confirms again that this pair is due to the DC-SLACK/Q type MMC2 converter connected at BUS4. To eliminate this pair it is necessary to act on MMC2 by modifying its equivalent impedance "seen" at its point of connection. In this case, we added a shunt capacitor to MMC2 to obtain stability. The eigenvalues loci of the power system after this intervention are shown in the bottom panel of Figure 7. The addition of shunt capacitors is clearly beneficial because all eigenvalues remain in the left-half plane regardless of the active power reference value.

To further check stability, we performed a transient stability analysis analogous to those of Section 3, whose results are reported in Figure 8.

As in the previously analysed scenarios, when the line between BUS2 and BUS5 is tripped at 200 s, a voltage drop occurs at all buses, including BUS4 and BUS5. This leads to a decrease in the load power absorbed by the constant impedance loads. In addition, the droop controller of MMC1 lowers its absorbed power in a similar way. From the same figure, one can notice that the long-term load voltage restoring action of the converter is similar to those of OLTCs. This was obtained through a proper tuning of the τ_L time constant in Equation (11).

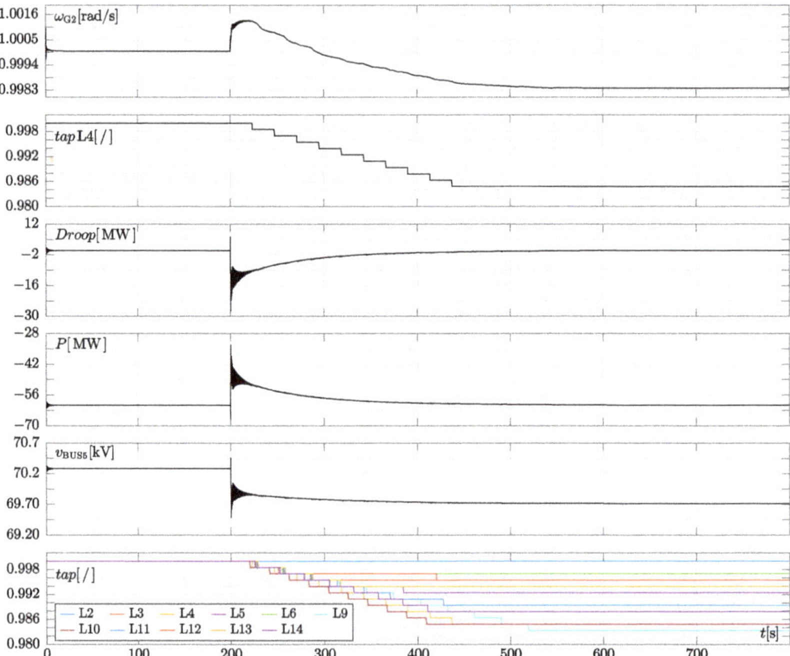

Figure 8. Simulation results of the modified IEEE14 test system shown in Figure 6. In this case, constant impedance loads behind OLTCs and dynamic line models are used. The MMC1 converter, which is equipped with an AC-voltage/active power droop, is set to exchange $P = 60.7\,\text{MW}$ and $Q = -13\,\text{MVAR}$ (this latter value is also exchanged by MMC2). The MMC2 converter is equipped with a shunt capacitor. Third panel from the top: MMC1 power exchange variation due to the implementation of the droop control. Fourth panel: overall power exchange of the MMC1 converter. The traces in the other panels have the same meaning as those in Figure 3.

4.2. Accurate MMC Model

In this section, we used the DCS1 test system by CIGRE as a detailed three-phase model of the MMC-based HVDC link. Its schematic, which is fully described in [27], is not reported here for space reasons. The original power rating of the DCS1 system is 800 MW, which is abundant for our scope: indeed, it is worth recalling that the HVDC link in Figure 6 is meant to transmit 60.7 MW. Therefore we reduced the power rating to 100 MW. The three-phase transformers of both MMCs were adapted to step-up voltage from 69 kV to 145 kV and 380 kV respectively to meet the original design requirements. The PLL of each MMC tracks the frequency at the point of connection and aligns to the q-axis of the DQ-frame.

To have some insight into the behaviour of the system and to understand the mutual interactions between the power system and the MMCs when a more accurate three-phase MMC model is used together with constant impedance loads, we computed the admittances of MMC1 and MMC2 at their points of connections and checked for possible stability issues [45–47]. Note that we deal with a hybrid model of the IEEE14 system, where the MMCs are simulated with detailed *three-phase* dynamic models and the IEEE14 with a *single-phase* equivalent model. A conventional eigenvalue analysis can not be performed with this hybrid power system.

At first, we chose the $P = 11.2\,\text{MW}$, $Q = \mp 2.6\,\text{MVAR}$ set-points and connected each step-up transformer to an infinite bus. This is the common configuration used to design an HVDC link and its converter stations that will be connected to AC systems. The adoption of such a configuration ensures that the HVDC link and its MMCs are designed to be stable *per se*. However, as explained in the following, this does not prevent possible undesired grid and converter interactions from occurring when the HVDC link is connected to a real power system, such as the IEEE14 one. The choice of the set-points is based on the top panel of Figure 7: since with these setpoints no eigenvalue has a positive real part, system stability should be ensured.

The numeric tool we used to compute admittances is fully described in [26] and implemented in our own simulator PAN [48,49]. In extreme synthesis, a small-signal tone with *variable frequency* is injected in one of the three phases of the step-up transformer (in our case the "a" phase) of each MMC. This tone superimposes the phase voltage (of large magnitude). We computed the corresponding phase current, extracted the small-signal current contribution due to the injected small signal, and computed the admittance (i.e., the current versus voltage small-signal transfer function).

The plots in Figure 9 depict the real and imaginary components of the admittance of the P/Q type MMC1.

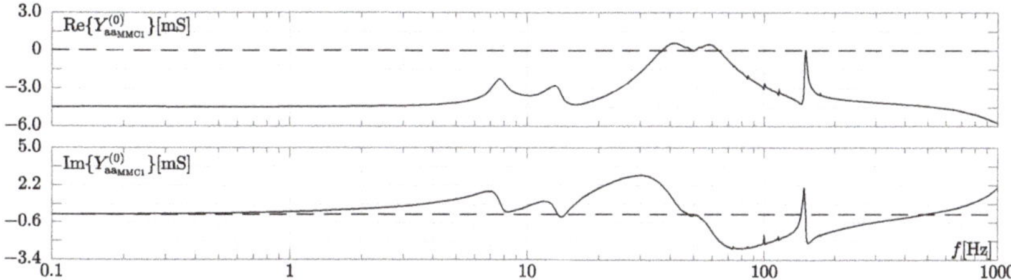

Figure 9. The admittances of the "a" phase of the P/Q type MMC1 when the accurate MMC model is used. Upper panel: real part of the admittance [mS]. Lower panel: imaginary part [mS]. X-axis: frequency [Hz]. In the panels, the dashed lines denote null real and imaginary parts.

The plots include several frequency intervals, both at low and high frequencies, where the real part of the admittance is *negative* and its imaginary part is *positive* (capacitive behaviour). In Figure 10 we report the admittance of MMC2.

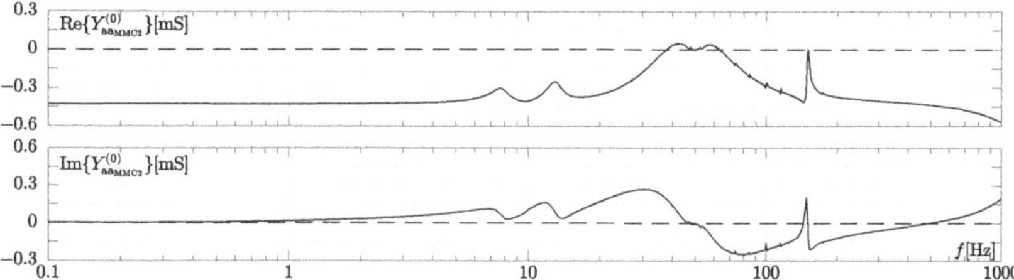

Figure 10. The admittances of the "a" phase of the DC-SLACK/Q type MMC2 when the accurate HVDC link model is used. Upper panel: real part of the admittance [mS]. Lower panel: imaginary part [mS]. X-axis: frequency [Hz]. In the panels, the dashed lines denote null real and imaginary parts.

The traces are similar to the corresponding ones in Figure 9, but their magnitudes are about ten times lower. Thus, since MMC1 potentially exhibits the largest negative conductance in both low and high-frequency intervals, it constitutes the most critical MMC from a system stability perspective. In particular, as suggested in the previous subsection, (i) MMC1 can form an *unstable* RLC resonator with the inductances of the lines and transformers (i.e., when dynamic models of passive elements are used). This phenomenon might occur at frequencies higher than those at which AC grids typically operate. In addition, (ii) MMC1 can even interact with the models of synchronous generators/condensers of the IEEE14 power system leading to poorly damped or unstable (electro-mechanical) *low-frequency* modes (less than 10 Hz).

To provide more details about the first comment (i), consider the high-frequency portion of the admittance of MMC1 in Figure 9. For example, at 1 kHz its susceptance is about

2.04 mS (corresponding to 0.324 µF equivalent capacitance), while its conductance is negative. If the line inductance is larger than 80 mH (a value attainable with an overhead line of sufficient length, such as 100 km), an unstable RLC equivalent circuit can originate [50]. This happens even at relatively low power levels since the detailed model of the MMC shows a much more complex admittance function with respect to its simplified counterpart. This kind of instability can be avoided at high frequency provided that the L inductance of the transformers and the lines is sufficiently low. It is also important to note that this instability does not occur when non-dynamic line and transformer models are adopted. Indeed, in this case, the inductance would be null, thereby preventing the unstable RLC resonator from originating.

Now, to further elaborate on the second comment (ii), consider the lower-frequency portion of the admittance in Figure 9. The admittance has a negative real part and positive imaginary part at frequencies less than 10 Hz. This means that undesired interactions among MMC1 and the electro-mechanical dynamics of synchronous generators and condensers can arise. These interactions are known as sub-synchronous oscillations (SSO) and are similar to sub-synchronous resonances [51–53]. Contrary to the previously mentioned instability issues at high frequencies, SSO cannot be effectively prevented by acting on L. Indeed, to do so, L should assume values impractical to obtain. In addition, as better shown in the following, SSOs depend on the MMC active power exchange.

To check the possible onset of an SSO we performed a long-lasting transient analysis of the IEEE14 benchmark, where at $t = 4\,$s the line between BUS2 and BUS5 was tripped as before (compared to other case studies, the time of occurrence of the disturbance was reduced to minimise the simulation effort, since the simulation of detailed MMC models is more CPU time consuming). We started with the same set-points used to compute impedances of MMCs. As previously mentioned, instabilities could be due to an unstable RLC resonator and/or SSOs. To confine the onset of possible unstable modes to only SSO, we did not consider any dynamics of passive elements (i.e., $L = 0$).

As it can be seen from the results in Figure 11, OLTCs intervene once again to restore load voltages and, thus, power. The voltage at BUS5, where the potentially critical MMC1 is connected, is also restored. Any stability problem was not evidenced at the current power level, as it was too low to trigger SSOs.

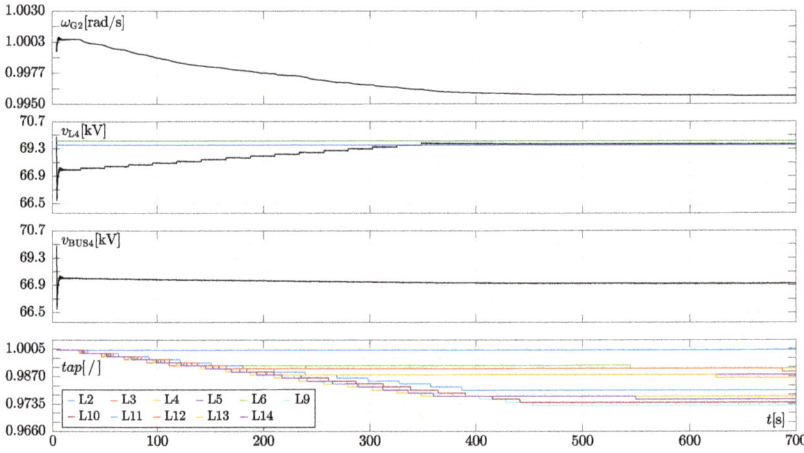

Figure 11. Simulation results of the modified IEEE14 test system shown in Figure 6 when the line between BUS2 and BUS5 is tripped at 4 s. In this case, an accurate MMC model is used for both MMC1 and MMC2. Constant impedance loads behind OLTCs and non-dynamic line and transformer models are used. The traces in each panel have the same meaning as those in Figure 3.

After having restored and stabilised bus voltages inside the dead-bands of OLTC, we started to slowly increase the power set point of the P/Q type MMC1. The slow increase of power is done to allow the system to adapt to the new working condition and to adequately identify the power level at which it becomes unstable. The corresponding results of the transient stability analysis are reported in Figure 12.

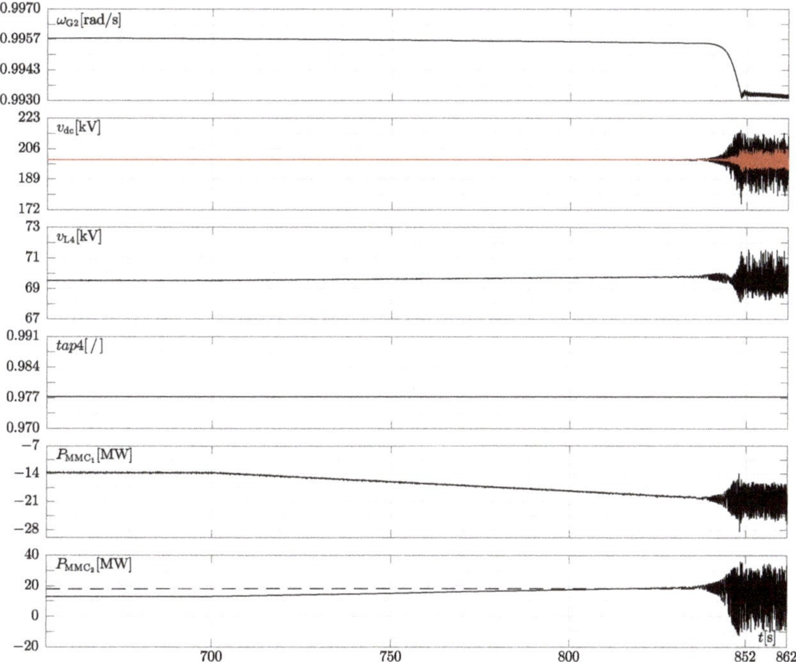

Figure 12. Simulation results of the modified IEEE14 test system shown in Figure 6 when the line between BUS2 and BUS5 is tripped at 4 s. The results shown in these plots are a sequel to those in Figure 11: MMC, load, line, and transformer models used are the same. In this case, the active power set-point of MMC1 changes linearly from 700 s and system instability arises when the power goes above 17 MW. Second panel from the top: pole-to-ground voltage of MMC1 (black trace) and MMC2 (red trace). Fifth and last panel from the top: active power exchange of MMC1 and MMC2. The dashed line in the last panels denotes the active power reference value above which the system becomes unstable (i.e., 17 MW). The traces in the other panels have the same meaning as those in Figure 3.

When the power goes above 17 MW (which is well below the final 60.7 MW target), instability occurs as shown by the divergence in the pole-to-ground voltage and power exchange of the MMCs. Since in this transient stability analysis we did not use dynamic models of lines and transformers, this unstable behaviour can only be ascribed to SSOs [51].

A possible countermeasure to prevent SSOs from originating (and, thus, damping the unstable electro-mechanical modes) consists of increasing the damping parameter of the synchronous generators and condensers. To validate this statement, we repeated the simulation and artificially increased the damping parameter of all the synchronous generators and condensers to 10.

The results obtained in this case are shown in Figure 13.

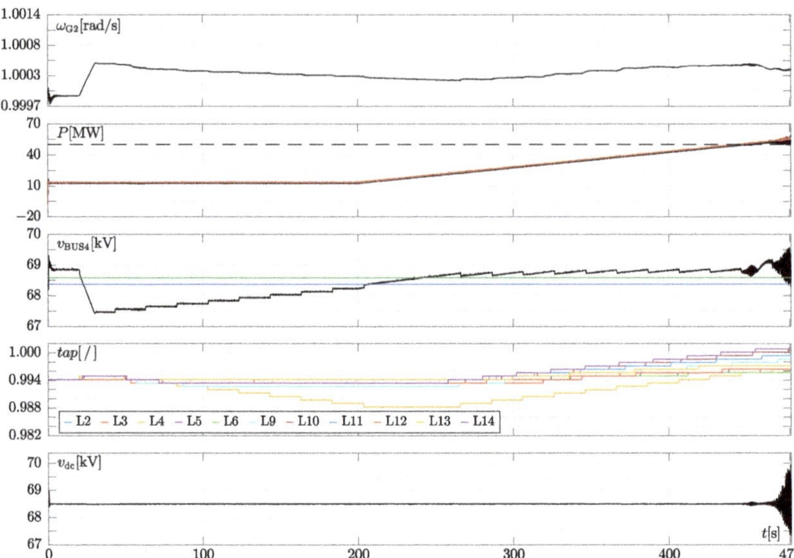

Figure 13. Simulation results of the modified IEEE14 test system shown in Figure 6 when the line between BUS2 and BUS5 is tripped at 4 s and the damping of all synchronous generators and condensers were set to 10. MMC, load, line, and transformer models used are the same as those of previous simulations. Analogously to Figure 12, the active power set-point of MMC1 changes linearly. The second panel from the top: active power exchange of MMC1 (black trace) and MMC2 (red trace). The dashed line denotes the active power reference value above which the system becomes unstable (i.e., 50 MW). Last panel: pole-to-pole voltage of MMC1. The traces in the other panels have the same meaning as those in Figure 3.

We see that during power ramp-up of MMC1 (and thus of MMC2) the OLTC at BUS4 increases its transformer ratio to bring voltage inside the dead band. Voltage increases since power voltage drops of transmission lines lower. The other OLTCs perform in a similar way. We see that stability is ensured up to $P = 50$ MW. If the power absorbed by the sending converter is further increased instability occurs, which means that increasing damping is effective only to a limited extent in preventing SSOs.

As a last comment, it is worth pointing out that all of these features would not be visible by adopting constant power loads. Moreover, also the model of the passive elements play a relevant role: indeed, if dynamic line and transformer representations were used, other instabilities than SSOs (i.e., attributable to an unstable RLC resonator) might arise.

5. Conclusions

Our analyses highlight that the well-established phasor analysis and single-phase equivalent models of power elements successfully used so far to study power system stability are no longer valid due to the ever-increasing penetration of IBRs (i.e., generation, load and energy conversion systems interfaced with the grid through power electronic converters). The simulated case studies suggest that a paradigm shift is necessary to study accurately modern electricity networks. In principle, detailed electro-magnetic transient (EMT) models of the entire system should be used to ensure accurate results are as adherent as possible to the real grid under study. However, this numerical approach is still impractical today, even on the most powerful computers, due to the scale and complexity of modern power grids, which require solving dynamic models composed of a very large number of equations and unknowns.

We believe that a more promising approach in this regard consists in suitably mixing single-phase dynamic models and accurate three-phase EMT models to perform hybrid phasor-EMT numerical analyses, thereby achieving a proper compromise between simulation speed and accuracy [54]. The former could represent non-critical parts of the grids (i.e., conventional power grids), while the latter could account for the critical ones (i.e., those comprising large shares of IBRs). However, the issue of how to efficiently partition the power grid in these two parts is still an open question and requires further analysis from our standpoint. According to our analyses, the coupling between conventional and modern grid elements lead to challenges in studying long-term dynamic stability. For example, despite being stable *per se* (i.e., their designs are correct and lead to stability if considered on their own), the connection of the IEEE14 and DCS1 HVDC benchmarks may lead to an overall unstable power system. In particular, this outcome also depends on the models of loads and passive components adopted, which need careful reconsideration due to the increasing presence of IBRs. This is the main challenge of modern and future IBR-dominated power grids.

Author Contributions: Conceptualization, A.M.B.; Formal analysis, D.d.G., F.B., S.G., D.L. and A.M.B.; Funding acquisition, S.G.; Investigation, A.M.B.; Methodology, A.M.B.; Software, A.M.B.; Validation, D.d.G., F.B., S.G., D.L. and A.M.B.; Visualization, D.d.G.; Writing—original draft, A.M.B.; Writing—review & editing, D.d.G., F.B., S.G., D.L. and A.M.B. All authors have read and agreed to the published version of the manuscript.

Funding: Italian MIUR project PRIN 2017K4JZEE_006 funded the work of S. Grillo (partially) and D. del Giudice (totally).

Data Availability Statement: Not applicable

Conflicts of Interest: The authors declare no conflict of interest.

References

1. Hatziargyriou, N.; Milanović, J.; Rahmann, C.; Ajjarapu, V.; Canizares, C.; Erlich, I.; Hill, D.; Hiskens, I.; Kamwa, I.; Pal, B.; et al. Definition and Classification of Power System Stability—Revisited & Extended. *IEEE Trans. Power Syst.* **2021**, *36*, 3271–3281. [CrossRef]
2. Musca, R.; Gonzalez-Longatt, F.; Gallego Sánchez, C.A. Power System Oscillations with Different Prevalence of Grid-Following and Grid-Forming Converters. *Energies* **2022**, *15*, 4273. [CrossRef]
3. Milanović, J.V.; Yamashita, K.; Martínez Villanueva, S.; Djokic, S.v.; Korunović, L.M. International Industry Practice on Power System Load Modeling. *IEEE Trans. Power Syst.* **2013**, *28*, 3038–3046. [CrossRef]
4. Korunović, L.M.; Milanović, J.V.; Djokic, S.Z.; Yamashita, K.; Villanueva, S.M.; Sterpu, S. Recommended Parameter Values and Ranges of Most Frequently Used Static Load Models. *IEEE Trans. Power Syst.* **2018**, *33*, 5923–5934. [CrossRef]
5. Pasiopoulou, I.; Kontis, E.; Papadopoulos, T.; Papagiannis, G. Effect of load modeling on power system stability studies. *Electr. Power Syst. Res.* **2022**, *207*, 107846. [CrossRef]
6. Overbye, T. Effects of load modelling on analysis of power system voltage stability. *Int. J. Electr. Power Energy Syst.* **1994**, *16*, 329–338. [CrossRef]
7. EL-Shimy, M.; Mostafa, N.; Afandi, A.; Sharaf, A.; Attia, M.A. Impact of load models on the static and dynamic performances of grid-connected wind power plants: A comparative analysis. *Math. Comput. Simul.* **2018**, *149*, 91–108. [CrossRef]
8. Zhang, X.P.; Chen, H. Analysis and selection of transmission line models used in power system transient simulations. *Int. J. Electr. Power Energy Syst.* **1995**, *17*, 239–246. [CrossRef]
9. Milano, F. *Power System Modelling and Scripting*; Power Systems; Springer: Berlin/Heidelberg, Germany, 2010.
10. Matevosyan, J.; MacDowell, J.; Miller, N.; Badrzadeh, B.; Ramasubramanian, D.; Isaacs, A.; Quint, R.; Quitmann, E.; Pfeiffer, R.; Urdal, H.; et al. A Future with Inverter-Based Resources: Finding Strength From Traditional Weakness. *IEEE Power Energy Mag.* **2021**, *19*, 18–28. [CrossRef]
11. Franquelo, L.G.; Rodriguez, J.; Leon, J.I.; Kouro, S.; Portillo, R.; Prats, M.A.M. The age of multilevel converters arrives. *IEEE Ind. Electron. Mag.* **2008**, *2*, 28–39. [CrossRef]
12. Lesnicar, A.; Marquardt, R. An innovative modular multilevel converter topology suitable for a wide power range. In Proceedings of the Power Tech Conference Proceedings, Bologna, Italy, 23–26 June 2003; Volume 3, pp. 6–10.
13. Dekka, A.; Wu, B.; Fuentes, R.L.; Perez, M.; Zargari, N.R. Evolution of topologies, modeling, control schemes, and applications of modular multilevel converters. *IEEE J. Emerg. Sel. Top. Power Electron.* **2017**, *5*, 1631–1656. [CrossRef]
14. Martinez-Rodrigo, F.; Ramirez, D.; Rey-Boue, A.; de Pablo, S.; Herrero-de Lucas, L. Modular Multilevel Converters: Control and Applications. *Energies* **2017**, *10*, 1709. [CrossRef]
15. Glover, J.D.; Sarma, M.S.; Overbye, T.J. *Power System Analysis and Design*; Brooks/Cole Publishing Co.: Pacific Grove, CA, USA, 2001.

16. Kundur, P. *Power System Stability and Control*; McGraw-Hill: New York, NY, USA, 1994.
17. Van Cutsem, R.; Papangelis, L. *Description, Modeling and Simulation Results of a Test System for Voltage Stability Analysis*; University of Liège: Liège, Belgium, 2013; pp. 1–16.
18. Milano, F. Hybrid Control Model of Under Load Tap Changers. *IEEE Trans. Power Deliv.* **2011**, *26*, 2837–2844. [CrossRef]
19. DIgSILENT. *PowerFactory User Manual*; DIgSILENT GmbH: Gomaringen, Germany, 2020.
20. Dommel, H.W. Digital Computer Solution of Electromagnetic Transients in Single-and Multiphase Networks. *IEEE Trans. Power Appar. Syst.* **1969**, *PAS-88*, 388–399. [CrossRef]
21. Murad, M.A.A.; Mele, F.M.; Milano, F. On the Impact of Stochastic Loads and Wind Generation on Under Load Tap Changers. In Proceedings of the 2018 IEEE Power Energy Society General Meeting (PESGM), Portland, OR, USA, 5–10 August 2018; pp. 1–5. [CrossRef]
22. Liu, M.; Bizzarri, F.; Brambilla, A.M.; Milano, F. On the Impact of the Dead-Band of Power System Stabilizers and Frequency Regulation on Power System Stability. *IEEE Trans. Power Syst.* **2019**, *34*, 3977–3979. [CrossRef]
23. Fudeh, H.; Ong, C.M. A Simple and Efficient AC-DC Load-Flow Method for Multiterminal DC Systems. *IEEE Trans. Power Appar. Syst.* **1981**, *PAS-100*, 4389–4396. [CrossRef]
24. Sanghavi, H.A.; Banerjee, S.K. Load flow analysis of integrated AC-DC power systems. In Proceedings of the Fourth IEEE Region 10 International Conference TENCON, Bombay, India, 22–24 November 1989; pp. 746–751.
25. Feng, W.; Yuan, C.; Shi, Q.; Dai, R.; Liu, G.; Wang, Z.; Li, F. Using virtual buses and optimal multipliers to converge the sequential AC/DC power flow under high load cases. *Electr. Power Syst. Res.* **2019**, *177*, 106015. [CrossRef]
26. del Giudice, D.; Brambilla, A.; Linaro, D.; Bizzarri, F. Modular Multilevel Converter Impedance Computation Based on Periodic Small-Signal Analysis and Vector Fitting. *IEEE Trans. Circuits Syst. I Regul. Pap.* **2022**, *69*, 1832–1842. [CrossRef]
27. Cigré Working Group B4.57. *Guide for the Development of Models for HVDC Converters in a HVDC Grid*; CIGRÉ (WG Brochure No. 604); CIGRÉ: Paris, France, 2014.
28. Wang, S.; Bao, D.; Gontijo, G.; Chaudhary, S.; Teodorescu, R. Modeling and Mitigation Control of the Submodule-Capacitor Voltage Ripple of a Modular Multilevel Converter under Unbalanced Grid Conditions. *Energies* **2021**, *14*, 651. [CrossRef]
29. Wu, H.; Wang, X. Dynamic Impact of Zero-Sequence Circulating Current on Modular Multilevel Converters: Complex-Valued AC Impedance Modeling and Analysis. *IEEE J. Emerg. Sel. Top. Power Electron.* **2020**, *8*, 1947–1963. [CrossRef]
30. Peralta, J.; Saad, H.; Dennetiere, S.; Mahseredjian, J.; Nguefeu, S. Detailed and Averaged Models for a 401-Level MMC–HVDC System. *IEEE Trans. Power Deliv.* **2012**, *27*, 1501–1508. [CrossRef]
31. Khan, S.; Tedeschi, E. Modeling of MMC for Fast and Accurate Simulation of Electromagnetic Transients: A Review. *Energies* **2017**, *10*, 1161. [CrossRef]
32. del Giudice, D.; Brambilla, A.; Linaro, D.; Bizzarri, F. Isomorphic Circuit Clustering for Fast and Accurate Electromagnetic Transient Simulations of MMCs. *IEEE Trans. Energy Convers.* **2022**, *37*, 800–810. [CrossRef]
33. Xu, J.; Ding, H.; Fan, S.; Gole, A.M.; Zhao, C.; Xiong, Y. Ultra-fast electromagnetic transient model of the modular multilevel converter for HVDC studies. In Proceedings of the 12th IET International Conference on AC and DC Power Transmission (ACDC 2016), Beijing, China, 28–29 May 2016; pp. 1–9. [CrossRef]
34. Meng, X.; Han, J.; Pfannschmidt, J.; Wang, L.; Li, W.; Zhang, F.; Belanger, J. Combining Detailed Equivalent Model with Switching-Function-Based Average Value Model for Fast and Accurate Simulation of MMCs. *IEEE Trans. Energy Convers.* **2020**, *35*, 484–496. [CrossRef]
35. Song, G.; Wang, T.; Huang, X.; Zhang, C. An improved averaged value model of MMC-HVDC for power system faults simulation. *Int. J. Electr. Power Energy Syst.* **2019**, *110*, 223–231. [CrossRef]
36. Freytes, J.; Papangelis, L.; Saad, H.; Rault, P.; Van Cutsem, T.; Guillaud, X. On the modeling of MMC for use in large scale dynamic simulations. In Proceedings of the 2016 Power Systems Computation Conference (PSCC), Genova, Italy, 20–24 June 2016; pp. 1–7. [CrossRef]
37. Zhu, S.; Liu, K.; Qin, L.; Ran, X.; Li, Y.; Huai, Q.; Liao, X.; Zhang, J. Reduced-Order Dynamic Model of Modular Multilevel Converter in Long Time Scale and Its Application in Power System Low-Frequency Oscillation Analysis. *IEEE Trans. Power Deliv.* **2019**, *34*, 2110–2122. [CrossRef]
38. Guo, D.; Rahman, M.H.; Ased, G.P.; Xu, L.; Emhemed, A.; Burt, G.; Audichya, Y. Detailed quantitative comparison of half-bridge modular multilevel converter modelling methods. *J. Eng.* **2019**, *2019*, 1292–1298. [CrossRef]
39. Bizzarri, F.; Giudice, D.D.; Linaro, D.; Brambilla, A. Partitioning-Based Unified Power Flow Algorithm for Mixed MTDC/AC Power Systems. *IEEE Trans. Power Syst.* **2021**, *36*, 3406–3415. [CrossRef]
40. Beerten, J.; Belmans, R. Development of an open source power flow software for high voltage direct current grids and hybrid AC/DC systems: MATACDC. *IET Gener. Transm. Distrib.* **2015**, *9*, 966–974. [CrossRef]
41. Aik, D.; Andersson, G. Use of participation factors in modal voltage stability analysis of multi-infeed HVDC systems. *IEEE Trans. Power Deliv.* **1998**, *13*, 203–211. [CrossRef]
42. Abed, E.H.; Lindsay, D.; Hashlamoun, W.A. On participation factors for linear systems. *Automatica* **2000**, *36*, 1489–1496. [CrossRef]
43. Perez-arriaga, I.J.; Verghese, G.C.; Schweppe, F.C. Selective Modal Analysis with Applications to Electric Power Systems, PART I: Heuristic Introduction. *IEEE Trans. Power Appar. Syst.* **1982**, *PAS-101*, 3117–3125. [CrossRef]

44. del Giudice, D.; Bizzarri, F.; Linaro, D.; Brambilla, A. Stability Analysis of MMC/MTDC Systems Considering DC-Link Dynamics. In Proceedings of the 2021 IEEE International Symposium on Circuits and Systems (ISCAS), Daegu, Korea, 22–28 May 2021; pp. 1–5. [CrossRef]
45. Wu, H.; Wang, X.; Kocewiak, L.H. Impedance-Based Stability Analysis of Voltage-Controlled MMCs Feeding Linear AC Systems. *IEEE J. Emerg. Sel. Top. Power Electron.* **2020**, *8*, 4060–4074. [CrossRef]
46. Zhu, S.; Qin, L.; Liu, K.; Ji, K.; Li, Y.; Huai, Q.; Liao, X.; Yang, S. Impedance Modeling of Modular Multilevel Converter in D-Q and Modified Sequence Domains. *IEEE J. Emerg. Sel. Top. Power Electron.* **2022**, *10*, 4361–4382. [CrossRef]
47. Zhu, S.; Liu, P.; Liao, X.; Qin, L.; Huai, Q.; Xu, Y.; Li, Y.; Wang, F. D-Q Frame Impedance Modeling of Modular Multilevel Converter and Its Application in High-Frequency Resonance Analysis. *IEEE Trans. Power Deliv.* **2021**, *36*, 1517–1530. [CrossRef]
48. Linaro, D.; del Giudice, D.; Bizzarri, F.; Brambilla, A. PanSuite: A free simulation environment for the analysis of hybrid electrical power systems. *Electr. Power Syst. Res.* **2022**, *212*, 108354. [CrossRef]
49. Bizzarri, F.; Brambilla, A. PAN and MPanSuite: Simulation Vehicles Towards the Analysis and Design of Heterogeneous Mixed Electrical Systems. In Proceedings of the NGCAS, Genova, Italy, 6–9 September 2017; pp. 1–4.
50. Zong, H.; Zhang, C.; Lyu, J.; Cai, X.; Molinas, M.; Rao, F. Generalized MIMO Sequence Impedance Modeling and Stability Analysis of MMC-HVDC with Wind Farm Considering Frequency Couplings. *IEEE Access* **2020**, *8*, 55602–55618. [CrossRef]
51. Damas, R.N.; Son, Y.; Yoon, M.; Kim, S.Y.; Choi, S. Subsynchronous Oscillation and Advanced Analysis: A Review. *IEEE Access* **2020**, *8*, 224020–224032. [CrossRef]
52. Ghaffarzadeh, H.; Mehrizi-Sani, A. Review of control techniques for wind energy systems. *Energies* **2020**, *13*, 6666. [CrossRef]
53. He, C.; Sun, D.; Song, L.; Ma, L. Analysis of subsynchronous resonance characteristics and influence factors in a series compensated transmission system. *Energies* **2019**, *12*, 3282. [CrossRef]
54. Linaro, D.; del Giudice, D.; Brambilla, A.; Bizzarri, F. Application of Envelope-Following Techniques to the Simulation of Hybrid Power Systems. *IEEE Trans. Circuits Syst. I Regul. Pap.* **2022**, *69*, 1800–1810. [CrossRef]

Article

Probabilistic Description of the State of Charge of Batteries Used for Primary Frequency Regulation

Elio Chiodo [1], Davide Lauria [1,*], Fabio Mottola [2], Daniela Proto [2], Domenico Villacci [1], Giorgio Maria Giannuzzi [3] and Cosimo Pisani [3]

[1] Department of Industrial Engineering, University of Naples Federico II, 80125 Naples, Italy
[2] Department of Electrical Engineering and Information Technology, University of Naples Federico II, 80125 Naples, Italy
[3] Terna Italian Transmission System Operator, 00156 Rome, Italy
* Correspondence: davide.lauria@unina.it

Abstract: Battery participation in the service of power system frequency regulation is universally recognized as a viable means for counteracting the dramatic impact of the increasing utilization of renewable energy sources. One of the most complex aspects, in both the planning and operation stage, is the adequate characterization of the dynamic variation of the state of charge of the battery in view of lifetime preservation as well as the adequate participation in the regulation task. Since the power system frequency, which is the input of the battery regulation service, is inherently of a stochastic nature, it is easy to argue that the most proper methodology for addressing this complex issue is that of the theory of stochastic processes. In the first part of the paper, a preliminary characterization of the power system frequency is presented by showing that with an optimal degree of approximation it can be regarded as an Ornstein–Uhlenbeck process. Some considerations for guaranteeing desirable performances of the control strategy are performed by assuming that the battery-regulating power depending on the frequency can be described by means of a Wiener process. In the second part of the paper, more realistically, the regulating power due to power system changes is described as an Ornstein–Uhlenbeck or an exponential shot noise process driven by a homogeneous Poisson process depending on the frequency response features requested of the battery. Because of that, the battery state of charge is modeled as the output of a dynamic filter having this exponential shot noise process as input and its characterization constitutes the central role for the correct characterization of the battery life. Numerical simulations are carried out for demonstrating the goodness and the applicability of the proposed probabilistic approach.

Keywords: primary frequency regulation; battery energy storage system; Ornstein–Uhlenbeck stochastic process; compound poisson stochastic process

Citation: Chiodo, E.; Lauria, D.; Mottola, F.; Proto, D.; Villacci, D.; Giannuzzi, G.M.; Pisani, C. Probabilistic Description of the State of Charge of Batteries Used for Primary Frequency Regulation. *Energies* **2022**, *15*, 6508. https://doi.org/10.3390/en15186508

Academic Editor: Alvaro Caballero

Received: 23 July 2022
Accepted: 3 September 2022
Published: 6 September 2022

Publisher's Note: MDPI stays neutral with regard to jurisdictional claims in published maps and institutional affiliations.

1. Introduction

The problem of continuously and instantaneously balancing generation and load is the most important requirement in power system operation. Traditionally, this service has been performed by power plants of high-rated active power (in Italy, for instance, by those whose rated power is greater than 10 MW) [1]. With the integration of renewable source generators which have priority dispatch, the assurance of stability of the power system is even more critical due to the intermittency typical of such systems [2,3]. Storage technologies seem to be ideal candidates for such service thanks to their ability to track rapid fluctuations encountered in power systems and to their relatively high-power capability, thus being even more appropriate than conventional generators. Conventional generators based on fossil fuels are slower to respond to the operator signal and a require higher and redundant frequency reserve compared to storage systems which are able to provide more accurate regulation. Another important aspect is related to the ability of electrical energy storage, unlike conventional power plants, to avoid the emission of greenhouse gases [4]. While there are many technologies which can be considered suitable for such operation, those based on electrochemical conversion, i.e., battery energy storage systems (BESS), seem to be interesting options thanks to their high ramp rates and their reliability [5].

The viability of BESSs for primary frequency regulation depends on a few issues mainly related to their economic feasibility as well as existing regulations. The U.S. has already moved towards the implementation of battery storage, motivated by the increased variable renewable energy due to state renewable portfolio standards, the falling costs of solar photovoltaic power, the increasing retail electricity rates and regulations that value and pay for fast frequency response [4]. Other challenging aspects related to the use of BESSs refer to technical tasks. The most challenging requirements of BESSs in providing primary frequency regulation are in maintaining their state of charge (SoC) at a level which guarantees capacity availability and preserving the battery lifetime. The considered SoC reestablishing strategy, in fact, significantly influences the lifetime of the battery.

In the technical literature, several proposals have been presented for the use of BESSs in primary frequency regulation aimed at preserving the battery life. In [1], a specific control strategy has been proposed where a variable-droop mode is considered, aimed at exploiting the fast response capability of the BESS when SoC is in a good state while working at a minimum when SoC is limited. In [6], a primary frequency modulation control strategy is proposed based on fuzzy control which also considers the SoC state self-recovery of energy storage. A control algorithm with adjustable SoC limits and the application of dissipation resistors is proposed in [7]. In [8], five strategies for delivering the primary frequency regulation and reestablishing the systems' SoC have been investigated from the battery lifetime perspective. The study in [9] proposes a control system model to simulate the operation of BESS, and the design of a controller whose strategy is based on model predictive control to ensure the optimal operation of the BESS. A frequency predictor based on the Grey model is also designed to improve the performance of the predictive controller. The prediction model involved can predict frequency multiple steps ahead, this information being used in the optimization part of the controller. An analysis of the incidence of the control law on the battery lifetime is proposed in [10] where a proper framework was tailored to take into account the stochastic nature of the system frequency. In [11], a two-stage stochastic control approach is proposed to optimize the charging of batteries on board the electric vehicles of a public charging station to provide frequency regulation and energy arbitrage. In [12], the profitability of the use of electric vehicles for the frequency regulation service is discussed together with a control strategy aimed at counteracting the battery lifetime degradation. An adaptive droop coefficient and SoC balance-based primary frequency modulation control strategy for energy storage is proposed in [13] which controls the SoC of the BESS and adaptively adjusts the depth of energy storage output to prevent the saturation or exhaustion of energy storage SoC. In [14], a droop control for the BESS is proposed which includes the SoC feedback with the aim of properly managing the SoC profile of multiple battery devices. An adaptive droop control method of a BESS is also proposed in [15] which allows for the recovery of the desired SoC level through a proper feedback action. Penalties function-based control is adopted in [16], where the management of the SoC allows access to potential reserves.

The description of the SoC is then a fundamental task when performing primary frequency control with BESSs. Proper control strategies are required in some approaches [13–16] to adaptively adjust the participation of the battery to the frequency regulation service while guaranteeing that the SoC falls within proper ranges. In that regard, the approach proposed in this paper allows for the regulation of the charging/discharging power of the battery according to the SoC of the battery through a gain parameter. The main novelty of this approach lies in the stochastic nature of this parameter's design, which is based on a stochastic analysis of the system frequency and of the corresponding regulating power requested by the primary frequency control. For this purpose, a novel approach is proposed to analyze the stochastic processes of the SoC based on the Ornstein–Uhlenbeck and compound Poisson processes. More in depth, a dynamic stochastic model is proposed for the BESS control where the regulating power is described as an Ornstein–Uhlenbeck or an exponential shot noise process driven by a homogeneous Poisson process depending on the frequency response features requested of the battery. The BESS SoC is modeled according to an Ornstein–Uhlenbeck process driven by an Ornstein–Uhlenbeck process. The accuracy of this approach has been investigated together with the definition of a gain parameter, which can be set appropriately to optimize the control of the battery for the balance between the availability of regulation service and battery lifetime degradation. Based on the derivation of this gain parameter, a BESS control strategy is proposed which allows for the control of the power exchanged with the grid to provide regulation while guaranteeing the proper profile of SoC according to the stochastic behavior of the frequency.

Compared to the literature, the main outcomes of this paper refer to the stochastic analysis of the BESS contribution to the primary frequency regulation which allows the identification of an estimation method for the SoC profiles as well as proper design of adaptive control techniques that properly manage the provision of the regulation service while keeping SoC within admissible and appropriate ranges. Possible applications of the proposed stochastic approach are multiple since it provides a tool (i) for the accurate identification of the battery lifetime, (ii) for the optimal design of control techniques, and (iii) for the optimal sizing of the BESS, despite the uncertainties affecting the frequency regulation problem.

The paper is organized as follows: Section 2 gives a few details on the use of BESSs for primary frequency regulation; in Section 3 the stochastic characterization of the power system frequency is carried out, while Section 4 describes the statistical characterization of BESS's SoC. The results of numerical applications are reported in Section 5. Our conclusions are drawn in Section 6.

2. Primary Frequency Regulation Employing Battery Energy Storage Systems

When perturbations occur, generation/load balance is not guaranteed, thus control of the frequency is required. This can be achieved by means of storage devices whose control has to be properly defined. In what follows, the problem of frequency regulation is introduced first, and then the use of BESS for such purpose is detailed.

2.1. The Problem of Frequency Regulation

Time evolution of frequency strictly depends on the degree of generation/load unbalance, on the system rotating reserve available when a disturbance occurred and on the features of the frequency regulation.

Regarding the available rotating reserve, R, an estimation can be provided by means of the following equation:

$$R = \sum_{j=1}^{n} \left(P_{nj} - P_j^0 \right) \tag{1}$$

where n is the number of groups in service after the perturbation, P_j^0 and P_{nj} are the power delivered by the jth group at the time of the perturbation and its nominal active power, respectively. It has to be noted that, typically, R assumes a value which is about 5% of the network nominal power, this resulting from economic considerations.

Based on the hypothesis of linearity, the frequency behavior depends on the step load perturbation ΔP_L, and, more specifically, it can be derived in terms of variations of equivalent load ΔP_L. In case of perturbations of the generated power (such as, for instance, large intermittences of renewable generation or disconnection of groups) or, more generally, in case of relevant imbalance, frequency reaches lower limits implying load disconnection. Control actions are then required to prevent frequency's inadmissible values.

The equation of the primary frequency regulation is given by the well-known transfer function of the regulator [17]:

$$\left.\frac{\Delta P_r}{\Delta f}\right|_{\Delta f_{ref}=0} = -E_P \frac{1+sT_2}{1+sT_1} \tag{2}$$

where Δf_{ref} is the variation of the frequency reference, ΔP_r is the regulating power, T_1 and T_2 are the time constants of the control, and E_P, which is referred to as permanent regulating energy (MW/Hz), is the ratio, at steady-state conditions, of the variation of regulating power and the corresponding variation of frequency, with the sign changed.

The frequency deviation $\left(\frac{\Delta f}{f_n}\right)$ can be derived from the power deviation $\left(\frac{\Delta P_L}{P_n}\right)$ through the transfer function, $G(s)$, as follows (Figure 1):

$$G(s) = \frac{1}{sT_a + \frac{1}{\sigma_c} + \frac{1}{\sigma_P}\frac{1+sT_2}{1+sT_1}} \tag{3}$$

where σ_P is the permanent droop consequent to the generation unit regulation, which is defined, in steady state condition, as the ratio between the relative variations of frequency and regulating power, with the sign changed; σ_c is the permanent load droop, which is related to the variation of load power consequent to frequency variation; T_a is the starting time of the unit, in seconds, linked to the inertia constant, H, from the simple relation $T_a = 2H$.

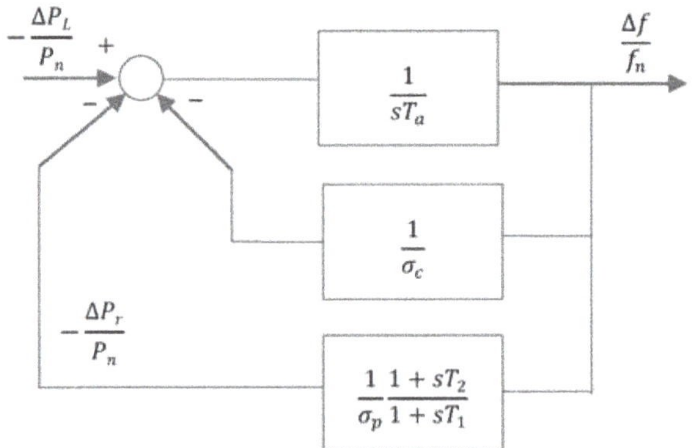

Figure 1. Block diagram for analyzing the frequency transients.

The characteristic equation of the transfer function (3) has complex and conjugated poles, resulting in oscillatory dynamics of frequency transients with pulsation ω_0 [17]. The frequency nadir corresponding to a step function of the load variation, $\Delta P_L = d/s$, can be evaluated as [17]:

$$f_{min} = f_n \left\{ 1 - \frac{\sigma_P}{1+\frac{\sigma_P}{\sigma_c}} d \left[1 + e^{-\frac{T_2+\sigma_P(T_a+\frac{T_1}{\sigma_c})}{2T_aT_1\sigma_P}t_p} \sqrt{\frac{T_1-T_2}{T_a\sigma_P}} \right] \right\} \tag{4}$$

where

$$t_p = \frac{1}{\omega_0}\left\{ \frac{\pi}{2} - arctg\left[\omega T_2 + \omega\sigma_P\left(\frac{T_1}{\sigma_c} - T_a\right) \right] \right\} \tag{5}$$

with $\omega_0 = \frac{1}{2\,T_aT_1\sigma_P\omega}$ and $\omega = \frac{1}{\sqrt{4\left(1+\frac{\sigma_P}{\sigma_c}\right)T_aT_1\sigma_P - \left[T_2+\sigma_P\left(T_a+\frac{T_1}{\sigma_c}\right)\right]^2}}$.

This value has to be maintained as close as possible to the frequency nominal value, thereby facilitating the action of the frequency regulators. Analogously, the frequency zenith value also must be maintained as close as possible to the nominal frequency value, this being a particularly critical issue due to the widespread presence of renewable generators in the power systems.

2.2. Use of the Battery Energy Storage System for Frequency Regulations

In this sub-section, the use of BESS for frequency regulation service is discussed. In particular, the fundamentals of the regulation service are described with respect to the requirements usually requested by the system operators. Then, a background of the BESS control for primary frequency regulation service is reported.

To keep the frequency deviation within admissible ranges, the charge and discharge of the battery can be controlled for the primary frequency regulation service. The regulation action of the storage device, which is shown in Figure 2, requires that the BESS be charged/discharged (absorbing/injecting power to the network) when frequency deviation exceeds the dead band $[f_a, f_b]$. The values of frequency requiring regulation are those lower than f_a, where BESS is required to be discharged (i.e., up-regulation), and those greater than f_b, where BESS is required to be charged (i.e., down-regulation). Regarding the rate at which the BESS has to be discharged and charged, it is required that the power be linearly increasing with frequency for values greater than f_{min} and lower than f_{max} and assumes its maximum value when the frequency exceeds the range $[f_{min}, f_{max}]$.

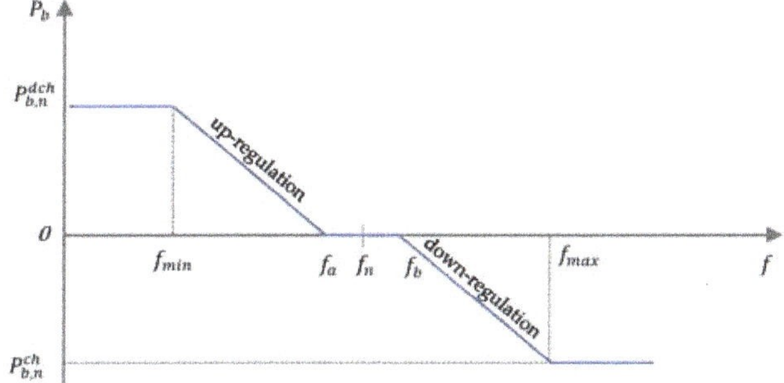

Figure 2. The action of the BESS for the primary frequency regulation service.

Typically, the battery control strategy is aimed at:

- controlling the charging/discharging power according to the up- and down-regulation,
- recovering the SOC according to a reference value (typically set at 0.5 p.u.), when frequency falls within the dead-band, and
- keeping the SOC within a range which avoids battery degradation (e.g., $[0.1, 0.9]$ p.u.).

The block diagram of the simulation set-up of a possible control strategy for the primary frequency regulation is shown in Figure 3 [10].

In Figure 3, two main parts can be identified, the former (lower part of the figure) which refers to the down-regulation which applies when $f < f_n$ and $|\Delta f| > \Delta f_a$ ($\Delta f_a = f_n - f_a$), the latter referring to the up-regulation which applies when $f > f_n$ and $|\Delta f| > \Delta f_b$ ($\Delta f_b = f_b - f_n$). The values of Δf_a and Δf_b are very important since they are related to both regulatory and control aspects [8,18,19].

It has to be noted that, to perform down-regulation, SoC must be lower than ΔSoC_{max} since, in this case, the BESS is requested to absorb power from the grid whose value depends on the difference between reference and actual frequencies and on the BESS setting droop σ_{batt} [20]. Similarly, for the up-regulation, SoC is required to be greater than ΔSoC_{min} since the BESS is requested to inject power to the network, whose value again depends on the difference between reference and actual frequencies and on the BESS setting droop, σ_{batt}.

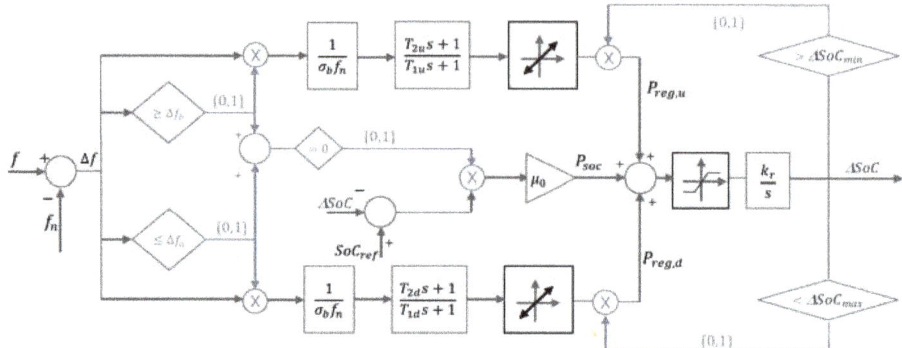

Figure 3. Block diagram of the BESS control strategy for primary frequency regulation.

The time constants of the transfer function of the BESS down-regulation are T_{2d} and T_{1d}, and those of the transfer function of the BESS up-regulation are T_{2u} and T_{1u}.

A limitation is imposed to down- and up-regulations due to the BESS rated power.

Regarding the control, in the case the frequency falls within the dead-band, the BESS can be charged or discharged to recover the reference value of the SoC. In this case, the transfer function can be seen as a proportional gain μ_0, which has a remarkable role. In this case, the transfer function can be seen as a proportional gain μ_0, which has a remarkable role. The value of the parameter μ_0, in fact, affects the speed of the SoC recovery and, in turn, the ability of the BESS to have enough stored energy to provide frequency regulation when required. On the other hand, the speed of SoC recovery affects the lifetime of the battery, since it implies a greater number of charging/discharging cycles. As a consequence of that, a trade-off between the regulation service availability of the BESS and the battery lifetime duration is required according to technical and/or economic issues [10].

In the control approaches such as that in Figure 3, the presence of the dead-band clearly influences the control model, since it implies the capability of the battery to recover the SoC when frequency falls within the dead-band. In this case, the dead-band width also influences the ability of SoC recovering, thus, in some cases it makes the SoC prone to exceeding the admissible range values and makes the battery unavailable to provide frequency regulation. The control based on the adaptive modification of the battery contribution based on the SoC, instead, is not affected by the presence of the dead-band if a correct identification of the SoC is carried out which allows optimization of the battery performance. At this aim, the control strategy for the use of BESS for frequency regulation can be summarized in Figure 4 where two block diagrams are shown, one referring to the regulation power P_{reg} provided by the battery for frequency regulation and the other referring to the power P_{SoC} requested to recover the reference value of SoC.

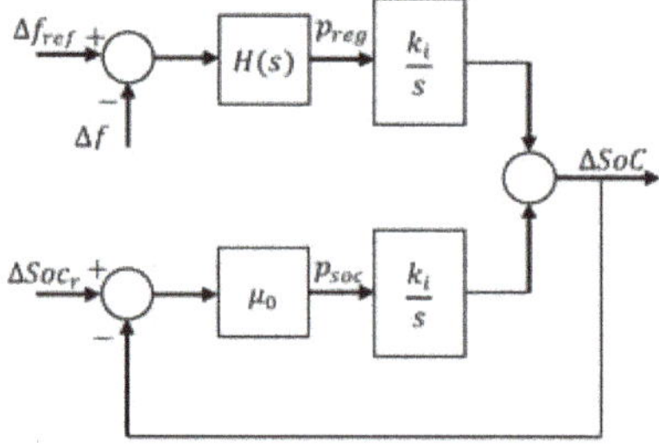

Figure 4. Block diagram of the battery control strategy.

The aforementioned power terms contribute to the total power provided by the battery, P_{bess}, which is then given by:

$$P_{bess}(t) = P_{reg}(t) + P_{SoC}(t) \tag{6}$$

and which determines the following value of SoC when the battery efficiency is neglected:

$$SoC(t) = SoC^0 - \frac{1}{E_n} \int_0^t P_{bess}(\tau)d\tau \tag{7}$$

being SoC^0 the value of SoC at $t = 0$ and E_n the rated capacity of the battery.

By substituting (6) in (7) and multiplying and dividing by P_n, the SoC value is derived as:

$$SoC(t) = SoC^0 - \frac{P_n}{E_n} \int_0^t \frac{P_{reg}(\tau) + P_{SoC}(\tau)}{P_n}d\tau \tag{8}$$

and its variation $\Delta SoC(t)$, with respect to the initial values SoC^0 is given by:

$$\Delta SoC(t) = k_i \int_0^t \left[p_{reg}(\tau) + p_{SoC}(\tau) \right]d\tau \tag{9}$$

where p_{reg} and p_{SoC} are the relative values of P_{reg} and P_{SoC} with respect to the rated power P_n, and $k_i = \frac{P_n}{E_n}$.

The Laplace transform of $\Delta SoC(t)$, $\Delta SoC(s)$, is:

$$\Delta SoC(s) = \frac{k_i}{s} p_{reg}(s) + \frac{k_i}{s} p_{SoC}(s) \tag{10}$$

with obvious meaning of the symbols.

According to the block diagram of Figure 4 the Laplace transforms $p_{reg}(s)$ and $p_{soc}(s)$ are:

$$p_{reg}(s) = H(s)\left(\Delta f_{ref} - \Delta f \right) \tag{11}$$

$$p_{SoC}(s) = R(s)\left(\Delta SoC_{ref} - \Delta SoC \right) \tag{12}$$

where $H(s)$ and $R(s)$ are the transfer functions of the regulators, $\Delta f_{ref} = 0$, and $\Delta SoC_{ref} = 0$. This last referring to the case in which the initial value of SoC is assumed to be equal to its reference value.

3. Stochastic Characterization of Power System Frequency

In this section, the fundamentals of the stochastic processes typically adopted to describe the power system frequency are reported and discussed.

To statistically analyze the output of the BESS control subject to the frequency variation, frequency's stochastic characterization is required based on available historical data. Typically, these data are provided in terms of discrete time series including sets of observations $f(t)$ referred to specified times $t \in T_0$ with T_0 the set of all the sample times. Two main approaches can be used to reproduce the most important statistical features of the available frequency data which proved to be characterized by a stochastic nature, i.e., logistic autoregressive and Ornstein–Uhlenbeck [10,21]. In what follows, details on these two approaches are given.

3.1. Logistic Autoregressive Mode

In [10], time series of available recorded data of the frequency deviations were analyzed which were represented by an appropriated autoregressive model, this being proven to be a solution to infer data from time series [22–24].

Starting from the more general autoregressive moving average process, the autoregressive model, in case of normal underlying distribution, can be described as follows:

$$\phi(B)Y_t = \theta(B)Z_t, \quad \{Z_t\} \sim WN\left(0, \sigma^2\right) \tag{13}$$

where Y_t is the stochastic process which in this case is related to the frequency deviation, B is the backward shift operator, $\phi = (\phi_1, \dots, \phi_p)$ and $\theta = (\theta_1, \dots, \theta_q)$ are the parameters of polynomials of orders p and q, these last being assumed to be known, and σ^2 is the variance of the white noise, WN, whose estimation is based on available time series.

In case of pure autoregressive models (i.e., with $\theta(z) \equiv 1$), the following expression of estimation of the process (13) can be used:

$$y_t - \hat{\phi}_1 y_{t-1} - \dots - \hat{\phi}_p y_{t-p} = z_t, \quad \{z_t\} \sim WN\left(0, \hat{\sigma}^2\right) \tag{14}$$

with $(\hat{\phi}_1, \dots, \hat{\phi}_p)$ and $\hat{\sigma}^2$ the estimators of $(\phi_1, \dots, \phi_p)$ and σ^2, respectively.

Based on accurate analysis of the actual frequency data, in [10], the logistic distribution has been evidenced to be more appropriate than Gaussian, as underlying distribution, and the following first order polynomial approximation has been proposed:

$$y_t - \hat{\phi} y_{t-1} = \hat{\mu} + \hat{a}_t \tag{15}$$

with $\hat{\mu}$ the estimation of the mean value μ of the frequency deviation and $\hat{a}_t$ distributed according to the symmetric logistic distribution, whose parameters are location (which is equal to zero being the distribution symmetric), and scale, σ_a.

By referring to the process (15), parameters ϕ and μ can be estimated through the solution of a system of equation which, in case of first order polynomial approximation, reads [25]:

$$\begin{bmatrix} \hat{\mu} \\ \hat{\phi} \end{bmatrix} = \begin{bmatrix} n_s & \sum y_{t-1} \\ \sum y_{t-1} & \sum y_t^2 \end{bmatrix}^{-1} \begin{bmatrix} \sum y_t \\ \sum y_t\, y_{t-1} \end{bmatrix} \tag{16}$$

with n_s the number of samples.

When referring to the process (15), where three parameters need to be evaluated, the parameters obtained by solving (16) provide a partial estimation of the process, since the scale factor σ_a is not included in (16). Although partial, this first estimation of μ and ϕ is useful to support classical statistic methods which allow obtaining the complete estimation of the process. At this purpose, the maximum likelihood estimation method can be adopted, which is based on the definition of z_t as:

$$z_t = (1/\sigma_a)(y_t - \phi y_{t-1}) - \mu. \tag{17}$$

The likelihood function of (15) is then [10]:

$$L(\phi, \mu, \sigma_a) = (1/\sigma_a)^{n_s} \prod_{t=1}^{n_s} \frac{e^{-z_t}}{(1 + e^{-z_t})^2}. \tag{18}$$

A set of equations in the unknown parameters μ, ϕ and σ_a can be then derived by imposing the partial derivatives of the log-likelihood function equal to zero:

$$\begin{cases} \dfrac{\partial \ln L(\phi,\mu,\sigma_a)}{\partial \mu} = 0 \\[2mm] \dfrac{\partial \ln L(\phi,\mu,\sigma_a)}{\partial \phi} = 0 \\[2mm] \dfrac{\partial \ln L(\phi,\mu,\sigma_a)}{\partial \sigma_a} = 0 \end{cases} \tag{19}$$

In particular, by substituting the expression (18) of $L(\phi, \mu, \sigma_a)$ in (19), the following system of equations is obtained [10]:

$$\begin{cases} n_s - 2 \sum\limits_{t=1}^{n_s} \dfrac{1}{1+e^{z_t}} = 0 \\[3mm] \sum\limits_{t=1}^{n_s} y_{t-1} - 2 \sum\limits_{t=1}^{n_s} y_{t-1} \dfrac{1}{1+e^{z_t}} = 0 \\[3mm] n_s - \sum\limits_{t=1}^{n_s} z_t + 2 \sum\limits_{t=1}^{n_s} z_t \dfrac{1}{1+e^{z_t}} = 0 \end{cases} \tag{20}$$

whose solution allows eventually us to obtain the estimation of the unknown parameters μ, ϕ and σ_a.

3.2. Ornstein–Uhlenbeck

In the literature, the Ornstein–Uhlenbeck stochastic process, has proven to be the most appropriate process for characterizing power system frequency random variations (e.g., [22]).

The stochastic differential equation which describes the Ornstein–Uhlenbeck process Y_t is:

$$dY_t = \tau_{ou}(\mu_{ou} - Y_t)dt + \sigma_{ou}dW_t \tag{21}$$

which applies for $t \geq 0$ and where μ_{ou} is the long-term mean, τ_{ou} is a parameter related to reversion speed, σ_{ou} is a parameter related to the process volatility, which is a measure of dispersion around the average of the random variable, and W_t is the Wiener process.

Regarding the Wiener processes, the well-known property of independent increments applies, that is $W_t - W_s \approx N(0, t - s)$ (for $0 \leq s < t$) where $N(0, t - s)$ refers to the Normal distribution with expected value and variance equal to 0 and $t - s$, respectively. It is worth noting that, since W_t is a Wiener process, $W_{t_1} - W_{s_1}$ and $W_{t_2} - W_{s_2}$ are independent random variables, in the case that $0 \leq s_1 < t_1 \leq s_2 < t_2$.

The solution of the stochastic differential Equation (21) is:

$$Y_t = Y_0 e^{-\tau_{ou}} + \mu_{ou}\left(1 - e^{-\tau_{ou}}\right) + \sigma_{ou} \int_0^t e^{-\tau_{ou}(t-s)}dW_t. \tag{22}$$

In the hypothesis that $Y_0 \sim N\left(\mu_{ou}, \dfrac{\sigma_{ou}^2}{2\tau}\right)$ and Y_0 and W_t uncorrelated, Y_t is a stationary Ornstein–Uhlenbeck process, with increments which are normally distributed, i.e.,

$$E[Y_t] = \mu_{ou}, var[Y_t] = \dfrac{\sigma_{ou}^2}{2\tau_{ou}}, cov[Y_t, Y_s] = \dfrac{\sigma_{ou}^2}{2\tau_{ou}}e^{-\tau_{ou}|t-s|}. \tag{23}$$

To estimate the parameters μ_{ou}, τ_{ou} and σ_{ou}^2 in (23), the method of moments can be applied to the discretized version of the process. More in detail, if the process observations Y_{t_i} are assumed to occur at times t_i such that $0 = t_0 < t_1 < \ldots t_{n_s} = T$, with $t_i - t_{i-1} = \frac{T}{n_s}$, then the estimates of parameters read [26]:

$$\hat{\mu}_{ou} = \dfrac{1}{n_s + 1} \sum_{i=0}^{n_s} Y_{t_i} \tag{24}$$

$$\hat{\tau}_{ou} = n_s \log \frac{\frac{1}{n_s} \sum_{i=0}^{n_s} \left(Y_{t_i} - \left(\frac{1}{n_s+1} \sum_{i=0}^{n_s} Y_{t_i} \right) \right)^2}{\frac{1}{n_s-1} \sum_{i=1}^{n_s} \left(Y_{t_i} - \left(\frac{1}{n_s+1} \sum_{i=0}^{n_s} Y_{t_i} \right) \right) \left(Y_{t_{i-1}} - \left(\frac{1}{n_s+1} \sum_{i=0}^{n_s} Y_{t_i} \right) \right)} \tag{25}$$

$$\hat{\sigma}_{ou}^2 = 2n_s \log \frac{1}{n_s} \sum_{i=0}^{n_s} \left(Y_{t_i} - \left(\frac{1}{n_s+1} \sum_{i=0}^{n_s} Y_{t_i} \right) \right)^2 \log \frac{\frac{1}{n_s} \sum_{i=0}^{n_s} \left(Y_{t_i} - \left(\frac{1}{n_s+1} \sum_{i=0}^{n_s} Y_{t_i} \right) \right)^2}{\frac{1}{n_s-1} \sum_{i=1}^{n_s} \left(Y_{t_i} - \left(\frac{1}{n_s+1} \sum_{i=0}^{n_s} Y_{t_i} \right) \right) \left(Y_{t_{i-1}} - \left(\frac{1}{n_s+1} \sum_{i=0}^{n_s} Y_{t_i} \right) \right)} \tag{26}$$

where, for ease of notation, it is assumed that $T = 1$.

By considering the presence of a noise D_{t_i} at time t_i, that in practical applications can affect the actual observations of the process, the process under study is modified as:

$$Z_{t_i} = Y_{t_i} + D_{t_i}. \tag{27}$$

With reference to the case of frequency, it can be assumed that the noise is Gaussian with mean equal to zero and variance v_{ou}^2 (i.e., $D \sim N(0, v_{ou}^2)$), thus obtaining the following expression for the expected value of Z_t:

$$E[Z_t] = \mu_{ou}, \tag{28}$$

$$var[Z_t] = \frac{\sigma_{ou}^2}{2\tau_{ou}} + v_{ou}^2 \tag{29}$$

and

$$cov\left[Z_{t_i}, Z_{t_j} \right] = \frac{\sigma_{ou}^2}{2\tau_{ou}} e^{-\tau_{ou}|t_i - t_j|} \qquad i \neq j \tag{30}$$

When the process (27) is considered, the parameters μ_{ou}, τ_{ou}, σ_{ou}^2 and v_{ou}^2 can be estimated as follows:

$$\hat{\mu}_{ou} = \frac{1}{n+1} \sum_{i=0}^{n} Y_{t_i} \tag{31}$$

$$\hat{\tau}_{ou} = n \log \frac{\frac{1}{n-1} \sum_{i=1}^{n} \left(Y_{t_i} - \left(\frac{1}{n+1} \sum_{i=0}^{n} Y_{t_i} \right) \right) \left(Y_{t_{i-1}} - \left(\frac{1}{n+1} \sum_{i=0}^{n} Y_{t_i} \right) \right)}{\frac{1}{n-2} \sum_{i=2}^{n} \left(Y_{t_i} - \left(\frac{1}{n+1} \sum_{i=0}^{n} Y_{t_i} \right) \right) \left(Y_{t_{i-1}} - \left(\frac{1}{n+1} \sum_{i=0}^{n} Z_{t_i} \right) \right)} \tag{32}$$

$$\hat{\sigma}_{ou}^2 = 2n \frac{\left(\frac{1}{n-1} \sum_{i=1}^{n} \left(Z_{t_i} - \left(\frac{1}{n+1} \sum_{i=0}^{n} Z_{t_i} \right) \right) \left(Z_{t_{i-1}} - \left(\frac{1}{n+1} \sum_{i=0}^{n} Z_{t_i} \right) \right) \right)^2}{\frac{1}{n-2} \sum_{i=2}^{n} \left(Z_{t_i} - \left(\frac{1}{n+1} \sum_{i=0}^{n} Z_{t_i} \right) \right) \left(Z_{t_{i-1}} - \left(\frac{1}{n+1} \sum_{i=0}^{n} Z_{t_i} \right) \right)} \cdot \log \frac{\frac{1}{n-1} \sum_{i=1}^{n} \left(Z_{t_i} - \left(\frac{1}{n+1} \sum_{i=0}^{n} Z_{t_i} \right) \right) \left(Z_{t_{i-1}} - \left(\frac{1}{n+1} \sum_{i=0}^{n} Z_{t_i} \right) \right)}{\frac{1}{n-2} \sum_{i=2}^{n} \left(Z_{t_i} - \left(\frac{1}{n+1} \sum_{i=0}^{n} Z_{t_i} \right) \right) \left(Z_{t_{i-1}} - \left(\frac{1}{n+1} \sum_{i=0}^{n} Z_{t_i} \right) \right)} \tag{33}$$

$$\hat{v}^2 = \frac{1}{n} \sum_{i=0}^{n} \left(Z_{t_i} - \left(\frac{1}{n+1} \sum_{i=0}^{n} Z_{t_i} \right) \right)^2 - \frac{\left(\frac{1}{n-1} \sum_{i=1}^{n} \left(Z_{t_i} - \left(\frac{1}{n+1} \sum_{i=0}^{n} Z_{t_i} \right) \right) \left(Z_{t_{i-1}} - \left(\frac{1}{n+1} \sum_{i=0}^{n} Z_{t_i} \right) \right) \right)^2}{\frac{1}{n-2} \sum_{i=2}^{n} \left(Z_{t_i} - \left(\frac{1}{n+1} \sum_{i=0}^{n} Z_{t_i} \right) \right) \left(Z_{t_{i-1}} - \left(\frac{1}{n+1} \sum_{i=0}^{n} Z_{t_i} \right) \right)}. \tag{34}$$

4. Stochastic Characterization of SoC

Prior to the study of stochasticity of SoC, the stochastic features of the regulation power p_{reg} need to be analyzed. In this regard, an accurate numerical analysis based on actual data has been proposed in the numerical application. By anticipating the results here, the analysis has evidenced that different stochastic interpretations can be provided based on the presence or absence of the dead-band of the frequency-power characteristic for the primary frequency response (Figure 2). The theoretical basis which proves these interpretations is discussed in the following sub-sections.

4.1. Stochastic Characterization of the Regulation Power in Absence of the Dead-Band

In the case of absence of the dead-band of Figure 2, the action requested to the BESS aimed at primary frequency regulation is that reported in Figure 5.

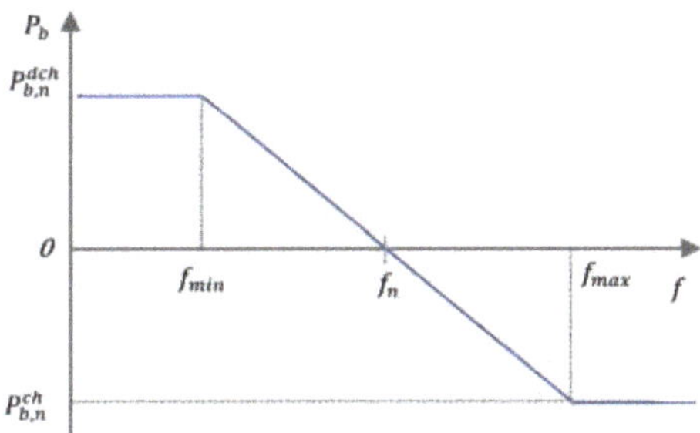

Figure 5. Action of the BESS for the primary frequency regulation service in the absence of the dead-band.

By analyzing the linear power-frequency relation of Figure 5, it is easy to argue that when the dead-band is not considered, the stochastic nature of p_{reg} strictly depends on the stochastic nature of the frequency. Thus, it can be accurately modelled according to an Ornstein–Uhlenbeck process. In Section 4.3 we adopt this model to characterize the dynamics of the battery SoC and to lead to the choice of the gain μ_0 of Figure 4, when the dead-band is not considered.

4.2. Stochastic Characterization of the Regulation Power in Presence of the Dead-Band

When the dead-band is considered, the analyses performed with respect to the various available data lead one to suggest that the regulation power consequent to frequency deviations can be regarded as a shot noise process X_t, with decay constant α, driven by a homogeneous Poisson process with intensity λ.

Based on the analysis of the actual data, it can be noted that the intensity of the homogeneous Poisson process, that is the peak values of the regulation power, can be accurately assumed as statistically distributed according to the truncated Laplace distribution. At this regard, starting from the probability density function (pdf) of the Laplace distribution:

$$f(x) = \frac{1}{2b_p} e^{-\frac{|x-\mu_p|}{b_p}}, \tag{35}$$

where x is the random variable (in this case, the amplitude of the regulation power), μ_p is the location parameter and b_p is the scale parameter distribution, the expression of the truncated Laplace pdf is:

$$f(x) = \begin{cases} \dfrac{1}{2b_p e^{-\frac{d}{b_p}}} e^{-\frac{|x|}{b_p}} & |x| \geq d \\ 0 & |x| < d \end{cases} \tag{36}$$

where d is the minimum absolute value of the amplitudes experienced and where, according to the typical frequency data, μ_p has been assumed equal to 0. An estimate $\hat{b}_p$ of the scale parameter b_p can be derived by the minimization of the log-likelihood function. Thus, being $L(b_p)$ the likelihood function of the truncated Laplace distribution:

$$L(b_p) = \prod_{i=1}^{n_s} \frac{1}{2b_p e^{-\frac{d}{b_p}}} e^{-\frac{|x_i|}{b_p}} \tag{37}$$

the estimate of $\hat{b}_p$ can be derived by solving the equation corresponding to $\frac{\partial \ln L\left(b_p\right)}{\partial b_p} = 0$, that is:

$$n_s b_p + n_s d - \frac{1}{n_s}\sum_{i=1}^{n_s}|x_i| = 0, \tag{38}$$

whose solution is

$$\hat{b}_p = \frac{1}{n_s}\sum_{i=1}^{n_s}|x_i| - d. \tag{39}$$

Regarding the statistical characterization of the time of occurrence of the regulation power, the analysis of the actual data obtained from the application of the control of Figure 3 allows us to identity the exponential distribution as an accurate interpretation. The pdf of the exponential distribution can be expressed as:

$$f(x) = \frac{1}{\mu_{tp}}e^{-\frac{x}{\mu_{tp}}} \tag{40}$$

where x is the random variable (in this case the time of occurrence of the regulation power) and μ_{tp} is its mean value.

Once the statistical analysis of the time of occurrence and amplitude of the peak values of the regulation power has been carried out, the stochastic process of the regulation power, X_t, can be analyzed. In this regard, the parameters of the process X_t are provided by the following relationships:

$$E[X_t] = \frac{\lambda E[Z]}{\alpha}\left(1 - e^{-\alpha t}\right) \tag{41}$$

$$var[X_t] = \frac{\lambda E[Z^2]}{2\alpha}\left(1 - e^{-2\alpha t}\right) \tag{42}$$

$$cov[X_r, X_s] = \frac{\lambda E[Z^2]}{2\alpha}e^{-\alpha(s-r)}\left(1 - e^{-2\alpha r}\right) \tag{43}$$

with $r < s$ and Z the random variable, which refers, in this case, to the power amplitude approximated by the truncated Laplace pdf.

It is not superfluous to highlight, as in [27], that the above parameters are the same as those relating to the Ornstein–Uhlenbeck process, Y_t, described by the following stochastic differential equation:

$$dY_t = (\lambda E[Z] - \alpha Y_t)dt + \sigma_{ou}dW_t \tag{44}$$

where $\sigma_{ou}dW_t$ is the differential Wiener diffusion process with standard deviation σ_{ou}, and where λ and α are the parameters of X_t (i.e., the parameters appearing in Equations (41)–(43)).

Of course, this does not mean that the two processes, X_t and Y_t, are the same, since the process Y_t is wholly characterized by the first two moments, being a Gaussian process. In what follows, we adopt this approximation for the purpose of characterizing the dynamics of the battery SoC and choosing the gain parameter μ_0, when the dead-band is considered.

4.3. Stochastic Characterization of the State of Charge

At the purpose of characterizing the battery SoC and of choosing the gain parameter μ_0, we can initially refer to the block diagram of Figure 4. Then, under the hypothesis discussed in the previous two sub-sections, the SoC control block diagram can be represented as that of Figure 6, where $w = \frac{k_i}{s}p_{reg}$, i.e., proportional to the integral (21). Indeed, in the case of absence of the dead-band, p_{reg} has been assumed as an Ornstein–Uhlenbeck process; in case of presence of the dead-band, the hypothesis to consider the shot noise process X_t equivalent to the Ornstein–Uhlenbeck process has been assumed

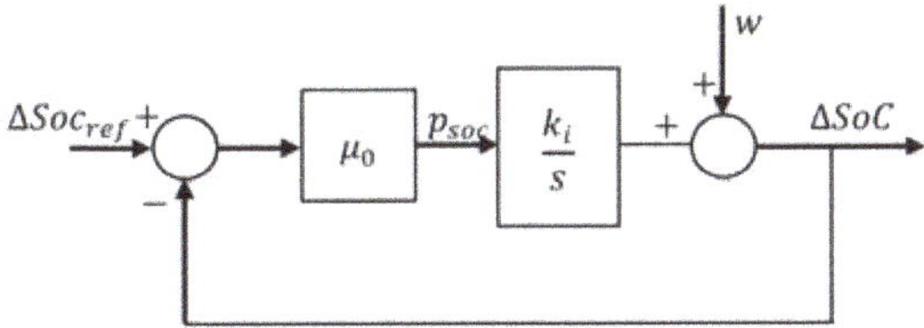

Figure 6. Block diagram of the SoC control scheme.

In both cases, it is not difficult to demonstrate that the variance of the process w asymptotically grows linearly in t. The linear growth of these parameters is a fundamental property of the Wiener process [27]. In other words, the integrated process p_{reg} exhibits similar features of the Wiener process. The SoC can be then statistically characterized by an Ornstein–Uhlenbeck process driven by an Ornstein–Uhlenbeck process, that is:

$$\frac{d\Delta SOC_t}{dt} = -\mu_0 \Delta SOC_t + Y_t \tag{45}$$

This process suggests the adoption of an adaptative control strategy of the BESS which modifies the regulation power provided by the BESS according to a proper choice of the parameter μ_0. The value of this parameter can be chosen to allow SoC variation to satisfy both the requirements of regulation service provision and battery lifetime preservation.

4.4. BESS Control Scheme

The above considerations suggest modifying the control scheme of Figure 3 in that shown in Figure 7. The proposed scheme of Figure 7 allows us to update the regulation power to the SoC level, thereby obtaining a balance between (i) the regulation power needed for the provision of the regulation service and (ii) the need of SoC recovery. This control scheme allows us to simplify the control of the BESS even preserving multiple complex requirements, provided that the correct choice of the gain parameter μ_0 is performed. The proposed approach allows us to consider the features of the frequency variations, the requirements imposed by the system operator in the service provided by the BESS, and the range of values within which the SoC must falls. This approach can be used for any battery technology.

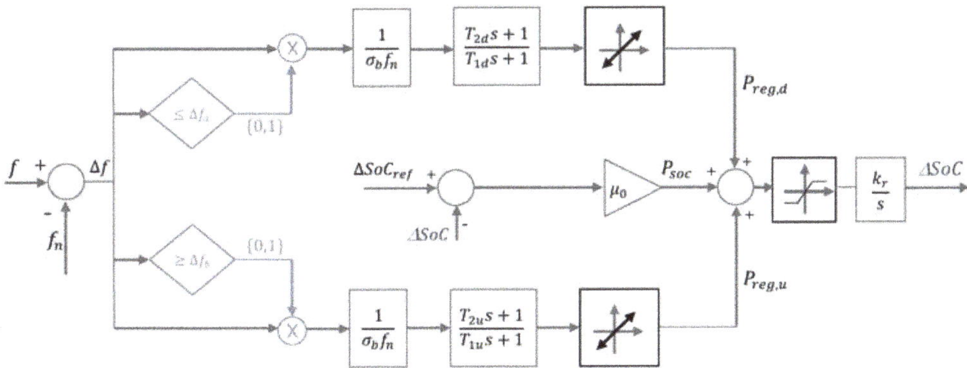

Figure 7. SoC control scheme.

In order to derive, that means to design the regulator, it is possible to refer to the control scheme of Figure 6, where the disturbance w has been previously identified as a random walk. The two control blocks above discussed identify two transfer functions:

- $G(s)$ which relates the output, i.e., ΔSoC, to the disturbance, i.e., w.
- $F(s)$ which relates the output, i.e, ΔSoC, to the input, i.e., ΔSoC_{ref}.

The two transfer functions can be identified as:

$$G(s) = \frac{s}{s + \mu_0 k_i} \tag{46}$$

$$F(s) = \frac{\mu_0 k_i}{s + \mu_0 k_i}. \tag{47}$$

$F(s)$ clearly identifies a low-pass filter, whose cut-off frequency is $\mu_0 k_i$. Thus, the correct evaluation of the proportional gain μ_0 must comply with a proper frequency bandwidth, which must be large enough to catch frequencies able to meet the requirements of both frequency regulation and SoC recovery. From one hand, highest cut-off frequency values can keep SoC values closest to its reference value by reducing the effect of the disturbance w. On the other hand, in the paper, the disturbance w is modelled to represent the effect of the battery power required for the regulation service. Thus, this reduction must not be too high, that is a proper value of the cut-off frequency must be assigned. In Figure 8, a schematic of the effect of the choice of the cut-off frequency is shown.

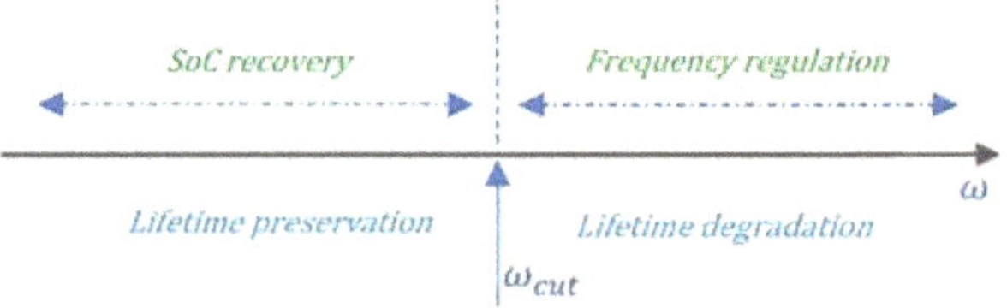

Figure 8. Cut-off frequency choice criterion.

The aim of the proposed control method is to define a correction to the BESS output power in the frequency domain which uses high frequency band to provide the desired regulation service and low frequency band to counteract the low frequency drift that takes SoC out of the limits. In order to choose the proper value of the cut-off frequency, the requirements of the power frequency regulation service should be considered. The primary regulation service time interval availability is assumed to be equal to 15 min [20], thus implying that the allowed boundary frequency f_w of the spectral signal w are those corresponding to $\tau = 900$ s, that is:

$$f_w = \frac{1}{900} = 1.1 \cdot 10^{-3} \text{ Hz} \tag{48}$$

which corresponds to the angular frequency, ω_w, given by:

$$\omega_w = 2\pi \cdot 1.1 \cdot 10^{-3} \cong 7 \cdot 10^{-3} \text{rad}/\text{s} \tag{49}$$

Eventually, the cut-off frequency of ω_{cut}, can be properly chosen as a decade lower, since ω_w allows us to obtain 3dB abatement, so obtaining:

$$\omega_{cut} = 7 \cdot 10^{-4} \text{rad}/\text{s}. \tag{50}$$

The value of the proportional gain μ_0 can finally be given by:

$$\mu_0 = \frac{\omega_{cut}}{k_i} \tag{51}$$

being

$$k_i = \frac{P_n}{E_n} \tag{52}$$

5. Numerical Applications

In this section, an analysis is performed on the frequency, the BESS regulation power and the BESS SoC with respect to the stochastic approaches described in Sections 3 and 4. For the sake of clarity, the results are reported in three sub-sections, related to the (i) analysis of frequency data, (ii) analysis of regulation power, and (iii) analysis of SoC, respectively. The size of the BESS used in the simulation corresponds to a power rating of 98.5 MW and an energy rating of 49.5 MWh [20].

5.1. Analysis of Frequency Data

The analysis of this section refers to the assessment of the stochastic assumptions on the frequency process discussed in Section 3. To test the logistic-autoregressive (logistic AR) and Ornstein–Uhlenbeck processes, actual frequency data were used, made available from the French Transmission operator RTE [28]. More specifically, one-year frequency values from April 2021 to March 2022 were used. In Figure 9 the frequency variation in the period 1–15 April is reported. In the same figure, a zoom of a portion of the April 6th is also reported.

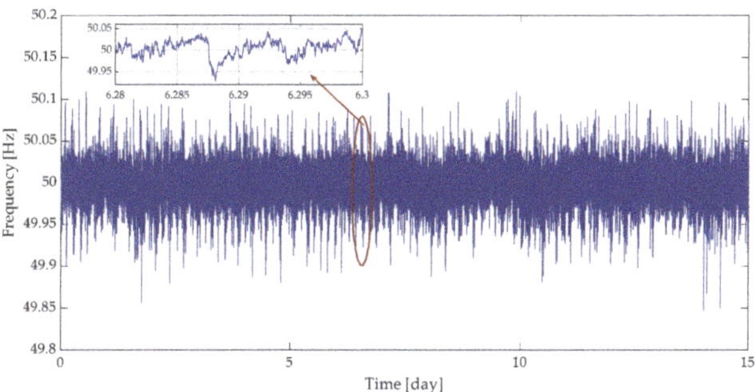

Figure 9. Actual frequency data from 1–15 April.

The data under consideration have been used for the stochastic characterization of the frequency with respect to the logistic-AR and Ornstein–Uhlenbeck processes. The effectiveness of the approaches has been analyzed through the method of the moments. At this purpose, the percentage error of the first and second moments evaluated by characterizing the frequency through the logistic-AR and Ornstein–Uhlenbeck processes are reported in Figures 10 and 11. Both the figures refer to the whole set of the available data and the errors are evaluated on a monthly interval basis.

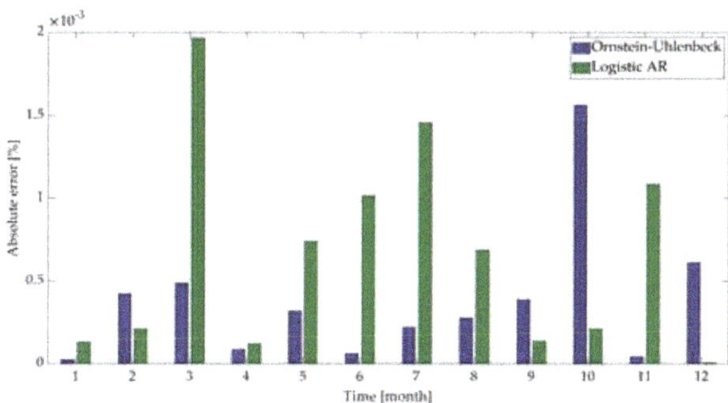

Figure 10. Errors of the mean evaluated with characterization by logistic–AR and Ornstein–Uhlenbeck processes.

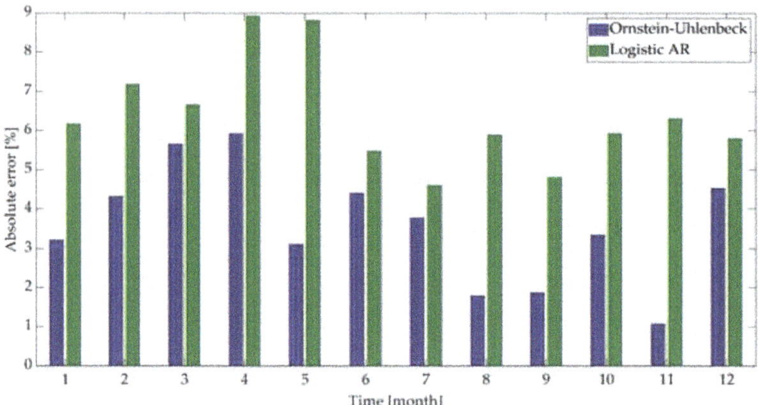

Figure 11. Errors of the second moments evaluated with characterization by logistic–AR and Ornstein–Uhlenbeck processes.

The analysis of Figures 10 and 11 shows that both the processes allow obtaining accurate characterization of the regulation power process. In particular, the Ornstein–Uhlenbeck process generally performs better than the logistic-AR process. Regarding the mean values (Figure 10), both the processes allow us to obtain quite a good estimation of the actual dataset with very low errors. The maximum absolute errors are lower than $1.6 \cdot 10^{-3}\%$ for the Ornstein-Uhlenbeck and $2 \cdot 10^{-3}\%$ for the logistic-AR. Regarding the second moment, the Ornstein-Uhlenbeck always has a lower error than that obtained through the logistic-autoregressive. The maximum absolute error of the Ornstein–Uhlenbeck is less than 6% and that of the logistic-AR is less than 9%. The better performance of the Ornstein–Uhlenbeck process is clearly reported in Table 1, where the mean absolute percentage error (MAPE) on the first and second moments of both processes are reported for the whole dataset. In the table, the MAPE is reported also with respect to the processes estimated on weakly and daily basis. In all the cases, the MAPE of Ornstein–Uhlenbeck is lower than the logistic-AR process.

Table 1. MAPE Errors.

Error⟋Process	Mean [%]	Variance [%]
Monthly		
Logistic Autoregressive	7.2×10^{-4}	6.2
Ornstein–Uhlenbeck	3.5×10^{-4}	3.0
Weekly		
Logistic Autoregressive	2.0×10^{-3}	8.8
Ornstein–Uhlenbeck	6.8×10^{-4}	3.3
Daily		
Logistic Autoregressive	6.7×10^{-3}	7.9
Ornstein–Uhlenbeck	1.8×10^{-3}	5.3

The fourth moments (kurtosis) of the two processes have also been compared. In this regard, the better performance of the logistic-AR process demonstrates that it can better reproduce the tails of the distribution of the time series values.

A synthetic profile estimated according to the Ornstein–Uhlenbeck process for seven days is reported in Figure 12. In the same figure, a zoom of a portion of the sixth day is also reported. A qualitative analysis of the comparison between the synthetic (Figure 12) and actual (Figure 9) profiles clearly confirms the accuracy of the Ornstein–Uhlenbeck approach.

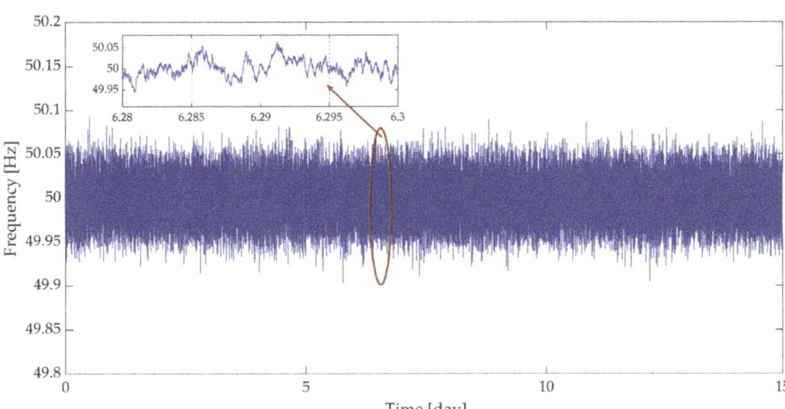

Figure 12. Frequency data estimated according to the Ornstein–Uhlenbeck process.

5.2. Analysis of Regulation Power

Based on the control scheme of Figure 3, the regulation power can be evaluated, starting from the real system frequency or that corresponding to the stochastic process modeled according to the Ornstein–Uhlenbeck processes. In what follows, three case studies are reported with respect to three possible scenarios involving the dead-bands:

- *Case 1*: the dead-band is not considered;
- *Case 2*: the dead-band is ±10 mHz;
- *Case 3*: the dead-band is ±20 mHz.

By neglecting the effect of the SoC recovery control, in Figure 13 an example of the regulation power for the same period of Figure 9 is reported. Case 1, Case 2, and Case 3 are referred to in Figures 13–15, respectively. By comparing the zooms of the three cases, the effect of the increasing dead-band range clearly appears.

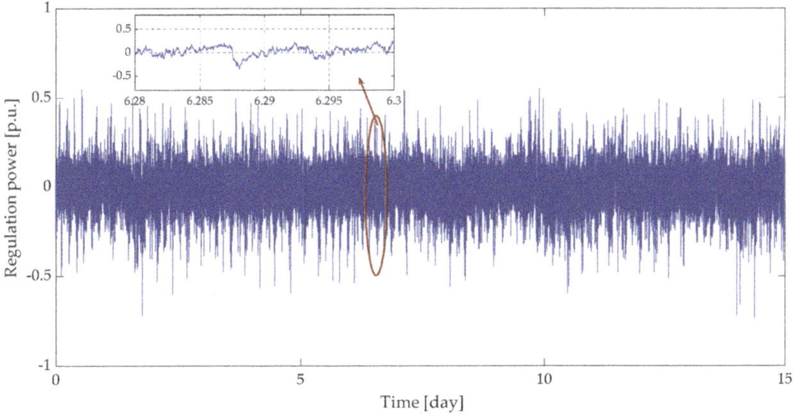

Figure 13. Regulation power in Case 1.

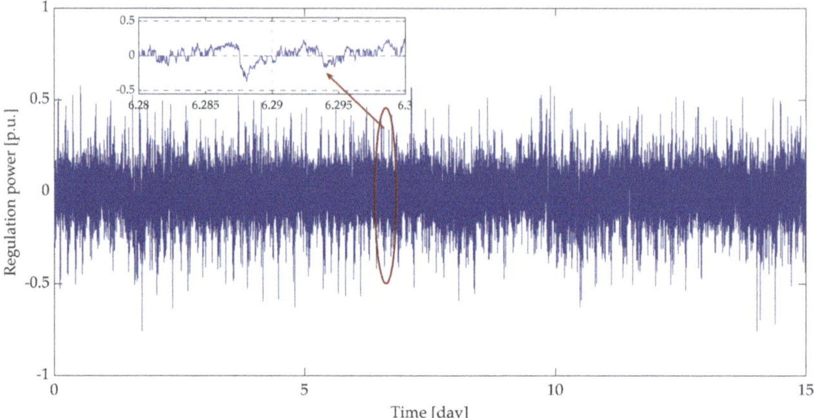

Figure 14. Regulation power in Case 2.

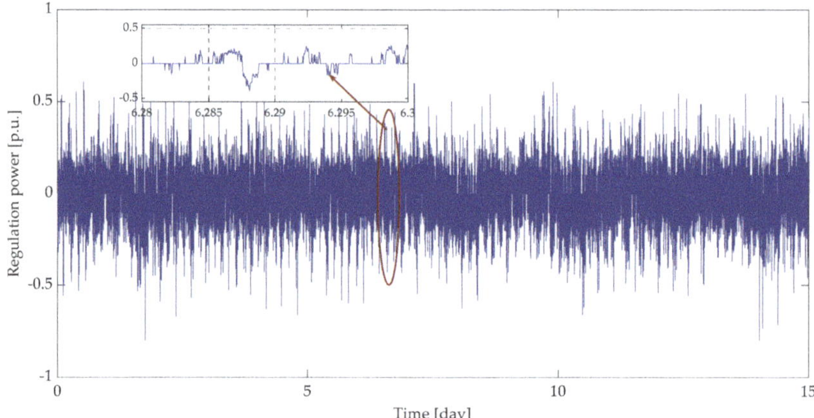

Figure 15. Regulation power in Case 3.

Based on the method of the moments, an analysis of the regulation power obtained in Case 1 similar to that reported in Section 5.1 (with reference to the frequency) was carried out, showing that it varies according to a Ornstein–Uhlenbeck process.

Regarding the other two cases, where it appeared that the regulation power can be approximated by a Poisson process, a statistical analysis of the amplitudes of the peak values of the regulation power and their time of occurrence was performed. In particular, the amplitude data were given in terms of non-parametric distributions and were compared to the truncated-Laplace distribution. These comparisons are shown in the Figure 16a,b), which correspond to Cases 2 and 3, respectively.

With reference to the data used in this application, the scale parameter of the truncated Laplace distribution is equal to $b_p = 0.0459$ in the case of a dead-band equal to ±10 mHz (Case 2), and $b_p = 0.0620$ in the case of a dead-band equal to ±20 mHz (Case 3); in both cases the location parameter is $\mu_p \cong 0$.

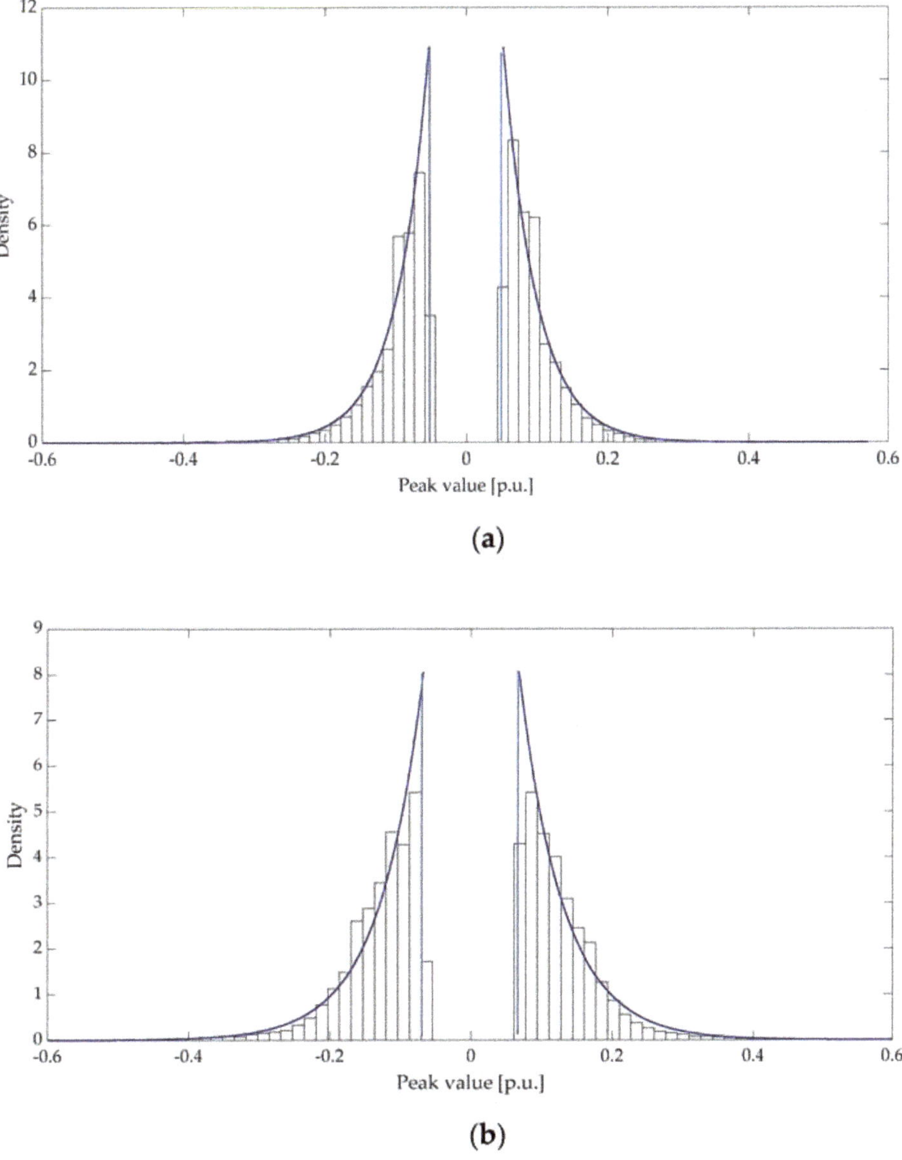

Figure 16. Approximation of the power peak values of the regulation power to a Laplace pdf in Case 2 (**a**) and Case 3 (**b**).

Regarding the time of occurrence of the regulation power, the analysis of the actual values is reported in Figure 17. In particular, Figure 17a refers to Case 2 and Figure 17b refers to the Case 3. These data are collected in terms of frequency of occurrence whose histograms are reported in the figure together with the fitting exponential distribution which provides an accurate approximation of the real distribution.

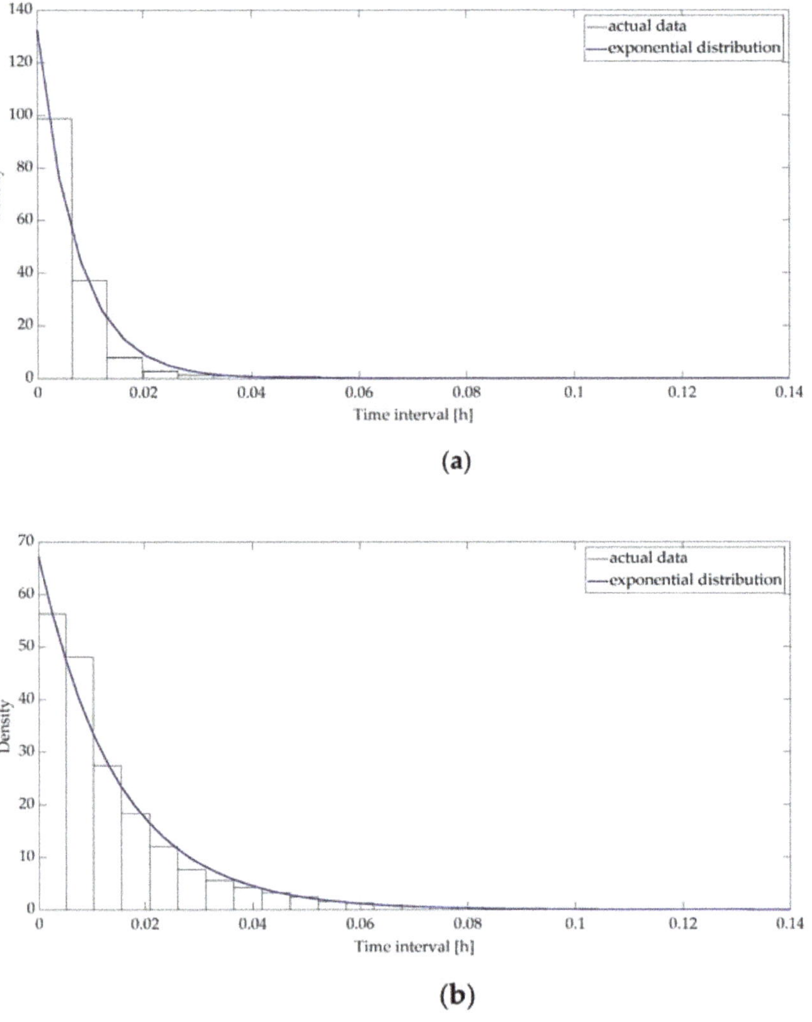

Figure 17. Approximation of the time occurrence of regulation power to an exponential pdf in Case 2 (**a**) and Case 3 (**b**).

In the two cases the mean value μ_t is equal to 0.0074 h (dead-band $\pm$10 mHz) and equal to 0.0148 h (dead-band $\pm$20 mHz).

5.3. Analysis of State of Charge

Various simulations have been performed for deriving different SoC profiles corresponding to the three case studies defined in the previous sub-sections, which refer to the different dead-bands. Regarding the gain parameter μ_0, the value estimated in Section 4 has been initially considered ($\mu_0{}^* = 1.4 \cdot 10^{-3}$). The effect of two further values have also been investigated, which are $\mu_0 = 10^{-3}\mu_0{}^*$ and $\mu_0 = 10^3\mu_0{}^*$. The SoC profiles obtained for each gain parameter are reported in Figure 18 for each case study: Figure 18a refers to Case 1, Figure 18b refers to Case 2, and Figure 18c refers to Case 3.

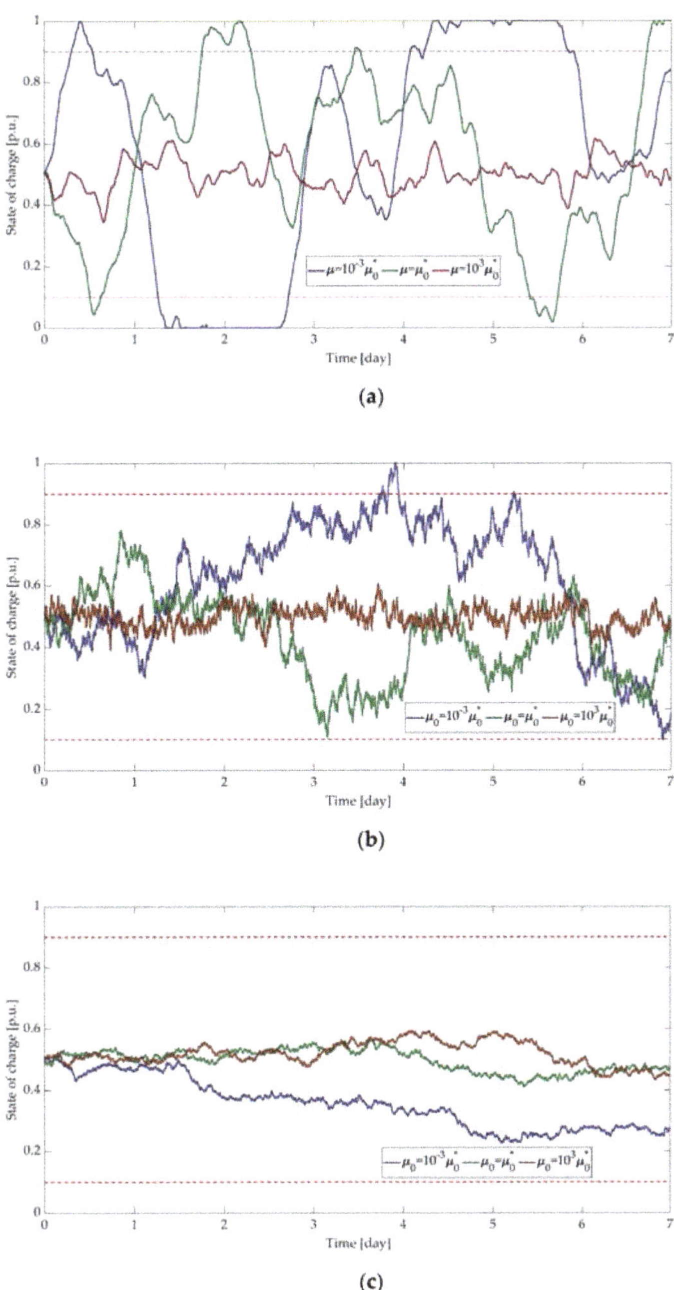

Figure 18. Examples of synthetic SoC profiles over a week in Case 1 (**a**), Case 2 (**b**), and Case 3 (**c**).

The profiles reported in Figure 18a clearly show that, in the absence of a dead-band, the BESS is continuously used for providing power to the grid for frequency regulation. This implies that during the considered time of observation the SoC profile often reaches its maximum and minimum capacity values. Particularly, this happens in the case of $\mu_0 = \mu_0^*$, for which the SoC exceeds, in some cases, the upper (0.9 p.u.) and lower (0.1 p.u.) limits typically adopted for the SoC to preserve the battery lifetime. As expected, the lower value of the gain parameter ($\mu_0 = 10^{-3}\mu_0{}^*$) implies an even higher use of the battery for frequency regulation thus resulting in a SoC profile that often exceeds the admissible limits. On the contrary, a greater value of the gain parameter ($\mu_0 = 10^3\mu_0{}^*$) implies less participation of the battery to the frequency regulation service, thus resulting in a SoC profile that is contained within a narrow range around the reference value of 0.5 p.u. This consideration is fundamental since it demonstrates that a proper choice of μ_0 is required to optimally design the control action of the BESS which guarantees the lifetime preservation.

The considerations made for Case 1 also apply to the case of Figure 18b which refers to a dead-band of ± 10 mHz (Case 2). However, the presence of the dead-band implies a lower participation of the BESS to the frequency regulation, thus implying that the SOC profile has lower variations than that of the previous case and the SOC limits are rarely reached. This behavior is stressed in the case of dead-band at ± 20 mHz (Case 3) for which the SoC always ranges in an interval very close to the reference value of 0.5 p.u. This value of the dead-band implies oversizing of the BESS.

In summary, the plots of Figure 18 show that, when the value of the parameter is optimally chosen, the BESS control is optimized, since it is controlled by maintaining the SoC within admissible ranges, thus optimizing the provision of the regulation service. When the parameter is lower than the optimal value, the SoC profile exceeds the admissible range; differently, when the parameter is greater than the optimal value, the SoC varies very slowly. As expected, the figures also show that the effect of the parameter value depends on the dead-band width.

By comparing the SoC profiles of Figure 18, it is clear that in case of Figure 18a the profile is smoother than that obtained in Figure 18b,c. This is obviously due to the nature of the stochastic process of the regulation power, which is an Ornstein–Uhlenbeck process in case of Figure 18a (absence of the dead-band), and is a compound Poisson process in case of Figure 18b,c (presence of the dead-band). The presence of jumps in the compound Poisson process, indeed, implies the presence of spikes in the profiles of SoC.

6. Conclusions

In this paper, a new stochastic approach has been applied to the study of the primary frequency regulation service provided by battery energy storage systems. The use of storage devices for frequency regulation in modern power systems is influenced by complexities due to the need to match the requirements imposed by the transmission system operators for the provision of the service, and the containment of the battery lifetime degradation, imposed by the economic issues. For this purpose, proper control strategies are requested to be designed which can be used despite the uncertainties typically affecting the frequency regulation service. In this application, a stochastic approach is proposed to provide accurate models for the time profiles of frequency, regulation power, and state of charge. In particular, the use of the Ornstein–Uhlenbeck stochastic process is studied for the frequency profile. Regarding the regulation power, it has been proven that it can still be modeled according to the Ornstein–Uhlenbeck process or the compound Poisson process, depending on the service performance requested of the storage device (more specifically, depending on the presence or absence of the dead-band). It has been demonstrated that, in both cases, the state of charge can be modeled according to an Ornstein–Uhlenbeck process driven by an Ornstein–Uhlenbeck process. The accuracy of this approach has been investigated in this paper together with the definition of a gain parameter, which can be set appropriately to optimize the control of the battery for the balance between availability of regulation service and battery lifetime degradation. In the numerical application it has been found that this problem can be studied as a hitting problem, for which more research efforts are needed in future works. The proposed approach also makes the primary frequency regulation service provided by battery energy storage systems prone to the use of stochastic control methods which inherently account for the existence of uncertainties in the evolution of the system frequency.

Author Contributions: Conceptualization, E.C., D.L., F.M., D.P., D.V., G.M.G. and C.P.; methodology, E.C., D.L., F.M., D.P. and D.V.; software, D.L., F.M. and D.P.; validation, E.C., D.L., F.M., D.P., D.V. and C.P.; formal analysis, E.C., D.L., F.M. and D.P.; investigation, E.C., D.L., F.M. and D.P.; data curation, F.M. and D.P.; writing—original draft preparation, D.L., F.M. and D.P.; writing—review and editing, E.C., D.L., F.M., D.P., D.V., G.M.G. and C.P.; supervision, D.L., D.V. and G.M.G. All authors have read and agreed to the published version of the manuscript.

Funding: This research received no external funding.

Institutional Review Board Statement: Not applicable.

Data Availability Statement: The data on power system frequency used in the numerical simulation have been taken from: RTE, Network Frequency. Available online: https://www.services-rte.com/en/view-data-published-by-rte.html (accessed on 4 July 2022).

Conflicts of Interest: The authors declare no conflict of interest.

References

1. Brivio, C.; Mandelli, S.; Merlo, M. Battery energy storage system for primary control reserve and energy arbitrage. *Sustain. Energy Grids Netw.* **2016**, *6*, 152–165. [CrossRef]
2. Guerra, K.; Haro, P.; Gutiérrez, R.E.; Gómez-Barea, A. Facing the high share of variable renewable energy in the power system: Flexibility and stability requirements. *Appl. Energy* **2022**, *310*, 118561. [CrossRef]
3. Carlini, E.M.; Del Pizzo, F.; Giannuzzi, G.M.; Lauria, D.; Mottola, F.; Pisani, C. Online analysis and prediction of the inertia in power systems with renewable power generation based on a minimum variance harmonic finite impulse response filter. *Int. J. Electr. Power Energy Syst.* **2021**, *131*, 107042. [CrossRef]
4. IRENA. Battery Storage for Renewables: Market Status and Technology Outlook. 2015. Available online: https://www.irena.org/publications/2015/Jan/Battery-Storage-for-Renewables-Market-Status-and-Technology-Outlook (accessed on 4 July 2022).
5. Thien, T.; Axelsen, H.; Merten, M.; Axelsen, H.; Merten, M.; Zurmùhlen, S.; Münderlein, J.; Leuthold, M.; Sauer, D.U. Planning of grid-scale battery energy storage systems: Lessons learned from a 5 MW hybrid battery storage project in Germany. In Proceedings of the Battcon International Stationary Battery Conference, Orlando, FL, USA, 12–14 May 2015.
6. Fu, H.; Tong, X.; Pan, Z.; Liu, F.; Wang, F.; Zhang, W. Research on BESS Participating in Power System Primary Frequency Regulation Control Strategy Considering State-of-Charge Recovery. In Proceedings of the 5th International Conference on Energy, Electrical and Power Engineering, Chongqing, China, 22–24 April 2022.
7. Oudalov, A.; Chartouni, D.; Ohler, C. Optimizing a Battery Energy Storage System for Primary Frequency Control. *IEEE Trans. Power Syst.* **2007**, *22*, 1259–1266. [CrossRef]
8. Stroe, D.; Knap, V.; Swierczynski, M.; Stroe, A.; Teodorescu, R. Operation of a Grid-Connected Lithium-Ion Battery Energy Storage System for Primary Frequency Regulation: A Battery Lifetime Perspective. *IEEE Trans. Ind. Appl.* **2017**, *53*, 430–438. [CrossRef]
9. Khalid, M.; Savkin, A.V. An optimal operation of wind energy storage system for frequency control based on model predictive control. *Renew. Energy* **2012**, *48*, 127–132. [CrossRef]
10. Andrenacci, N.; Chiodo, E.; Lauria, D.; Mottola, F. Life Cycle Estimation of Battery Energy Storage Systems for Primary Frequency Regulation. *Energies* **2018**, *11*, 3320. [CrossRef]
11. Wu, F.; Sioshansi, R. A stochastic operational model for controlling electric vehicle charging to provide frequency regulation. *Transp. Res. Part D Transp. Environ.* **2019**, *67*, 475–490. [CrossRef]
12. Scarabaggio, P.; Carli, R.; Cavone, G.; Dotoli, M. Smart Control Strategies for Primary Frequency Regulation through Electric Vehicles: A Battery Degradation Perspective. *Energies* **2020**, *13*, 4586. [CrossRef]
13. Meng, G.; Lu, Y.; Liu, H.; Ye, Y.; Sun, Y.; Tan, W. Adaptive Droop Coefficient and SOC Equalization-Based Primary Frequency Modulation Control Strategy of Energy Storage. *Electronics* **2021**, *10*, 2645. [CrossRef]
14. Tan, Z.; Li, X.; He, L.; Li, Y.; Huang, J. Primary frequency control with BESS considering adaptive SoC recovery. *Int. J. Electr. Power Energy Syst.* **2020**, *117*, 105588. [CrossRef]
15. Shim, J.W.; Verbič, G.; Kim, H.; Hur, K. On Droop Control of Energy-Constrained Battery Energy Storage Systems for Grid Frequency Regulation. *IEEE Access* **2019**, *7*, 166353–166364. [CrossRef]
16. Dang, J.; Seuss, J.; Suneja, L.; Harley, R.G. SOC feedback control for wind and ESS hybrid power system frequency regulation. In Proceedings of the IEEE Power Electronics and Machines in Wind Applications Conference, Denver, CO, USA, 16–18 July 2012.
17. Marconato, R. *Electric Power Systems Vol. 2: Steady-State Behaviour Controls, Short Circuits and Protection Systems*, 2nd ed.; CEI: Milan, Italy, 2004.
18. Vorobev, P.; Greenwood, D.M.; Bell, J.H.; Bialek, J.W.; Taylor, P.C.; Turitsyn, K. Deadbands, Droop, and Inertia Impact on Power System Frequency Distribution. *IEEE Trans. Power Syst.* **2019**, *34*, 3098–3108. [CrossRef]

19. Quint, R.; Ramasubramanian, D. Impacts of droop and deadband on generator performance and frequency control. In Proceedings of the IEEE Power and Energy Society General Meeting, Chicago, IL, USA, 16–20 July 2017.

20. Knap, V.; Chaudhary, S.K.; Stroe, D.I.; Swierczynski, M.; Craciun, B.I.; Teodorescu, R. Sizing of an Energy Storage System for Grid Inertial Response and Primary Frequency Reserve. *IEEE Trans. Power Syst.* **2016**, *31*, 3447–3456. [CrossRef]

21. del Giudice, D.; Brambilla, A.; Grillo, S.; Bizzarri, F. Effects of inertia, load damping and dead-bands on frequency histograms and frequency control of power systems. *Int. J. Electr. Power Energy Syst.* **2021**, *129*, 106842. [CrossRef]

22. Brockwell, P.J.; Davis, R.A. *Introduction to Time Series and Forecasting*, 2nd ed.; Springer: New York, NY, USA, 2002.

23. Kallas, M.; Honeine, P.; Richard, C.; Francis, C.; Amoud, H. Prediction of time series using Yule-Walker equations with kernels. In Proceedings of the IEEE International Conference on Acoustics, Speech and Signal Processing (ICASSP), Kyoto, Japan, 25–30 March 2012.

24. Hassanzadeh, M.; Evrenosoğlu, C.Y.; Mili, L. A Short-term nodal voltage phasor forecasting method using temporal and spatial correlation. *IEEE Trans. Power Syst.* **2016**, *31*, 3881–3890. [CrossRef]

25. Wong, W.K.; Bian, G. Estimating parameters in autoregressive models with asymmetric innovations. *Stat. Probab. Lett.* **2005**, *71*, 61–70. [CrossRef]

26. Holý, V.; Tomanová, P. Estimation of Ornstein-Uhlenbeck Process Using Ultra-High-Frequency Data with Application to Intraday Pairs Trading Strategy. *arXiv* **2019**, arXiv:1811.09312v2.

27. Smith, P.L. From Poisson shot noise to the integrated Ornstein–Uhlenbeck process: Neurally principled models of information accumulation in decision-making and response time. *J. Math. Psychol.* **2010**, *54*, 266–283. [CrossRef]

28. RTE. Network Frequency. Available online: https://www.services-rte.com/en/view-data-published-by-rte.html (accessed on 4 July 2022).

Article

Development and Performance Verification of Frequency Control Algorithm and Hardware Controller Using Real-Time Cyber Physical System Simulator

Tae-Hwan Jin [1,2], Ki-Yeol Shin [1,*], Mo Chung [1,3] and Geon-Pyo Lim [4,*]

[1] School of Mechanical Engineering, Yeungnam University, Gyeongsan 38541, Korea
[2] Energy System Research Center, Korea Textile Machinery Convergence Research Institute, Gyeongsan 38542, Korea
[3] A1 Engineering, Inc., Gyeongsan 38542, Korea
[4] Korea Electric Power Research Institute, Korea Electric Power Corporation, Naju 58277, Korea
*   Correspondence: shinky@ynu.ac.kr (K.-Y.S.); gplim16@kepco.co.kr (G.-P.L.); Tel.: +82-53-810-3060 (K.-Y.S.); +82-42-865-5601 (G.-P.L.)

Abstract: Frequency stability is a critical factor in maintaining the quality of the power grid system. A battery energy storage system (BESS) with quick response and flexibility has recently been used as a primary frequency control (PFC) resource, and many studies on its control algorithms have been conducted. The cyber physic system (CPS) simulator, which can perform virtual physical modelling and verification of many hardware systems connected to the network, is an optimal solution for the performance verification of control algorithms and hardware systems. This study introduces a large-scale real-time dynamic simulator that includes the national power system. This simulator comprises a power grid model, an energy management system (EMS) model, a BESS system model, and a communication model. It performs the control algorithm performance evaluation and the hardware controller's response performance evaluation. The performance of the control algorithm was evaluated by tracking the power system's characteristic trajectory in the transient state based on the physical response delay time between the output instruction of the frequency regulation controller (FRC), a hardware controller, and the output response of the BESS. Based on this, we examined the response performance evaluation results by linking them to the optimally designed actual FRC. As a result, we present an analysis of the BESS's characteristic trajectories in the transient state, such as frequency, power system inertia, and power grid constant, and provide FRC response performance evaluation results at a level of 163 ms, by connecting the BESS installed at the actual site with the CPS simulator.

Keywords: real-time dynamic simulation; national power grid; cyber physical system (CPS); co-simulation; battery energy storage system (BESS); frequency control; energy management system (EMS)

Citation: Jin, T.-H.; Shin, K.-Y.; Chung, M.; Lim, G.-P. Development and Performance Verification of Frequency Control Algorithm and Hardware Controller Using Real-Time Cyber Physical System Simulator. *Energies* **2022**, *15*, 5722. https://doi.org/10.3390/en15155722

Academic Editors: Cosimo Pisani and Giorgio Maria Giannuzzi

Received: 27 June 2022
Accepted: 28 July 2022
Published: 6 August 2022

1. Introduction

Recently, power grid systems have been operated based on conventional fossil fuel-oriented centralised power plants (CPPs) and also distributed power sources of renewable energy. This is due to the global eco-friendly energy policy stance and emerging countries' efforts to improve electrification ratio. This phenomenon has advantages such as carbon emission reduction and power generation cost reduction, but there are problems in the frequency control part due to the phenomena of the system's complexity, intermittent power generation, and system inertia decline. Power generation increased as the proportion of distributed power source installations using renewable energy sources increased, but it also brought problems, such as a decrease in system inertia and frequency changes due to power output changes owing to intermittent outputs. A battery energy storage system (BESS) with high flexibility and fast responsiveness is gaining traction as a solution to this problem. The BESS can perform the roles of power generation and load because it can charge and discharge. Therefore, it can alleviate the power output changes of intermittent distributed power sources and can also be used in frequency controls since it responds quickly in a low-inertia system. As a result, it is widely used as a resource for primary frequency control (PFC). The main areas of research related to this can be classified into research on the development of physical models and research on optimal operation strategies for PFC. The physical model domain of the BESS is divided into dynamic models for a single battery [1–3], dynamic models at the battery pack level [4,5], and dynamic models at the energy storage system level [6–8]. Studies on physical models form the basis of development research on optimal operation strategies for PFC that reflect the properties of batteries. Based on this, simulation studies have been conducted for optimal operation strategies. The main research areas are research on the effectiveness and feasibility assessment of BESS for PFC [9–15], optimal SOC operation strategies of BESS for PFC [16,17], and optimal droop control and inertia control strategies of BESS for PFC [18–22]. As such, virtual physical models of BESS have been used in PFC simulations to analyse the characteristics of the BESS operated by PFC or the characteristics of the power system, resulting in the development of optimal operation strategies.

However, most previous studies did not consider the physical response delay time that occurs between the PFC hardware controller for the BESS and the power conditioning system (PCS). If a frequency transient state occurs in a low-inertia system, the rate of change of frequency (RoCoF) decreases rapidly in the initial stage. In this case, the inhibition of RoCoF may vary depending on how quickly the output of BESS is put in. To examine this, it is necessary to track the power system's characteristic trajectories in a short time, considering the response delay time of the BESS and the amount of PFC response provided by the CPP. Because of this, these characteristics cannot be examined in detail when closed-loop-type simulations or power system equivalent models are used, as in the aforementioned previous studies. As a result, the concept of cyber physical system (CPS) has recently emerged, which connects many hardware systems linked by a network with virtual physical systems. In general, the target CPS is a hybrid system because it includes both analogue continuous elements (such as physical elements and electricity) and digital discrete elements (such as control algorithms). In a power grid system, therefore, analogue continuous models, such as power generation, power transmission, and load, should be implemented to construct CPS-type systems, and digital discrete models, such as control algorithms and hardware controllers, should be distinguished and developed. Furthermore, the simulation-based model verification techniques, software in the loop simulation (SILS) and hardware in the loop simulation (HILS), should be constructed to provide a basis for verifying the target model's control algorithms and hardware controllers. As can be seen in recent research cases, the following studies have been conducted: performance verification of Volt-Var optimization engine using IEC61850t-Var [23], co-simulation verification of power and communication systems for dynamic analysis of a micro-grid [24], verification of wide-area communication platform for power grid monitoring system [25], efficiency verification of wireless communication technology in a power system [26,27], and investigation of the impacts of cyber contingency on power system operation [28]. As such, CPS simulations can implement new power operation systems and solutions in simulations to evaluate the impacts and validity of the system's control algorithms and communication systems, which are to be adopted.

As a result, this study employs CPS simulations to obtain real-time characteristic trajectory analysis results based on the effects of the BESS response delay time for PFC, as well as to implement hierarchical multi-systems of hardware as virtual physical systems to provide results evaluating the response performance of the hardware controllers.

Section 2 introduces A1GridSim, named as advanced real-time dynamic power grid simulator for CPS developed using ProTRAX software. The concept design began development in 2014, and the first prototype and its results were published in 2016 [29].

It consists of a power grid model that includes power generation, power transmission, power transformation, and load; a BESS model that includes batteries, PCS, and FRC; and a communication model that includes the protocols of DNP3.0 and Modbus. Two types of verification and evaluation studies were conducted to test the performance and various functions of the developed simulator. Section 3 shows the results of SILS for the comparative verification of the dynamic frequency model of A1GridSim based on the transient state frequency history data; characteristic trajectory analysis in the transient state of the BESS for PFC; and the impact of the BESS and the characteristic trajectory analysis of the power system according to the physical response delay time (PRDT) between the PCS and FRC. Section 4 constructs the hierarchical multi-system of the actual hardware system and performs co-simulation with the hardware controller installed on-site to show the results of HILS for the performance evaluation.

2. Development of Co-Simulation "A1GridSim"

The name of the CPS simulator developed in this study is "A1GridSim", and it was developed using the ProTRAX software, which provides a real-time dynamic simulation environment [30]. ProTRAX provides high-quality, high-fidelity virtual physical components, such as a boiler, combine cycle, circulating fluidized bed (CFB), the balance of plant, AC/DC electric power, and renewable energy. It is used as simulation software for CPS and digital-twins by providing programmable malfunctions that can simulate the transient state of each physical component; an emulator function for a distributed control system; and an application programming interface (API) function for the linkage with external solvers.

As shown in Figure 1, "A1GridSim" shows the configuration of the national power system, targeting the power system of South Korea. The national power grid model connects the mainland power grid of 97 GW and the Jeju island power grid of 0.8 GW with a total of four HVDC lines. The main components consist of 177 CPPs and power transmission and transformation facilities of the 154, 345, and 765 kV classes.

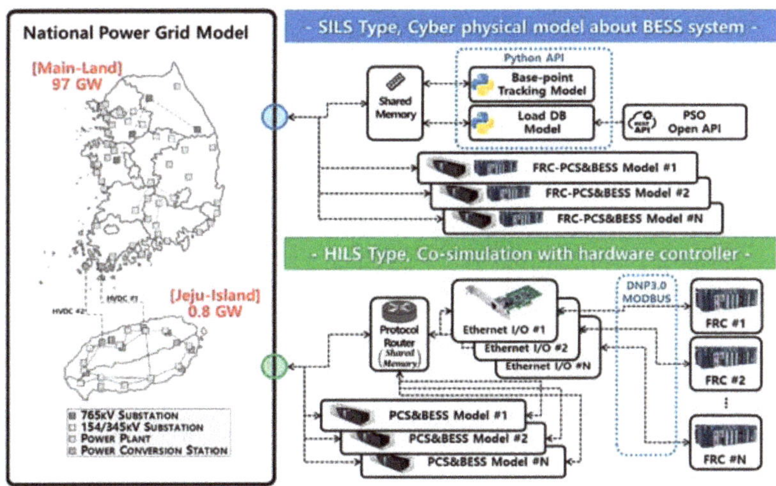

Figure 1. National power grid model and configuration of A1GridSim simulation modes.

The simulation mode can be selectable into the SILS or HILS type. The SILS type is used to analyse PFC operation algorithms by modelling the BESS for PFC as a virtual physical model. In addition, the HILS type is used to evaluate the performance of hardware equipment by building batteries and PCS as virtual physical systems among hierarchical multi-systems and configuring a communication model that can link up with the actual

FRC using DNP3.0 and Modbus protocols. As a result, this section introduces the main components of A1GridSim, which are a CPP model, an EMS model, a dynamic frequency model, and a BESS system model.

2.1. Centralised Power Plant (CPP) Model

The CPP models consist of turbine and generator modules. There are four model types, as shown in Figure 2: nuclear power plant (NP), thermal power plant (TP), combined-cycle power plant (CC), and pumped-storage hydroelectricity plant (PSH). NP and TP consist of IEEG1; CC has GAST as the gas turbine and CIGRE HRSG/ST as the steam turbine; and PSHP has the HYGOV module as a turbine module [31–33]. They are linked to ELGEN, which is a synchronous generator module of ProTRAX [34]. They are dynamic models, in which the turbine and the generator are synchronised; when the turbine and generator are synchronised and connected to the power system, the phase control of the electro-motive force (EMF) phasor is adjusted so that the generator's electrical output will be the same as the output of the shaft rotating by the turbine; the generator's free electric power and voltage change according to the change in the EMF vector size; a deviation between the mechanical output and electrical output occurs according to the angular velocity variation.

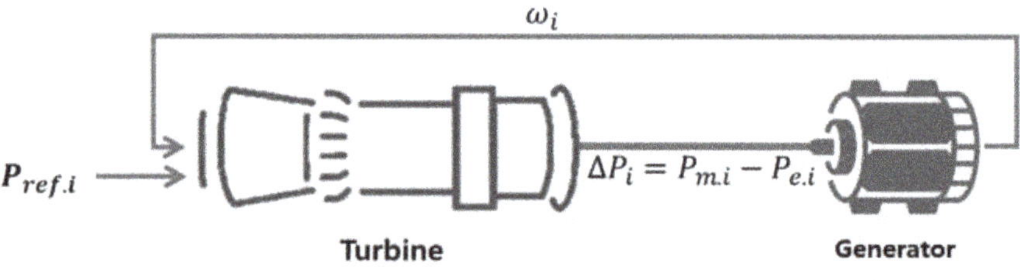

Plant Type		Turbine	Generator
Nuclear Power Plant	(NP)	IEEEG1	ELGEN
Thermal Power Plant	(TP)	IEEEG1	ELGEN
Combined-Cycle Power Plant	(CC)	GAST, CIGRE HRSG/ST	ELGEN
Pumped-Storage Hydroelectricity Plant	(PSH)	HYGOV	ELGEN

Figure 2. Configuration of CPP model and detail module about turbine and generator.

Table 1 shows the typical parameters of the CPP models. The parameters were setuped from the KPX open-API information based on 177 CPP generators in South Korea. The fuel types include nuclear, coal, heavy oil, LNG, and water. In the total capacity of 98.12 GW, NP accounts for 23.7%, TP accounts for 38.68% (Coal-35.74%, Heavy oil-2.95%), CC accounts for 32.48%, and PSH accounts for 3.7%.

Table 1. Typical parameter of CPP models [35–37].

	NP	TP	TP	CC	PSH
Fuel Type	Nuclear	Coal	Heavy Oil	LNG	Water
Rated capacity (GW)	23.25	35.74	2.95	32.48	3.7
Number (EA)	24	65	13	63	12
Efficiency (%)	33.0~39.0	33.0~42.0	32.0~33.0	42.0~54.0	76.0~82.0
Inertia Constant (s)	6.0~9.0	4.0~7.0	4.0~5.0	3.0~6.0	2.0~4.0
Turbine speed (RPM)	1800	3600	3600	3600	300~600
Droop (%)	-	5~9	4~6	6~8	3~4
Dead band (Hz)	-	±0.0~±0.038	±0.01~±0.03	±0~±0.06	±0.02~±0.033
Ramp rate (%MW/min)	-	0.7~3.1	1.1~2.3	2.6~16.0	22.0~32.0
Heat rate (MBtu/MWh)	9200~10,700	8000~10,900	11,000~11,200	6400~8500	-

To synchronise the turbine's speed with the frequency of the system measured at the generator, the CPP model performs the governor free (GF), which is a PFC service. The response $P_{PFC.i}$ is calculated using the rated power $P_{rated.i}$, the droop coefficient K_{droop}, the generator's rotating speed ω_i and rated rotating speed $\omega_{rated.i}$, as shown in Equation (1).

$$P_{PFC.i} = \frac{P_{rated.i}(\omega_{rated.i} - \omega_i)}{\omega_{rated.i}K_{droop}}100 \ [\text{MW}] \tag{1}$$

K_{droop} is configured to be 5% to 9% for TP (coal), 4% to 6% for TP (heavy oil), 6% to 8% for CC, and 3% to 4% for PSH. The CPP's PFC response section refers to the operation outside the frequency dead-band region, which is configured to be $\pm0\sim0.038$ Hz for TP (coal), $\pm0.01\sim0.03$ Hz for TP (heavy oil), $\pm0\sim0.06$ Hz for CC, $\pm0.02\sim0.033$ Hz for PSH. The output relationship between the turbine and the generator according to the angular velocity is defined as a swing equation, as shown in Equation (2). The deviation between the turbine output $P_{Mech.i}$ and the generator output $P_{Elec.i}$ is caused by the inertia constant H_i, rated capacity $S_{rated.i}$, and angular acceleration, which is the differential of the angular velocity ω_i.

$$2H_i S_{rated.i}\frac{d(\omega_i/\omega_{rated.i})}{dt} = P_{Mech.i} - P_{Elec.i}[\text{MW}] \tag{2}$$

The inertia constant H_i is given in seconds and represents the time it takes to fix the rotating generator at the synchronous speed in a state that mechanical power is not supplied, and the rated output is extracted. H_i is 6.0 to 9.0 for NP, 4.0 to 7.0 for TP (coal), 4.0 to 5.0 for TP (heavy oil), 3.0 to 6.0 for CC, and 2.0 to 4.0 for PSH. Therefore, an output deviation occurs depending on the angular speed of the kinetic energy stored in the turbine-generator, which is the product of the inertial constant H_i and the rated capacity $S_{rated.i}$. Because the rotation speed of the synchronised turbine and generator does not vary significantly, it is changed according to the power system's frequency change, and if a transient state frequency occurs, insufficient power is transferred from the stored kinetic energy, resulting in a decrease in the turbine's speed.

2.2. Energy Management System (EMS) Model

The EMS Model performs the secondary frequency control (SFC) and the tertiary frequency control (TFC) to maintain the power system's power and frequency balance, as shown in Figure 3. SFC is carried out by the AGC module, which estimates the power generation of the power grid model's CPP model and the load of the transmission/substation model and the load model. TFC is performed through the base-point tracking module. The area AGC module that provides the SFC service performs the role of generating a frequency control response every four seconds by measuring the frequency change of a short period in 2-s intervals. The area control error, ACE is calculated through the frequency deviation Δf and the system control gain B. By calculating the integral of ACE, the power system's total AGC demand P_{AGC_Demand} is calculated, as shown in Equation (3):

$$P_{AGC\_Demand} = \int ACE = \int 10B(\Delta f)[\text{MW}] \tag{3}$$

The AGC participation rate of each CPP is calculated through the ramp rate $K_{ramp.i}$ of the CPP model currently incorporated into the system, and as shown in Equation (4), the AGC response $P_{AGC.i}$ is delivered every four seconds.

$$P_{AGC.i} = P_{AGC\_Demand}\left(\frac{K_{ramp.i}}{\sum_{i=0}^{n}K_{ramp.i}}\right) \ [\text{MW}] \tag{4}$$

The base point tracking module, which provides the TFC service, responds to the load change over a long period to estimate the load and select a start generator every five minutes and delivers the base point $P_{base.i}$ every one minute. Here, the methods of selecting a start generator are divided into economic load dispatch (EcoLD) and environmental load dispatch (EnvLD) methods. First, the heat rate equation for each CPP model in Table 1 can be shown as Equation (5):

$$H_i(P_{base.i}) = \frac{a_i}{P_{base.i}} + b_i + c_i P_{base.i} \; [\text{MBtu/MWh}] \tag{5}$$

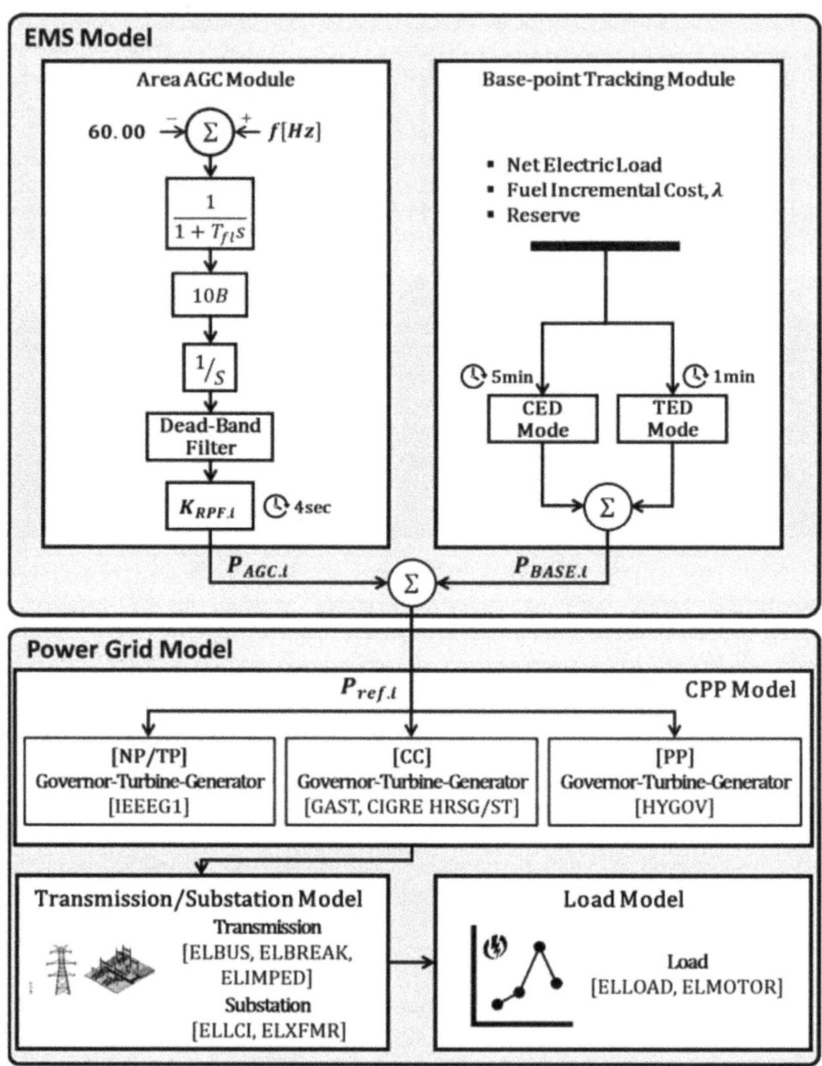

Figure 3. Main component and process of EMS model.

Here, a fuel cost equation can be created by multiplying the fuel cost FC_i and the basepoint $P_{base.i}$. Equation (6) is the fuel cost equation F_{Eco} of EcoLD, and Equation (7) is the fuel cost equation F_{Env} of EnvLD, considering the CO_2 trading price C_{Co2} and the CO_2 emission coefficient K_{emi}:

$$F_{Eco}(P_{base.i}) = \left(c_i P_{base.i}{}^2 + b_i P_{base.i} + a_i \right) \times FC_i \ [\$/h] \tag{6}$$

$$F_{Env}(P_{base.i}) = \left(c_i P_{base.i}{}^2 + b_i P_{base.i} + a_i \right) \times FC_i + (C_{Co2} \cdot K_{emi}) \times P_{base.i} \ [\$/h] \tag{7}$$

Through the fuel cost equations, we can find the incremental fuel cost λ_i, which shows the slope of the power output and fuel cost characteristics, as shown in Equation (8):

$$\lambda_i = \frac{dF(P_{base.i})}{dP_{base.i}} = 2\alpha_i P_{base.i} + \beta_i \ [\$/MWh] \tag{8}$$

In the case of the CPP model operating with a certain power output, λ_i represents the incremental fuel cost per unit of time required when the output is increased by 1 MW in this operating state. Therefore, start generators are connected to the power system according to the load in order of the lowest λ_i. In general, in plant power costs, plant operating costs, maintenance and repair costs, and labour costs are all included, but this model only considers pure fuel costs. The cost for each CPP fuel and emission factor were considered according to the EcoLD and EnvLD operation methods for this process, as shown in Table 2, based on which plans of start generators were simulated. Figure 4 shows the results. NP, which has the lowest fuel cost, is responsible for the base load, and the operation start up order is determined in the order of TP (coal), CC (LNG), and TP (heavy oil). In the results of the start up plans, the difference between the two power supply methods is that the incremental fuel cost λ_i is about 10 to 20 \$/MWh higher overall in the EnvLD method than in the EcoLD method. It is found that in EnvLD, some CCs, which have good efficiency because of emission factor adjustments, begin to output, starting from a power load of 40 GW.

Table 2. Average fuel cost and CO_2 emission factor by fuel type.

	NP	TP Bituminous	TP Anthracite	CC LNG	TP Heavy Oil
Fuel Cost (\$/MWh)	4.96	44.17	53.35	72.46	141.11
Emission Factor	-	0.8867	0.8867	0.3889	0.6588

Based on the power generation start up plan, the generator's reference output $P_{base.i}$ is calculated in the base point tracking module. This process can satisfy all power loads estimated for the CPPs that have been set up with the start up plan every five minutes, and the control economic/environmental dispatch (CED), which undergoes the minimisation process of the incremental fuel cost λ_i, is performed. After performing CED every minute, the tracking economic/environmental dispatch (TED), which calculates the power output adjustment to respond to the changed load, is performed. This process was implemented using the cvxpy optimization library through the Python API and configured according to Theorem 1. In the CED, constraints are set up through variables and parameters, such as the CPP's maximum power output $P_{max.i}$ and minimum power output $P_{min.i}$, the incremental fuel cost equation's coefficients α_i and β_i, and the start up state On_i based on the generator's operation and stop time constraints. Furthermore, $P_{base.i}$ is allocated by performing optimization to minimize the maximum λ_i among all operating CPPs. The TED performs the power output adjustment according to the load at which the sum of $P_{base.i}$ calculated in CED is estimated. Based on this process, we performed a one-day simulation that allocates $P_{base.i}$ in A1GridSim.

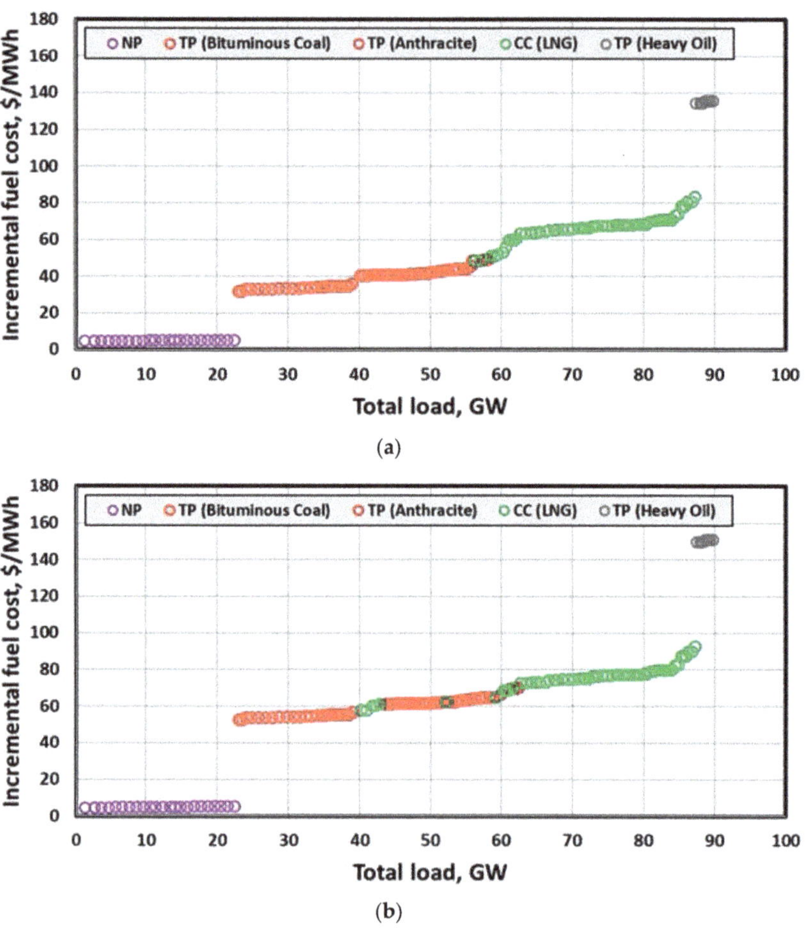

Figure 4. Optimum generator unit commitment curve by load; (**a**) Economic load dispatch (EcoLD); (**b**) Environmental load dispatch (EnvLD).

Figure 5 shows the one-day simulation result based on the base-point tracking mode. It shows the power generation situations layered in the start up order defined by the power generation start up plan. In the overall trend, it is found that power is generated in the order of NP, TP (coal), CC, and TP (heavy oil). The biggest difference between EcoLD and EnvLD, as shown in the start up plan, is that NP and TP (coal) account for the majority of EcoLD below 55 GW, whereas CCs, which have high efficiency due to the emission factor, are operated for base power generation starting at 40 GW in EnvLD. As such, the EMS model performs the SFC in the area AGC module in a short period and performs the TFC in the base-point tracking module in a long period to balance the power generation demands and maintain the quality of the power frequency. This means that the EMS model includes the centralised frequency control system that operates in the actual EMS. Furthermore, when it is performed according to the power generation start up plan and the base-point tracking mode, considering the dynamic characteristics of CPP, it shows the characteristics of the national power grid, in which the CPP models that have different inertia characteristics depending on the changing load characteristics are connected to the system and operated.

Theorem 1. *Process of optimization for CED and TED by python API.*

1. ***Control Economic/Environmental Dispatch***

   Import cvxpy as cp
   Import numpy as np
   ***def Control_ED**(n, P_{load}, P_{max}, P_{min}, α, β, On):*

   $P_{CED.i} = cp.variable(shape{=}n, nonneg{=}True)$
   $\lambda_i = cp.variable(shape{=}n, nonneg{=}True)$
   $P_{max.i} = cp.parameter(shape{=}n)$
   $P_{min.i} = cp.parameter(shape{=}n)$
   $\alpha_i = cp.parameter(shape{=}n)$
   $\beta_i = cp.parameter(shape{=}n)$
   $On_i = cp.parameter(shape{=}n)$
   $P_{max.i}.value = np.array(P_{max})$
   $P_{min.i}.value = np.array(P_{min})$
   $\alpha_i.value = np.array(\alpha)$
   $\beta_i.value = np.array(\beta)$
   $On_i.value = np.array(On)$

   Constraints = list()

   $Constraints.append(\sum_{i=0}^{n} P_{CED.i} = P_{load})$
   $Constraints.append(P_{CED.i} \leq P_{max.i} \times On_i)$
   $Constraints.append(P_{CED.i} \geq P_{min.i} \times On_i)$
   $Constraints.append(\lambda_i \leq 2\alpha_i P_{CED.i} + \beta_i)$

   $Object = cp.Minimize(max(\lambda_1, \lambda_2, \cdots, \lambda_{n-1}, \lambda_n))$
   Prob = cp.Problem(Object, Constraints)
   Prob.solve()

   $Return\ \lambda_i.value,\ P_{CED.i}.value$

2. ***Tracking Economic/Environmental Dispatch***

   ***def Traking_ED**(P_{load}, \lambda_i, \sum_{i=0}^{n} P_{CED.i}):*

   $If\ P_{load} - \sum_{i=0}^{n} P_{CED.i} > 0:$

   $$P_{TED.i} = \left(P_{load} - \sum_{i=0}^{n} P_{CED.i} \right) \times \frac{\lambda_i}{\sum_{i=0}^{n} \lambda_i}\, elseif\ P_{load} - \sum_{i=0}^{n} P_{CED.i} < 0:$$

   $$P_{TED.i} = \left(P_{load} - \sum_{i=0}^{n} P_{CED.i} \right) \times \frac{\frac{1}{\lambda_i}}{\sum_{i=0}^{n} \frac{1}{\lambda_i}}\, Return\ P_{TED.i}.value$$

2.3. Dynamic Frequency Model

The dynamic frequency model is a frequency model that has dynamic characteristics according to the power generation and demand. It is possible to approximate the relationship between the CPP model's turbine and generator by creating a kinetic equation of all parallel generators in the current power system, such as Equation (2). It can be represented approximately as Equation (9):

$$\sum_{i=0}^{n} \frac{2H_i S_i}{f_{rated}} \frac{df_{sys}}{dt} = \sum_{i=0}^{n} P_{Mech.i} - \sum_{i=0}^{n} P_{Elec.i} [\text{MW}] \tag{9}$$

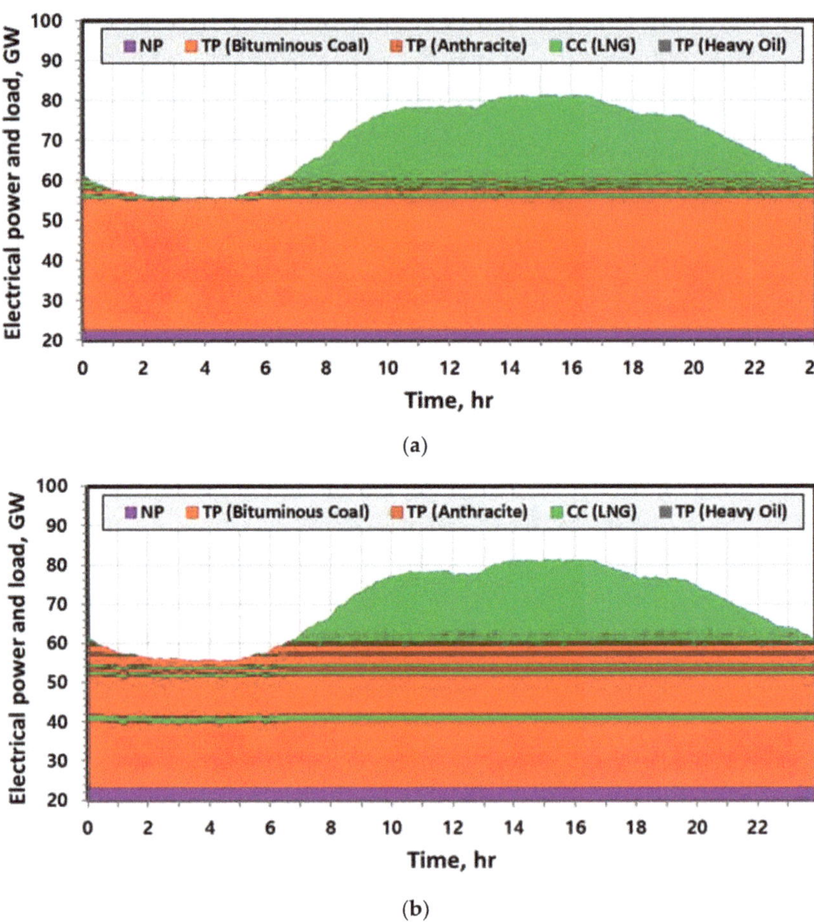

Figure 5. The 24-h Simulation result of electric power supply and demand by base-point tracking mode (December 2019); (**a**) Economic load dispatch (EcoLD); (**b**) Environmental load dispatch (EnvLD).

Equation (9) shows the deviation between $\sum\limits_{i=0}^{n} P_{Mech.i}$ (the sum of $P_{Mech.i}$, which is the output of the turbine) and $\sum\limits_{i=0}^{n} P_{Elec.i}$ (the sum of $P_{Elec.i}$, which is the output of the generator) causes a change in the power system frequency f_{sys}. Furthermore, general resistance load is not related to frequency but has the self-control characteristics, in which the power consumption changes according to the change in frequency in proportion to the rotational speed at the rotator load, as shown in Equation (10):

$$K_L \Delta f = \frac{\%K_L \cdot P_{Load}}{10} \Delta f [\text{MW}] \tag{10}$$

It is obtained using the load constant K_L [MW/0.1 Hz] and the frequency deviation Δf. K_L is obtained using $\%K_L$ of a 0.2–0.6% range and the total load P_{Load}. Δf is obtained as the deviation between f_{sys} and f_{rated}(60 Hz).

Therefore, Equation (11) can represent the characteristics of the power frequency based on the CPP Model's inertia characteristics and the load's self-control characteristics.

$$\sum_{i=0}^{n} \frac{2H_i S_{rated.i}}{f_{rated}} \frac{df_{sys}}{dt} + K_L \Delta f = \Delta P_{sys} \, [\text{MW}] \tag{11}$$

It can be seen that the power deviation occurs due to the deviation of the turbine-generator output caused by the frequency change and the load's self-control characteristics. If Laplace transform is applied to this, the relationship between the power deviation δ and frequency deviation φ can be represented by Equation (12):

$$\frac{\varphi}{\delta} = \left(\sum_{i=0}^{n} \frac{2H_i S_{rated.i}}{f_{rated}} s + K_L \right)^{-1} \, [\text{Hz/MW}] \tag{12}$$

where the power system's proportional gain K_F and the frequency time constant T_F can be obtained by Equations (13) and (14):

$$K_F = \frac{1}{K_L} \, [\text{Hz/MW}] \tag{13}$$

$$T_F = \sum_{i=0}^{n} \frac{2H_i S_{rated.i}}{f_{rated} K_L} \, [\text{s}] \tag{14}$$

The power system's proportional gain K_F is determined by the change in the power load according to the load constant K_L. The frequency time constant T_F is determined by the load constant K_L and the kinetic energy stored in the CPP model incorporated according to the generator start up plan.

If 24-h simulation of the dynamic frequency model is performed using the CPP model, EMS model, and power grid model examined so far, we can obtain results such as those shown in Figure 6. Figure 6 shows the results of performing the 24-h simulation using power load data from December 2019. For loads of 55.401 GW minimum and 81.535 GW maximum, the power generation output is operated from 55.393 GW minimum to 81.454 GW maximum. In general, the base load is shown between 00:00 and 06:00, and the peak load is shown between 12:00 and 18:00. Accordingly, the frequency deviation is −0.012 to +0.037 Hz in the base load period and −0.062 to +0.049 in the peak load period.

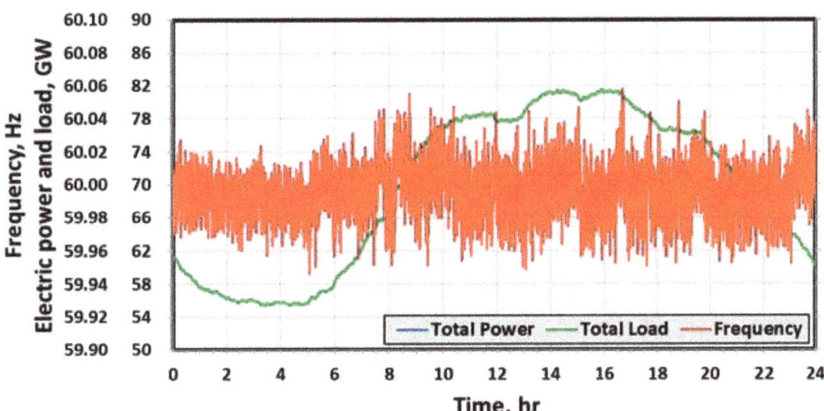

Figure 6. The 24-h simulation result of dynamic frequency model in A1GridSim.

2.4. BESS System Model

The BESS model is constructed based on the control algorithm and hardware control system developed in the "Frequency regulation control for BESS" project led by the Korea Electric Power Corporation [38–40], and its target was the BESS of 380 MW/111 MWh installed in substations across the country. Figure 7 shows the status of BESS for PFC installed on A1GridSim. In total, 376 MW PCS and 103 MWh batteries have been installed at 13 substations on the mainland, and 4 MW PCS and 8 MWh batteries have been installed on Jeju Island. A typical BESS consists of one FRC and four PCS/ESS sets, though this varies by substation. The hardware hierarchical multi-system of the BESS for PFC has a structure of the master power management system (MPMS), Local power management system (LPMS), Frequency regulation control master (FRCM), FRC, PCS, and BESS. The FRC and FRCM are referred to as the LPMS, and their role is to coordinate the FRC's participation rate. MPMS is a hierarchical layer for the integrated operation of LPMS at the substation level. The virtual physical model's PCS/BESS is physically linked to the power grid model, and the discrete models, such as the operation sequence of PCS, are constructed as emulation models in linkage with FRC. The FRC's control algorithm is simulated in linkage with the PCS/ESS model in the SILS mode. In the HILS mode, the simulation is performed by linking the actual FRC hardware controller with a communication model, such as DNP3.0 or Modbus.

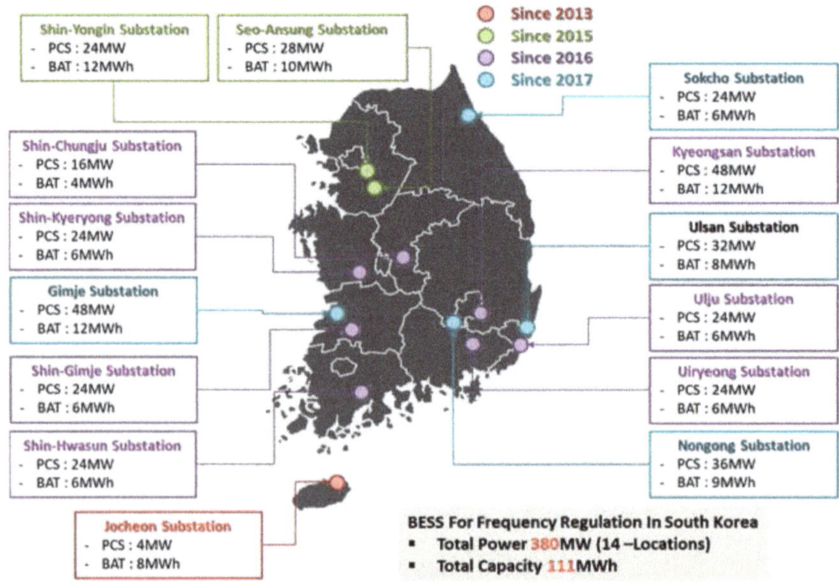

Figure 7. Installation status of BESS System for PFC in South Korea power grid (2013~2017).

3. Software in the Loop Simulation for BESS System

In this section, we will examine the performance of the PFC algorithm considering PRDT, based on the SILS mode of A1GridSim. To this end, Section 3.1 shows the results comparatively verifying the actual transient state frequency history data and the A1GridSim's transient state simulation frequency. Section 3.2 performs transient state simulations to compare the case of no BESS and the cases of PRDT being 140 ms, 160 ms, and 180 ms and shows the results of analysing the effects of operating PFC and analysing the trajectories of the system characteristics by physical delay time. Finally, Section 3.3 analyses the system characteristic trajectories for each PRDT from 140 ms to 640 ms in 20 ms increments and shows the results of the comparative analysis of the major characteristics.

3.1. Verification of Dynamic Frequency Model

To verify the dynamic frequency model of A1GridSim, a real-time dynamic simulation that includes the national power grid model, we conducted a comparative analysis using the frequency transient state history data. The frequency transient state history data target is the data of the frequency transient state caused by the halting incident of one 950 MW nuclear power generator during the peak load time at 3:00 p.m. in August 2016. In the simulation environment, too, the same load data were entered to simulate the halting incident of one 950 MW nuclear power generator where the CPP model is operated by the EMS model. Figure 8 shows the actual historical data and the simulation results. Real refers to the actual frequency history data, and History refers to the history of actual situations.

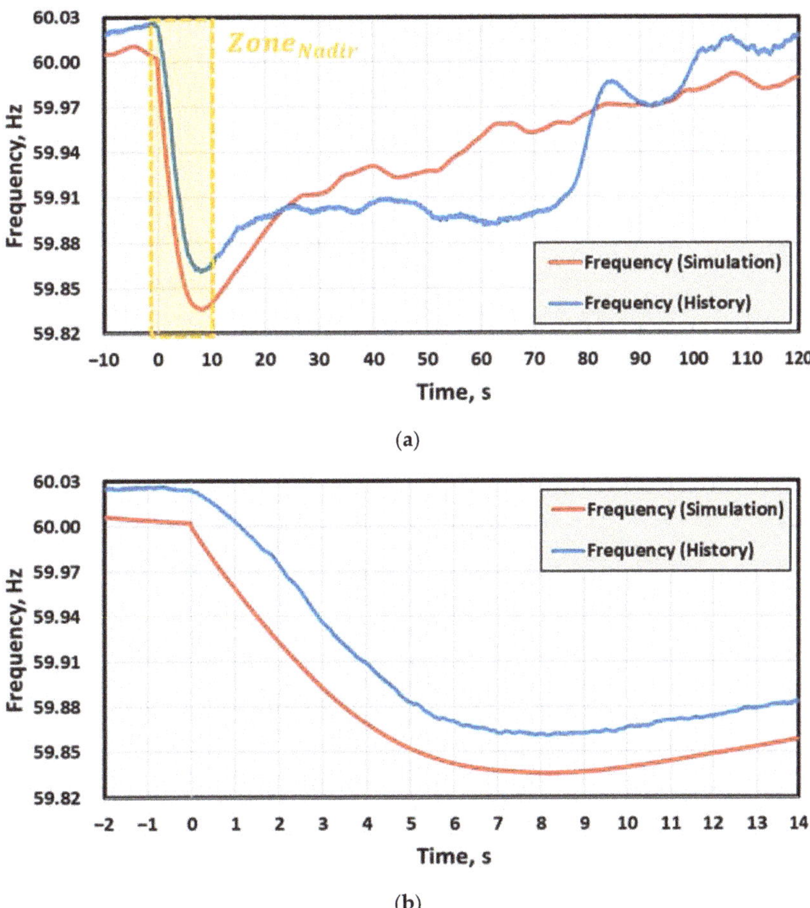

Figure 8. Comparison of history and simulation frequency for verification; (**a**) Frequency transient state until recovering to rated frequency; (**b**) Frequency transient state until nadir frequency point.

In history, a frequency drop of 0.1632 Hz occurred, and it took 8.16 s to reach the nadir frequency point. In the simulation, a frequency drop of 0.1648 Hz occurred, and it took 7.88 s to reach the nadir frequency point. When we examined the average RoCoF up to the nadir point in the state just before the incident, we found that it was 0.0207 Hz/s in Real and 0.0202 Hz/s in Sim. When the error rates were examined, it was found that the error rate was 0.9% in the frequency drop, 3.5% in the time consumed to reach the nadir point, and 2.48% in the average RoCoF, showing there is a high degree of similarity in the transient state.

3.2. Performance Verification of the BESS System for PFC

In this subsection, to examine the performance of the BESS for PFC, we simulate a halting incident of one 950 MW generator in a situation where the frequency change is the largest in the low-inertia system. To this end, we conducted case studies by configuring a case where the transient state is recovered with the CPP model alone without deploying the BESS and cases in which the PRDT is 140 ms, 160 ms, and 180 ms. Figure 9 shows the results of case studies in terms of RoCoF, BESS power, and CPP's GF response.

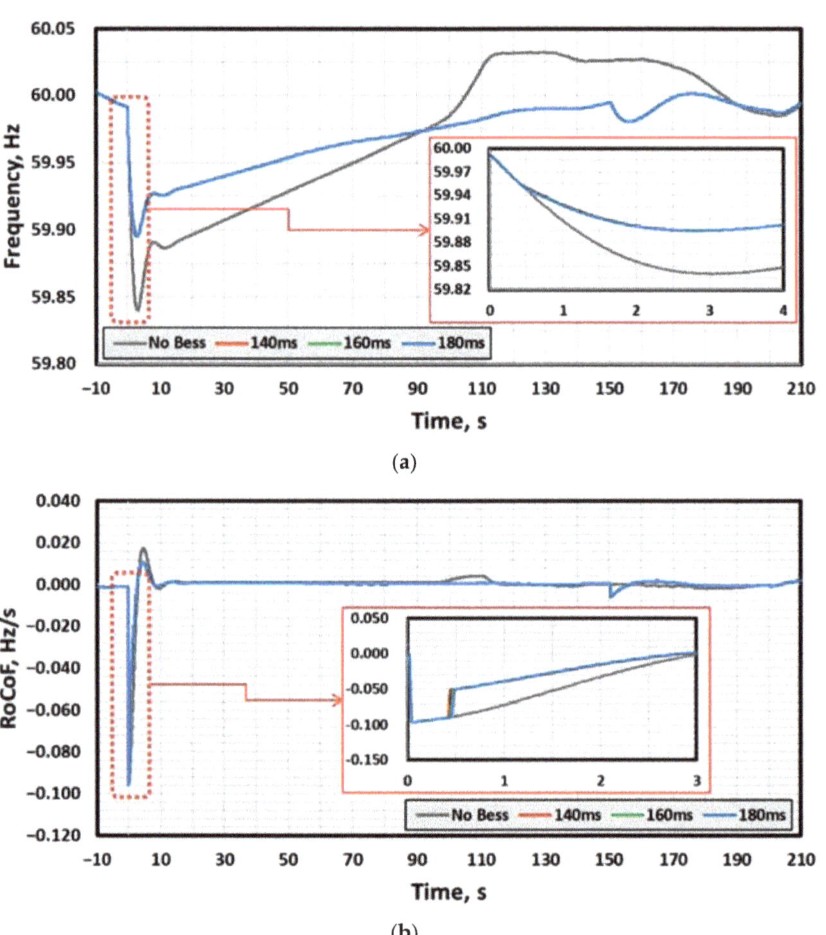

(a)

(b)

Figure 9. *Cont.*

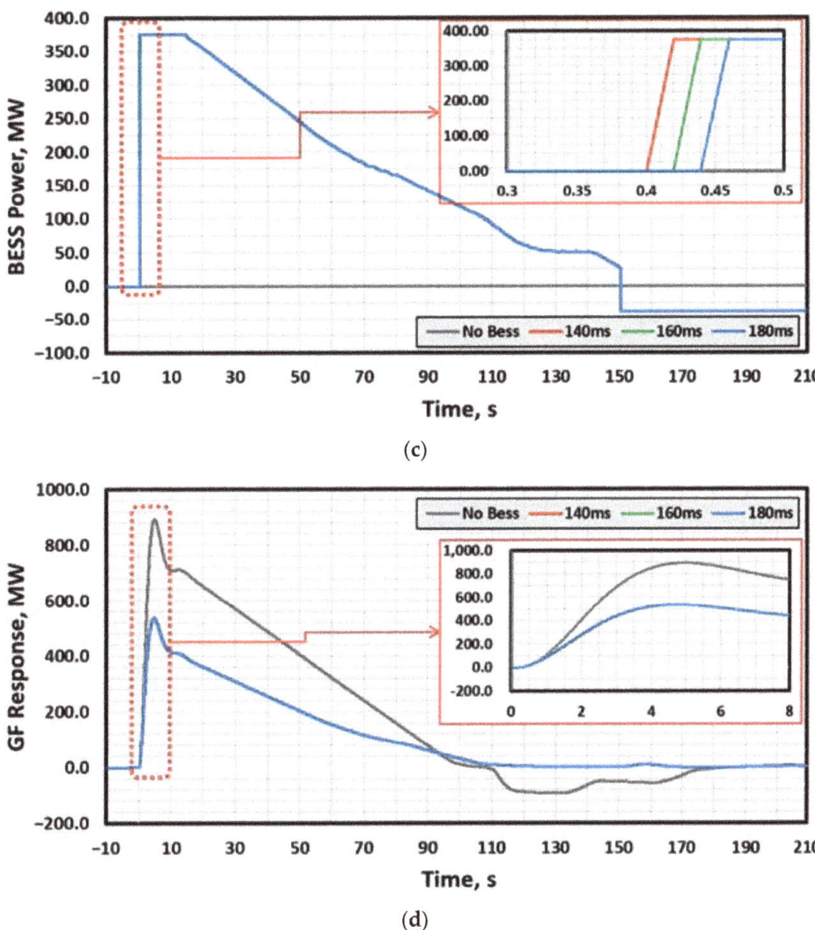

Figure 9. Result of frequency transient state simulation by case study; (**a**) Frequency; (**b**) RoCoF; (**c**) BESS Power; (**d**) GF response.

In Figure 9, (a) shows the frequency, (b) shows RoCoF, (c) shows the BESS power, and (d) shows the GF response. The main results of Figure 9 are summarised in Table 3. When the results of BESS cases are compared to the case of No BESS as a reference case, the time for reaching the nadir frequency point, T_{Nadir}, decreased by 6.58%, 7.24%, and 7.89% in the cases, respectively. Meanwhile, the frequency deviation Δf_{Nadir} decreased by 34.58%, 34.46%, and 34.40%, respectively. When the average RoCoF is calculated using Δf_{Nadir} and T_{Nadir}, the result is 0.0526 Hz/s, 0.0368 Hz/s, 0.0372 Hz/s, and 0.0375 Hz/s in Cases 1 to 4, respectively. As many previous studies have shown, it is found that Δf_{Nadir} and the average RoCoF are greatly improved in the cases of having the BESS for PFC compared to the case of No BESS. Furthermore, as shown in the simulation results of BESS power in Figure 9c, Δf_{Nadir} and the average RoCoF improve as the PRDT becomes faster. Finally, the time for recovering to the rated frequency f_{rated} is 30% faster for No BESS, but in the case of No BESS, an overshooting phenomenon occurs in the process of recovering to f_{rated}, showing that it takes about 20% longer time compared to the case of having BESS.

Table 3. Summary of simulation results from the several case studies.

	No BESS (Case1)	140 ms (Case2)	160 ms (Case3)	180 ms (Case4)
$f_{initial}$	59.9914	59.9914	59.9914	59.9914
T_{nadir}	3.04	2.84	2.82	2.80
f_{nadir}	59.8401	59.8954	59.8952	59.8951
Δf_{nadir}	0.1599	0.1046	0.1048	0.1049
$RoCoF_{average}$	0.05259	0.0368	0.0372	0.0375
$RoCoF_{BESS}$	-	−0.0510	−0.0505	−0.0500
GFR_{max}	893.86	535.02	538.72	539.11
$T_{recovery}$	105.1	150.2	150.2	150.2

When the instantaneous RoCoF in the transient state is examined, it is found that the minimum RoCoF to reach f_{Nadir} is −0.0958 Hz/s in every case. The reason is that it is the result before the BESS system responds. Therefore, when it is examined at the respective response time of the BESS (Cases 2 to 4), it is −0.051 Hz/s, −0.0505 Hz/s, and −0.05 Hz/s, respectively, which means the BESS responds faster as the RoCoF decreases. Lastly, GFR_{max}, the CPP's GR response value, has decreased by 40.14%, 37.73%, and 39.69% in the cases of having BESS compared to the case of No BESS.

The results of the case studies examined so far were analysed through the system characteristic trajectory according to each case. In the transient state of the entire system, the trajectory was tracked with the relationship between the power deviation ΔP_{trn} and the frequency deviation Δf_{trn}. ΔP_{trn} and Δf_{trn} were calculated by Equation (15) and (16):

$$\Delta P_{trn} = \left(\frac{P_{Gen}^{\,t} - P_{Gen}^{\,t=0}}{P_{Gen}^{\,t=0}} \right) - \left(\frac{P_{Load}^{\,t} - P_{Load}^{\,t=0}}{P_{Load}^{\,t=0}} \right) \ [\mathrm{pu}] \tag{15}$$

$$\Delta f_{trn} = \frac{f_{sys}^{\,t} - f_{sys}^{\,t=0}}{f_{sys}^{\,t=0}} \ [\mathrm{pu}] \tag{16}$$

Figure 10 shows the relationship between the power deviation and the frequency deviation in the frequency transient state, and shows the trajectory of the power grid constant. The power grid constant can be shown by Equation (17) according to the relationship between the two:

$$K_{sys} = \frac{\Delta P_{trn}}{\Delta f_{trn}} \tag{17}$$

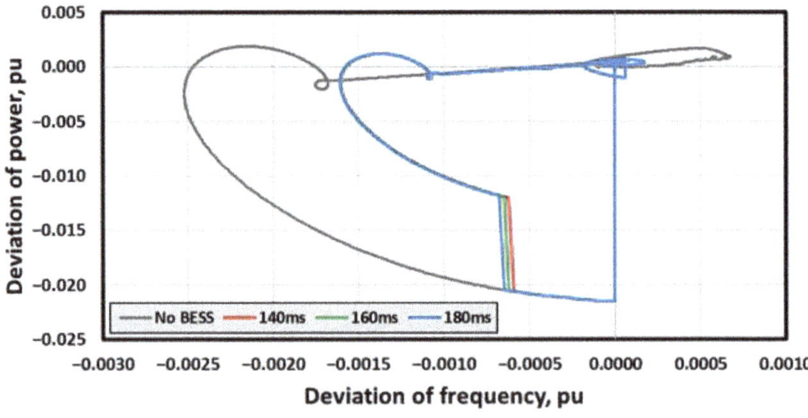

(a)

Figure 10. *Cont.*

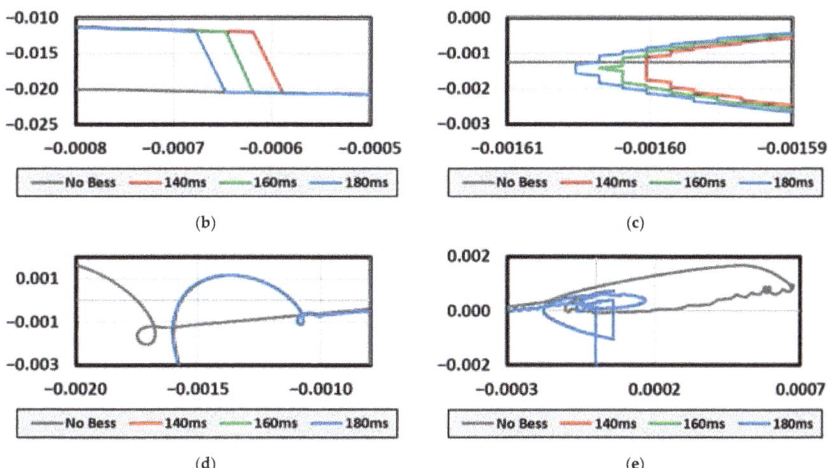

Figure 10. Curve about deviation of power and frequency by case study; (**a**) Power grid constant at entire period; (**b**) BESS response point; (**c**) Nadir frequency point; (**d**) Quasi−steady state point; (**e**) Overshooting point.

In Figure 10, (a) shows the power grid constant over the entire period, (b) shows the response point of the BESS, (c) shows the nadir frequency point, (d) Quasi-steady state point, and (e) shows the overshooting region of the frequency control in the region where the rated frequency is reached. If the generator halts at a point where the power distribution and frequency deviation are both 0 pu, the power deviation decreases sharply to −0.0215 pu, after which a frequency deviation occurs. As shown in Figure 9, every case follows the same power grid constant before the BESS responds. At the point where the BESS responds, the decrease in the power and frequency deviations is temporarily reduced, as shown in Figure 10b. After this, it proceeds to the nadir frequency point, and as shown in Figure 10c, it reaches the peak point of frequency deviation. Here, the power grid constant is 0.946, 0.891, 0.893, and 0.898 in Cases 1 to 4, respectively. These points are Points where the rated frequency is recovered from the nadir frequency, and as the value of the power grid constant decreases, the recovering ability increases. The BESS is operating at its maximum power output of 376 MW, and from the point of recovery by the CPP's GF response, the next point, the quasi-steady state, can be reached with a relatively smaller GF response as the power grid constant decreases. This is the reason why GFR_{max} is small in the case of the BESS with a fast PRDT. Next, Figure 10d shows the region for reaching the quasi-steady state through the nadir frequency point. In the presence of a power deviation, the power deviation is positive because the power generation exceeds the load, and the frequency change in the transient state is converged to 0, allowing the quasi-steady state point to be reached. The power grid constant at the quasi-steady state point is 0.935, 0.784, 0.782, and 0.779, respectively. In this region, a transient state occurs, and a stable state is maintained primarily. As the power grid constant increases, it can be shown that the stability increases in the frequency change. Here, the case of No BESS has the highest value because it is in a state where very high GF and AGC responses are incorporated into the power grid, which later causes the overshooting at the transient state recovery point. Finally, Figure 10e shows that the recovery has been achieved, as it was before the transient state occurred. In the case of No BESS, overshooting occurs, but in the case of applying the BESS, the overshooting phenomenon is inhibited through immediate charging control.

3.3. Analysis of Impact on Physical Response Delay Time between FRC and PCS

We have examined the performance verification results of the BESS for PFC above. Here, we have found that the faster the PRDT, the better the transient state recovery performance. As shown in the results of Figures 9 and 10, there is no large difference between the cases of 140 ms and 180 ms in the case studies. As a result, in this subsection, we will examine the main results by analysing the trajectories, as in Section 3.2, for the results from 140 ms, where the system can respond most quickly due to hardware characteristics, to 640 ms in 20 ms intervals.

Figure 11 shows the main results of analysing the trajectory at the transient state frequency according to the PRDT. It shows (a) the power grid constant at the nadir frequency point, (b) nadir frequency, (c) average RoCoF, (d) GF response when the BESS responds, and (e) GFR_{max}. At the point where the frequency is recovered after the maximum drop of the frequency at the nadir frequency point, the quasi-steady state can be recovered more stably as the power grid constant decreases. In case of PRDT of 200 ms, the power grid Constant is 0.86198, which is 4% lower than that of the case where the PRDT is faster than 180 ms. As explained earlier, as the PRDT increases, the power grid constant at the nadir frequency point decreases, in which case the quasi-steady state can be stably reached. However, this must be checked with the trajectories of frequency, RoCoF, and GF response. As confirmed earlier, the nadir frequency and the average RoCoF decrease as the PRDT increases. When the GF response is examined, it is found that the GF response at the BESS response point increases after 200 ms, and finally, GFR_{max} also increases. Therefore, the reason why the power grid constant at the nadir frequency point starts to decrease at 200 ms is that the power deviation decreases since the GR response rate for the BESS response is larger than that of the case faster than 200 ms. This means that the BESS responds faster than the CPP's PFC, reducing the CPP's GF response, and the goal of quickly inhibiting the frequency drop is not met due to the slow response time. This means that the PRDT faster than 200 ms must be maintained to use the performance of the BESS as much as possible.

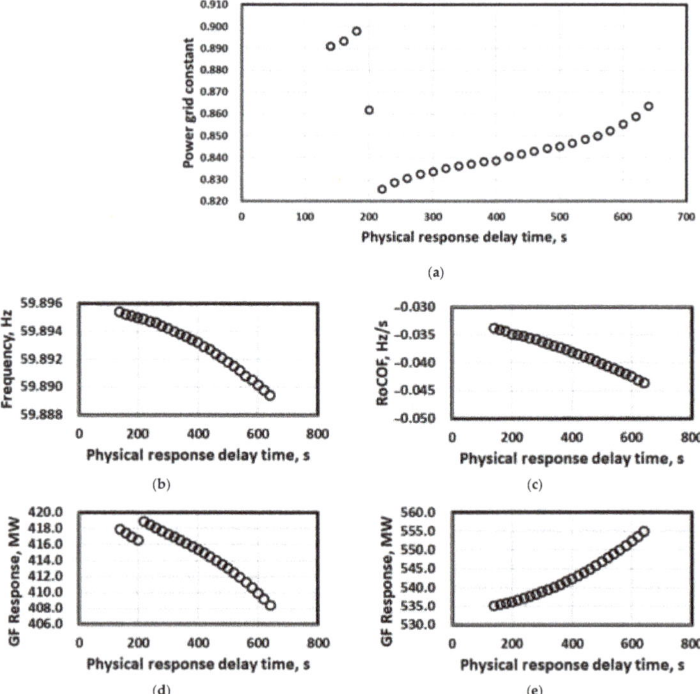

Figure 11. Main results of transient characteristic trajectory analysis considering the effect of physical response delay time; (**a**) Power grid constant at nadir frequency point; (**b**) Nadir frequency; (**c**) Average RoCoF; (**d**) GF Response at starting BESS; (**e**) Maximum GF Response.

4. Hardware in the Loop Simulation for BESS System

4.1. Configuration of the HILS System for Response Performance Evaluation

For HILS, we constructed the PCS-BESS in the BESS for PFC as a cyber physical model and linked it to the real FRC to evaluate the optimal PRDT performance. Figure 12 shows the configuration and characteristics of the actual and virtual BESS. The actual BESS measures the frequency of the power grid in 16.67 ms cycles through the frequency measurement unit (FMU) and sends it to the FRC. It takes 7 ms to generate the FRC demand based on this frequency and 150 ms to generate the BESS power through the demand received from the FRC. Through this system, the PRDT of 180 ms or less should be satisfied, as the result of SILS. To implement such a real hardware hierarchical multi-system as a CPS, we constructed the PCS, BESS, and power grid systems as physical models and linked them to the real FRC. The communication model was configured with MODBUS, a protocol that links the real FRC and the PCS, and a shared memory was configured so that the respective PCS-BESS processes can access the FRC simultaneously. Here, for the electric power grid model and the BESS model, the calculations were performed in time-steps of 16.625 ms so that the response level would be the same as that of the real facility. These configurations were implemented with portable test equipment that can check the response performance of the FRC installed at the substations, and the performance test was conducted.

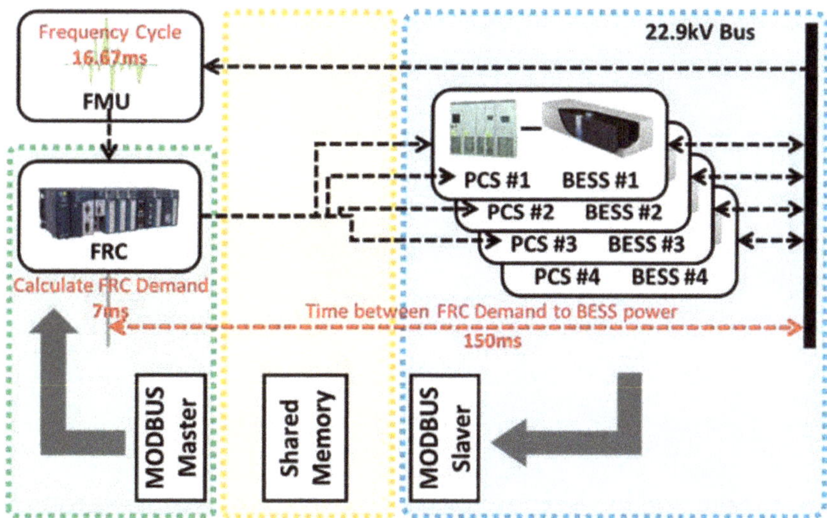

Figure 12. Configuration and characteristic of BESS System for actual and cyber physical system.

4.2. Result of BESS's Response Performance Evaluation by HILS

Figure 13 shows the response experiment results of the FRC at Non-gong Substation, which were checked through the Iba analyser, a process data acquisition system. For this, we constructed virtual physical models for the 36 MW PCS and 9 MWh batteries of Non-gong Substation to conduct the response performance test. As shown in Figure 13a, the frequency transient state simulation was performed for about 90 s, and the nadir frequency became 59.8501 Hz at 8.91 s. To gain a more detailed view of the time that the BESS responded, we examined the BESS response point, as shown in Figure 13b. At 8.45650 s, the frequency dead-band was deviated, and the FRC determined the frequency transient state. At 8.47928 s, an FRC demand was generated, and at 8.61935 s, the BESS power responded.

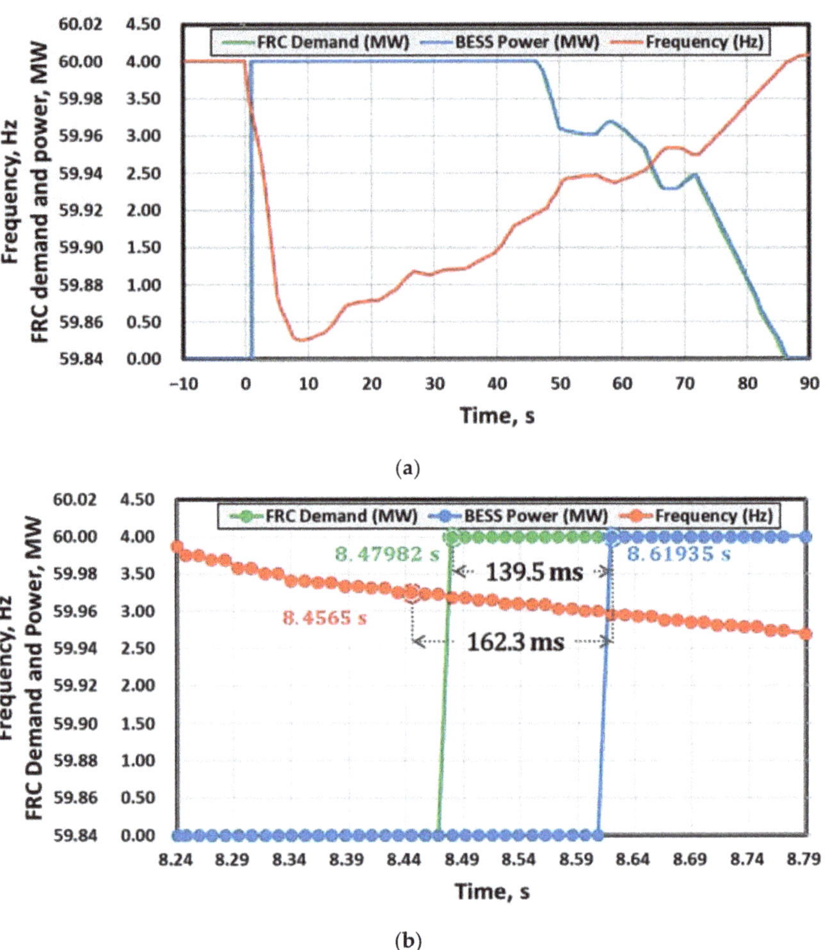

Figure 13. Result of FRC's response performance evaluation by Iba Analyzer in Non-gong substation; (**a**) Entire period; (**b**) BESS response starting point.

Therefore, it took 23.32 ms to generate the FRC demand according to the frequency change, and it took 139.5 ms for the BESS power to respond after the generation of the FRC demand. As a result, 162.3 ms was required for the PRDT, which was the time taken while the FRC determined the frequency, and finally, the BESS responded. As such, through the HILS, it was confirmed that the PRDT response performance was sufficiently satisfied for the FRC of the on-site BESS.

5. Summary Results

In this study, we used the SILS and HILS to provide the performance verification and evaluation results of the BESS. In the SILS type, compared to the frequency history data, the error rate in the maximum drop frequency was 0.9%, the error rate in the time consumed to reach the nadir frequency was 3.5%, and the average RoCoF was 2.45, confirming a high level of reliability. Furthermore, comparative analysis was performed between the cases of recovering without the BESS for PFC in a frequency transient state caused by a halting incident of a 950 MW generator and the case of PRDT of 140 ms, 160 ms, and 180 ms in a 376 MW BESS. The results showed that the maximum frequency drop was reduced by

34.40 to 34.58% in the case of having the BESS for PFC compared to the case of No BESS. Furthermore, the time taken to reach the nadir frequency point was reduced by 6.58 to 7.89%, and the average RoCoF was reduced by 28.70 to 30.03%. Therefore, it was confirmed that the frequency drop inhibition in the transient state was improved. Meanwhile, the CPP's GF response decreased by 37.73 to 40.14%. Furthermore, the power grid's constant trajectory was tracked in the frequency transient state. In the case of adding the BESS for PFC, the power grid constant at the nadir frequency point was reduced by 5.07 to 5.81%, and the ability to recover to the quasi-steady state was greatly improved. In addition, we conducted the trajectory analysis with a PRDT of 140 ms to 640 ms to examine the power grid constant, nadir frequency, RoCoF, and GF response at the nadir frequency point. In the results, as the PRDT decreased, the nadir frequency decreased, and the RoCoF's decrease rate increased, while the power grid constant started to decrease greatly at 200 ms. This is because the GF response trend at the BESS response point starts to decrease at 200 ms, and as the maximum GF response's increase rate increases, the change in the power deviation decreases relative to the frequency deviation, and the power grid constant decreases, thus reducing the inhibition of the frequency drop. Therefore, when examined comprehensively, it is found that when a PRDT faster than 180 ms is satisfied, the GF response reduction effect increases, improving the power system's stability. The HILS-type simulation was designed to mitigate the communication load with the FRC through the shared memory to verify the optimal response performance of the BESS installed at actual substations, and the performance test results showed that the PRDT was 162.3 ms, which satisfied the requirement.

6. Conclusions

This study introduced a novel power grid simulation with the BESS model for PFC. It is a real-time dynamic simulator that has a national power grid model, based on which real-time simulations can be performed in time-steps ranging from several ms to tens of ms. The continuous characteristics of the models can be analysed in real-time by analysing the characteristic trajectories of the power system at the national power grid level, based on the dynamic CPP model, the EMS models of PFC, SFC, and TFC, and the dynamic frequency model according to the power generation type, which are the key models. We presented the results of verifying the performance of the BESS algorithm for PFC and evaluating the response performance of the hardware controller based on these characteristics.

As result, it was confirmed that when BESS was used in the frequency transient state, the maximum frequency drop, the RoCoF, and the GF response of CPP were improved by about 30% than No BESS. In addition, the PRDT also confirmed that performance can be satisfied only when it reacts faster than up to 180 ms, and that the response performance of HILS' hardware controller is satisfied at 162.3 ms.

These functions of the CPS simulator can be used as an evaluation platform for solutions, such as a virtual power plant consisting of hardware hierarchical multi-systems. To this end, we need to develop reliable models for distributed power sources consisting of various energy sources and variable types of communication protocol models to ensure scalability.

Author Contributions: Conceptualization, G.-P.L. and K.-Y.S.; methodology, K.-Y.S.; software, T.-H.J. and K.-Y.S.; validation, T.-H.J., K.-Y.S. and G.-P.L.; formal analysis, T.-H.J.; investigation, T.-H.J. and K.-Y.S.; resources, M.C. and G.-P.L.; data curation, K.-Y.S.; writing—original draft preparation, T.-H.J.; writing—review and editing, K.-Y.S.; visualization, T.-H.J. and K.-Y.S.; project administration, K.-Y.S.; funding acquisition, G.-P.L. and M.C. All authors have read and agreed to the published version of the manuscript.

Funding: This research was performed as self-funding by A1 Engineering with technical support from KEPCO Research Institute.

Institutional Review Board Statement: Not applicable.

Informed Consent Statement: Not applicable.

Conflicts of Interest: The authors declare no conflict of interest.

Abbreviations

$P_{PFC.i}$	Power demand of generator about primary frequency control	MW
$P_{rated.i}$	Rated power of generator	MW
$\omega_{rated.i}$	Rated rotating speed of generator	rad/s
ω_i	Rotating speed of generator	rad/s
K_{droop}	Droop coefficient of generator	%/100
H_i	Inertia constant of generator	s
$S_{rated.i}$	Rated capacity of generator	MVA
$P_{Mech.i}$	Mechanical power of generator	MW
$P_{Elec.i}$	Electrical power of generator	MW
P_{AGC_Demand}	Total AGC demand in power system	MW
ACE	Area control error	MW
B	Control gain of AGC	MW/Hz
Δf	Deviation of frequency	Hz
$P_{AGC.i}$	AGC demand of generator	MW
$K_{ramp.i}$	Ramp rate of generator	%/100
$P_{base.i}$	Base point of generator by tertiary frequency control	MW
$H_i(P_{base.i})$	Heat rate equation of generator	MBtu/MWh
$F_{Eco}(P_{base.i})$	Fuel cost equation of generator by EcoLD mode	$/h
$F_{Env}(P_{base.i})$	Fuel cost equation of generator by EnvLD mode	$/h
FC_i	Cost of fuel used by generator	$
C_{Co2}	CO_2 trading price	$
K_{emi}	CO_2 emission coefficient	%/100
λ_i	Incremental fuel cost equation of generator	$ /MWh
$P_{max.i}$	Maximum power of generator	MW
$P_{min.i}$	Minimum power of generator	MW
On_i	Start up state of generator	-
P_i	Power of generator	MW
$P_{CED.i}$	Power demand of generator by CED	MW
$P_{TED.i}$	Power demand of generator by TED	MW
f_{rated}	Rated frequency of power system	Hz
f_{sys}	Frequency of power system	Hz
K_L	Load constant of power system	MW/0.1Hz
P_{Load}	Power consumption	MW
φ	Change of frequency	Hz
δ	Change of electric power	MW
K_F	Proportional gain of power system	Hz/MW
T_F	Frequency time constant	s
T_{Nadir}	Time for reaching the nadir frequency at frequency transient state	s
$f_{initial}$	Initial frequency before frequency transient state	Hz
f_{nadir}	Nadir frequency at frequency transient state	Hz
Δf_{Nadir}	Frequency deviation between initial frequency and nadir frequency	Hz
$RoCoF_{average}$	Average rate of change of frequency	Hz/s
$RoCoF_{BESS}$	Rate of change of frequency at BESS response point	Hz/s
GFR_{max}	Maximum governor free response	MW
$T_{recovery}$	Time to recover from transient state	s
ΔP_{trn}	Deviation of electric power in transient state	pu
Δf_{trn}	Deviation of electric frequency in transient state	pu
P_{Gen}^{t}	Total generation power in power system at time t	MW
$P_{Gen}^{t=0}$	Total generation power in power system at initial time	MW
P_{Load}^{t}	Total load in power system at time t	MW
$P_{Load}^{t=0}$	Total load in power system at initial time	MW
f_{sys}^{t}	Frequency at time t	Hz
$f_{sys}^{t=0}$	Frequency at initial time	Hz
K_{sys}	Power grid constant	pu

References

1. Tremblay, O.; Dessaint, L.-A. Experimental validation of a battery dynamic model for EV applications. *World Electr. Veh. J.* **2009**, *3*, 289–298. [CrossRef]
2. Zhang, Y.; Lyden, S.; de la Barra, B.L.; Haque, M.E. Optimization of Tremblay's battery model parameters for plug-in hybrid electric vehicle applications. In Proceedings of the 2017 Australasian Universities Power Engineering Conference (AUPEC), Melbourne, VIC, Australia, 19–22 November 2017; pp. 1–6.
3. Surya, S.; Channegowda, J.; Datar, S.D.; Jha, A.S.; Victor, A. Accurate battery modeling based on pulse charging using MATLAB/Simulink. In Proceedings of the 2020 IEEE International Conference on Power Electronics, Drives and Energy Systems (PEDES), Jaipur, India, 16–19 December 2020; pp. 1–3.
4. Xia, Q.; Yang, D.; Wang, Z.; Ren, Y.; Sun, B.; Feng, Q.; Qian, C. Multiphysical modeling for life analysis of lithium-ion battery pack in electric vehicles. *Renew. Sustain. Energy Rev.* **2020**, *131*, 109993. [CrossRef]
5. Fan, X.; Zhang, W.; Wang, Z.; An, F.; Li, H.; Jiang, J. Simplified battery pack modeling considering inconsistency and evolution of current distribution. *IEEE Trans. Intell. Transp. Syst.* **2020**, *22*, 630–639. [CrossRef]
6. Dubarry, M.; Baure, G.; Pastor-Fernández, C.; Yu, T.F.; Widanage, W.D.; Marco, J. Battery energy storage system modeling: A combined comprehensive approach. *J. Energy Storage* **2019**, *21*, 172–185. [CrossRef]
7. Tan, R.H.; Tinakaran, G.K. Development of battery energy storage system model in MATLAB/Simulink. *Int. J. Smart Grid Clean Energy* **2020**, *9*, 180–188. [CrossRef]
8. Rancilio, G.; Lucas, A.; Kotsakis, E.; Fulli, G.; Merlo, M.; Delfanti, M.; Masera, M. Modeling a large-scale battery energy storage system for power grid application analysis. *Energies* **2019**, *12*, 3312. [CrossRef]
9. Jo, H.; Choi, J.; Agyeman, K.A.; Han, S. Development of frequency control performance evaluation criteria of BESS for ancillary service: A case study of frequency regulation by KEPCO. In Proceedings of the 2017 IEEE Innovative Smart Grid Technologies–Asia (ISGT–Asia), Auckland, New Zealand, 4–7 December 2017; pp. 1–5.
10. Sanduleac, M.; Toma, L.; Eremia, M.; Boicea, V.A.; Sidea, D.; Mandis, A. Primary frequency control in a power system with battery energy storage systems. In Proceedings of the 2018 IEEE International Conference on Environment and Electrical Engineering and IEEE Industrial and Commercial Power Systems Europe (EEEIC/I&CPS Europe), Palermo, Italy, 12–15 June 2018; pp. 1–5.
11. Stein, K.; Tun, M.; Matsuura, M.; Rocheleau, R. Characterization of a fast battery energy storage system for primary frequency response. *Energies* **2018**, *11*, 3358. [CrossRef]
12. Bang, H.; Aryani, D.R.; Song, H. Application of battery energy storage systems for relief of generation curtailment in terms of transient stability. *Energies* **2021**, *14*, 3898. [CrossRef]
13. Meng, Y.; Li, X.; Liu, X.; Cui, X.; Xu, P.; Li, S. A Control Strategy for Battery Energy Storage Systems Participating in Primary Frequency Control Considering the Disturbance Type. *IEEE Access* **2021**, *9*, 102004–102018. [CrossRef]
14. Izadkhast, S.; Cossent, R.; Frías, P.; García-González, P.; Rodríguez-Calvo, A. Performance Evaluation of a BESS Unit for Black Start and Seamless Islanding Operation. *Energies* **2022**, *15*, 1736. [CrossRef]
15. Abayateye, J.; Corigliano, S.; Merlo, M.; Zimmerle, D. BESS Primary Frequency Control Strategies for the West Africa Power Pool. *Energies* **2022**, *15*, 990. [CrossRef]
16. Schiapparelli, G.P.; Massucco, S.; Namor, E.; Sossan, F.; Cherkaoui, R.; Paolone, M. Quantification of primary frequency control provision from battery energy storage systems connected to active distribution networks. In Proceedings of the 2018 Power Systems Computation Conference (PSCC), Dublin, Ireland, 11–15 June 2018; pp. 1–7.
17. Zhu, D.; Zhang, Y.-J.A. Optimal coordinated control of multiple battery energy storage systems for primary frequency regulation. *IEEE Trans. Power Syst.* **2018**, *34*, 555–565. [CrossRef]
18. Toma, L.; Sanduleac, M.; Baltac, S.A.; Arrigo, F.; Mazza, A.; Bompard, E.; Musa, A.; Monti, A. On the virtual inertia provision by BESS in low inertia power systems. In Proceedings of the 2018 IEEE International Energy Conference (ENERGYCON), Limassol, Cyprus, 3–7 June 2018; pp. 1–6.
19. Choi, W.Y.; Kook, K.S.; Yu, G.R. Control strategy of BESS for providing both virtual inertia and primary frequency response in the Korean Power System. *Energies* **2019**, *12*, 4060. [CrossRef]
20. Pinthurat, W.; Hredzak, B. Decentralized Frequency Control of Battery Energy Storage Systems Distributed in Isolated Microgrid. *Energies* **2020**, *13*, 3026. [CrossRef]
21. Obaid, Z.A.; Cipcigan, L.; Muhssin, M.T.; Sami, S.S. Control of a population of battery energy storage systems for frequency response. *Int. J. Electr. Power Energy Syst.* **2020**, *115*, 105463. [CrossRef]
22. Amin, M.R.; Negnevitsky, M.; Franklin, E.; Alam, K.S.; Naderi, S.B. Application of battery energy storage systems for primary frequency control in power systems with high renewable energy penetration. *Energies* **2021**, *14*, 1379. [CrossRef]
23. Manbachi, M.; Sadu, A.; Farhangi, H.; Monti, A.; Palizban, A.; Ponci, F.; Arzanpour, S. Real-time co-simulation platform for smart grid volt-VAR optimization using IEC 61850. *IEEE Trans. Ind. Inform.* **2016**, *12*, 1392–1402. [CrossRef]
24. Mana, P.T.; Schneider, K.P.; Du, W.; Mukherjee, M.; Hardy, T.; Tuffner, F.K. Study of microgrid resilience through co-simulation of power system dynamics and communication systems. *IEEE Trans. Ind. Inform.* **2020**, *17*, 1905–1915.
25. Adewole, A.C.; Tzoneva, R. Co-simulation platform for integrated real-time power system emulation and wide area communication. *IET Gener. Transm. Distrib.* **2017**, *11*, 3019–3029. [CrossRef]
26. Garau, M.; Ghiani, E.; Celli, G.; Pilo, F.; Corti, S. Co-simulation of smart distribution network fault management and reconfiguration with lte communication. *Energies* **2018**, *11*, 1332. [CrossRef]

27. Suzuki, A.; Masutomi, K.; Ono, I.; Ishii, H.; Onoda, T. Cps-sim: Co-simulation for cyber-physical systems with accurate time synchronization. *IFAC Pap.* **2018**, *51*, 70–75. [CrossRef]

28. Cao, Y.; Shi, X.; Li, Y.; Tan, Y.; Shahidehpour, M.; Shi, S. A simplified co-simulation model for investigating impacts of cyber-contingency on power system operations. *IEEE Trans. Smart Grid* **2017**, *9*, 4893–4905. [CrossRef]

29. Jin, T.-H.; Chung, M.; Shin, K.-Y.; Park, H.; Lim, G.-P. Real-time dynamic simulation of Korean power grid for frequency regulation control by MW battery energy storage system. *J. Sustain. Dev. Energy Water Environ. Syst.* **2016**, *4*, 392–407. [CrossRef]

30. ProTRAX, Energy Solutions. Available online: https://energy.traxintl.com/protrax (accessed on 27 June 2022).

31. *Turbine-Governor Models: Standard Dynamic Turbine-Govorner System in NEPLAN Power System Analysis Tool*; NEPLAN AG: Küsnacht, Switzerland, 2015.

32. Araújo, P. Dynamic Simulations in Realistic-Size Networks. Master's Thesis, Universidade Técnica De Lisboa, Lisboa, Portugal, 2010.

33. *PSS/E 34 Model Library*; Siemens, PTI: New York, NY, USA, 2015.

34. *ProTRAX Analyst's Instruction Manual*; Version 7.2; TRAX International: Virginia, VA, USA, 2014.

35. Korea Power Exchanger, Electric Power Statistics Information System: Generator Detail Contents. Available online: http://epsis.kpx.or.kr/epsisnew/selectEkfaFclDtlChart.do?menuId=020600 (accessed on 10 June 2021).

36. Kundur, P. Power system stability. In *Power System Stability and Control*; Grigsby, L.L., Ed.; CRC Press: Boca Raton, FL, USA, 2007.

37. Kothari, D.P.; Nagrath, I. *Modern Power System Analysis*; Tata McGraw-Hill Education: New York, NY, USA, 2003.

38. Lim, G.-P.; Han, H.-G.; Chang, B.-H.; Yang, S.-K.; Yoon, Y.-B. Demonstration to operate and control frequency regulation of power system by 4MW energy storage system. *Trans. Korean Inst. Electr. Eng. P* **2014**, *63*, 169–177. [CrossRef]

39. Han, J.B.; Kook, K.S.; Chang, B. A study on the criteria for setting the dynamic control mode of battery energy storage system in power systems. *Trans. Korean Inst. Electr. Eng.* **2013**, *62*, 444–450. [CrossRef]

40. Yun, J.Y.; Yu, G.; Kook, K.S.; Rho, D.H.; Chang, B.H. SOC-based control strategy of battery energy storage system for power system frequency regulation. *Trans. Korean Inst. Electr. Eng.* **2014**, *63*, 622–628. [CrossRef]

Article

Power Hardware-in-the-Loop Test of a Low-Cost Synthetic Inertia Controller for Battery Energy Storage System

Sergio Bruno, Giovanni Giannoccaro, Cosimo Iurlaro, Massimo La Scala * and Carmine Rodio

Department of Electrical and Information Engineering, Politecnico di Bari, Via E. Orabona, 4, 70125 Bari, Italy; sergio.bruno@poliba.it (S.B.); giovanni.giannoccaro@poliba.it (G.G.); cosimo.iurlaro@poliba.it (C.I.); carmine.rodio@poliba.it (C.R.)
* Correspondence: massimo.lascala@poliba.it

Abstract: In the last years, the overall system inertia is decreasing due to the growing amount of energy resources connected to the grid by means of power inverters. As a consequence, reduced levels of inertia can affect the power system stability since slight variations of power generation or load may cause wider frequency deviations and higher rate of change of frequency (RoCoF) values. To mitigate this trouble, end-user distributed energy resources (DERs) interfaced through grid-following inverters, if opportunely controlled, can provide additional inertia. This paper investigated the possibility of improving the control law implemented by a low-cost controller on remotely controllable legacy DERs to provide synthetic inertia (SI) contributions. With this aim, power hardware-in-the-loop simulations were carried out to test the capability of the proposed controller to autonomously measure frequency and RoCoF and provide SI actions by controlling an actual battery energy storage system.

Keywords: virtual inertia; fast frequency measurement; fast frequency regulation; distributed energy resources; microgrids; ancillary services; power hardware-in-the-loop; legacy resources

Citation: Bruno, S.; Giannoccaro, G.; Iurlaro, C.; La Scala, M.; Rodio, C. Power Hardware-in-the-Loop Test of a Low-Cost Synthetic Inertia Controller for Battery Energy Storage System. *Energies* **2022**, *15*, 3016. https://doi.org/10.3390/en15093016

Academic Editors: Cosimo Pisani and Giorgio Maria Giannuzzi

Received: 31 March 2022
Accepted: 18 April 2022
Published: 20 April 2022

Publisher's Note: MDPI stays neutral with regard to jurisdictional claims in published maps and institutional affiliations.

1. Introduction

The electric power system is facing new technical challenges due to the progressive integration of alternative energy sources in transmission and distribution networks. Massive efforts have been made in Europe to promote the employment of renewable energy sources (RES) and further steps will be taken to reach the energy transition targets of 2030 [1]. Nevertheless, the replacement of conventional power plants with RES is affecting the power system stability. Resources such as photovoltaics and wind turbines are interfaced with the grid by means of power converters, and are therefore not equipped with rotating masses that can release or absorb mechanical energy. As a result, high penetration levels of RES will reduce the total system inertia (TSI), and slight variations of generation or demand will cause wider frequency deviations, affecting power system security [2]. With reduced TSI, severe frequency fluctuations can result in undesired tripping of protections or load/generation units disconnection, or even instability [2,3].

Therefore, in the next years, new countermeasures must be adopted to limit the values of frequency *nadir* and rate of change of frequency (RoCoF) following a disturbance, in order to preserve power system safety and stability [4]. This is especially true in the case of smaller electrical systems such as isolated microgrids or non-synchronous islands, which are often interconnected to the mainland only through high-voltage direct current connections (such as, for example, in the case of Northern Europe [3]). Theoretically, as observed in [5], additional synchronous capacitors may be installed to improve the total system inertia. However, since these systems could result as expensive and complicated to be implemented, alternative solutions based on already installed power inverters, able to provide provide virtual or synthetic inertia (SI) by emulating the inertial behavior of synchronous generators, have been proposed.

Recent developments have shown that prosumers and active end-users at the distribution level are theoretically able to manage their own generation/load resources to provide ancillary services, such as congestion management, frequency restoration reserve [1,6], or inertia support to system operators. With this goal, an example of peer-to-peer cyberphysical infrastructure aimed at optimizing the inertial response of distributed energy resources (DERs) in an energy community was presented in [7]. The possibility to obtain distributed SI contributions by means of widespread distributed generation (DG) and DERs was also proposed in [8–10]. Even if storage systems are considered the best source to provide fast frequency control services [11–14], several studies have also demonstrated that is possible to generate SI by controlling domestic loads, such as refrigerators and boilers [15,16], or single-phase electric vehicles [17]. In this sense, the possibility to provide fast frequency regulation support by means of public LED lighting systems has been also investigated in [18].

Although the idea of using DERs for fast frequency regulation and SI support is generally accepted, few practical implementations can be found in the literature, and the actual controllability of legacy distributed resources was never addressed. In [19], the authors preliminarily investigated the possibility to develop a low-cost controller able to autonomously measure frequency and RoCoF, and implement an SI control law on the management system of remotely controllable DERs. Such a controller can enable SI response for any distributed component that possesses the ability to receive a remote control signal on a fast communication channel, without the need of reprogramming its management system or inverter.

In this paper, the studies on the low-cost SI controller have been further extended with more extensive power hardware-in-the-loop (PHIL) tests on the control of a battery energy storage system (BESS) as provider of system inertia support. The tests carried out in this paper were aimed to test the implementation of the SI controller on a single-board computer with more advanced computation capabilities than the one used in [19], which permitted to overcome some observed frequency measurement issues and improve the overall frequency response. Power hardware-in-the-loop tests allowed us to analyze the impacts that real-time fast frequency and RoCoF measurements have on the actual feasibility of SI control, and to tune control parameters such as RoCoF dead-band, frequency smoothing factor, and measurement reporting time. The performances of the real controller are compared to the PHIL simulation of an ideal controller with negligible RoCoF measurement error and reporting time. Moreover, several PHIL tests were aimed to address the issue of RoCoF error and study the impacts of a filtering stage in terms of stability and response delay. The possibility of increasing frequency measurement reporting time to reduce RoCoF error was also investigated, demonstrating how the controller has enough idle time to extend the number of controlled devices and control functions without affecting the efficacy of the control.

2. A Low-Cost Controller for Distributed Synthetic Inertia

In the coming years, ancillary services for power system management and stability will be also provided by distributed energy resources located at end-user level [20]. Several research projects investigated the capability of DERs such as battery energy storage systems, distributed generation, and loads to provide specific flexibility services, including synthetic inertia [11,21]. The provision of synthetic inertia, or fast frequency regulation, requires that a power device must be able to detect and respond to frequency variations very rapidly, in a few hundred milliseconds from the beginning of the transient event. As demonstrated in [11], a battery energy storage system represents one of the best candidates to provide frequency ancillary services thanks to its technical characteristics, such as long discharge time, high ramping rate, and high voltage/frequency control capability of its inverter.

The efficacy of SI control is severely affected by the quality of frequency and RoCoF measurements. The delays introduced by computation and communication processes can negatively affect the virtual inertia response, as shown in [13], even when power converter-based resources have the capability to adjust their power output within few cycles. Obtaining fast and reliable frequency and RoCoF measurements at field level represents a crucial aspect to be considered, since, unfortunately, speed and accuracy are two characteristics that tend to be mutually exclusive. In addition, the typical presence of unbalanced voltages, noise, and distortions in the distribution grid compromises the reliability of frequency measurements at end-user level. Authors in [22] even suggested that the assumption of having reliable RoCoF signals at distribution level may be unrealistic.

In this paper, the capability of a low-cost controller aimed to provide real-time synthetic inertia control of a BESS is assessed. The main characteristics of the proposed controller are that it must be autonomous (no external frequency/RoCoF signal should be needed) and based on the use of low-cost technology. More advanced frequency measurement can be provided using intelligent electronic device (IED) or phasor measurement unit (PMU) technologies, which are clearly not suitable for end-user applications due to their high costs [22,23]. The controller studied in this paper is instead based on a very low-cost architecture (below USD 100), being based on the use of an off-the-shelf single-board computer. In our implementation, a Raspberry Pi 4 Model B was used, but clearly any other similar processing unit could be adopted. This device has the following technical characteristics: 64-bit quad-core processor, 4 GB of RAM, dual-band 2.4/5.0 GHz wireless LAN, Bluetooth 5.0, Gigabit Ethernet, USB 3.0, power over Ethernet capability, and a standard 40-pin general-purpose input/output (GPIO) header.

The main purpose of this controller is to estimate RoCoF variations and generate SI control of a BESS, accordingly. SI control is obtained by changing the battery's power output set-points with an additional control signal proportional to RoCoF variations. The SI control can happen even without having to reprogram the BMS, with particular advantages in the case of legacy devices, whose internal control schemes and logic, based on proprietary languages and codes, cannot be modified.

The proposed controller, whose scheme is shown in Figure 1, is able to acquire and process grid voltage signals, calculate both frequency and RoCoF, and communicate with other external devices through Ethernet (LAN), Bluetooth, and USB. The GPIO interface of the single-board computer is used to acquire the voltages. The GPIO works with digital signals only, and therefore a voltage transducer and an analog to digital converter (ADC) are required. The transducer used in out tests is a high-voltage differential probe, with a signal attenuation of $200\times$, coupled with a DC source used to add a 1.65 V offset in order to adapt the voltage waveform to the ADC-shield input voltage (i.e., 0–3.3 V). A 10-bit ADC, with a cost of about \$10, equipped with serial peripheral interface (SPI) is also mounted on a prototype shield and connected to the GPIO's pins.

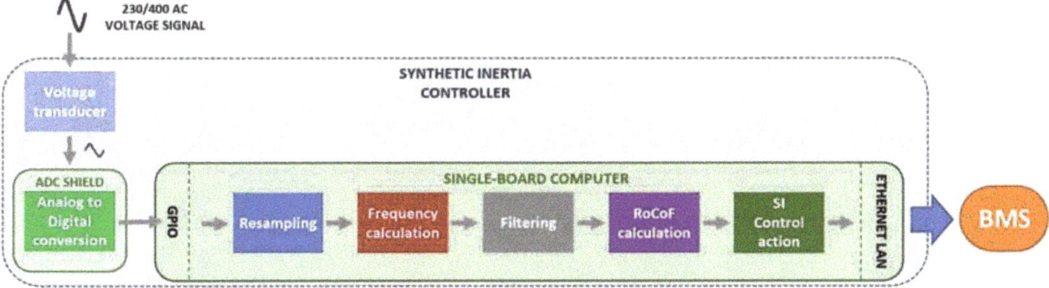

Figure 1. Scheme of the investigated SI controller.

The digital voltage signals are sampled by the single-board computer and are used to calculate the system frequency using the algorithm described in Section 4. This algorithm assumes that the acquired samples are equally spaced. In practical implementations on a single-board computer, this requirement is hardly satisfied without introducing idle times on samples or reducing the sampling rate. For this reason, a resampling process is performed to interpolate the acquired samples and obtain regularly-spaced input data (in the tested application the sampling time is 50 µs).

As a next step, the frequency values are employed to calculate the RoCoF and generate the corresponding SI control law. Nevertheless, as described in Section 4, due to the presence of noise and distortions on the acquired voltage signal, a filtering process is added between frequency and RoCoF calculation. This filtering stage is also recommended by ENTSO-E in a document discussing frequency and RoCoF measurement requirements [24]. As specified in this report, a wide range of different filter algorithms are available, such as Bessel, Butterworth, Chebyshev, elliptic, etc. However, these filters must be tuned in such a way to minimize the presence of noise without introducing excessive delays. An application case of frequency filtering is presented in [25], in which an appropriate feedback and feedforward filter was applied in order to remove noise as much as possible without affecting the RoCoF behavior. The algorithm is based on a combined version of "comb" filter, which acts by adding a delayed version of the signal to itself and causing constructive and destructive inference that can attenuate the specified frequency signal and its harmonics [26]. In this paper, a less sophisticated filter based on exponential moving average is employed in order to limit the necessary computational burden. This algorithm is able to reduce lags introduced from the average process by applying more weight to the recent samples than the older ones, and, therefore, results as particularly sensitive to recent signal changes [27]. According to the measured RoCoF values, the controller generates an SI control law and communicates it to the battery management system (BMS) of the BESS. In our tests, the control law is transferred to the BMS through the LAN and using Modbus TCP/IP protocol.

3. Experimental Tests of Synthetic Inertia Control by Means of a Battery Energy Storage System

In order to assess the performances of the proposed SI controller, power hardware-in-the-loop (PHIL) tests were performed by means of a microgrid facility located at LabZERO laboratory of Politecnico di Bari [28,29]. This facility permits to implement, in a real-time simulating environment, the interaction between the power system and real power devices installed in the microgrid.

The very first step of our analysis was to test the behavior of the 5 kVA LiFePO4 BESS currently installed in the LabZERO microgrid when it is controlled to provide synthetic inertia during a simulated frequency event. The controlled physical battery interacts with a power system model that reproduces, in real time, the electromechanical transient following a sudden power imbalance. This first PHIL simulation was used as a benchmark to compare, in the following tests, the SI response of the BESS when it is regulated by the proposed controller. In these additional PHIL tests, the SI control law sent to the BMS is directly generated by the Raspberry controller, which autonomously measures frequency and RoCoF, and controls the battery. These tests will also be aimed to demonstrate how, by moving the evaluation of the SI control at field level, it is possible to reduce the delays due to communication and control processes.

The power hardware-in-the-loop test bench is shown in Figure 2. The real-time simulator OPAL RT5600 (OPAL-RT Technologies, Montreal, QC, Canada) is used to simulate the electromechanical response of a generic power system, whereas the programmable power source Triphase PM15A30F60 (Triphase, Holsbeek, Belgium) is programmed to control the entire microgrid in grid-forming mode. Through synchronous communication, based on a fiber optics channel, the real-time frequency of the simulated system is sent by the real-time simulator (RTS) to the programmable source. This frequency value is imposed in real time on the actual microgrid and, therefore, on all its components. The frequency excursions following load/generation imbalances, which are controlled through the primary frequency regulation of the simulated system, are applied to the microgrid and to the BESS under study.

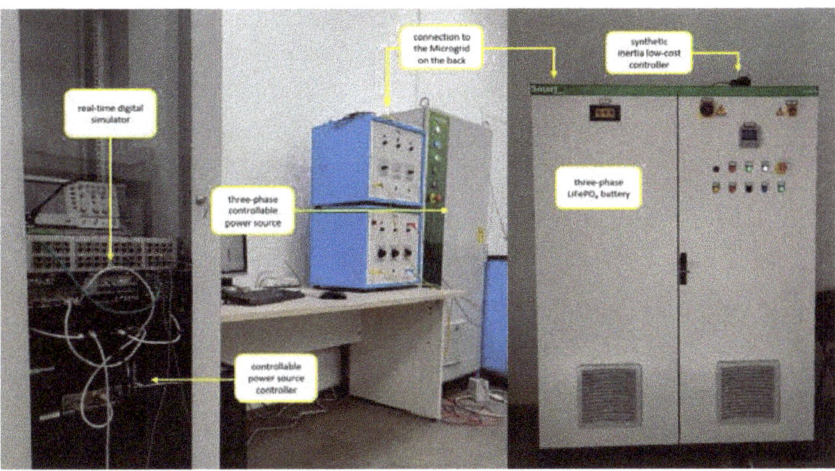

Figure 2. Power hardware-in-the-loop test equipment.

3.1. System Model and Frequency Response without Synthetic Inertia

In Figure 3, the block diagram used to build an equivalent model of the power system is shown. The term ΔP_L is the active power load variation applied to generate a power imbalance, whereas ΔP_{SI} represents the active power exchanged by BESS to provide SI, and ΔP_G is the primary frequency regulation following the disturbance. Table 1 summarizes the values assigned to the droop R and to the delays $T_G, T_{T1}, T_{T2}, T_{T3}$ of the governor and reheat turbine models. The limits set for the governor model consider the valve opening constraints, whereas limitations imposed on the turbine model take into account the active power limits. A substandard level of system inertia H (3 seconds) was assumed to simulate the arising of reduced TSI conditions due to the high penetration of inverter-based RES. This assumption was made in accordance with [30], where the authors estimated system inertia of the Italian transmission system, highlighting how a TSI even lower than 3 seconds has been experienced during specific real-time conditions.

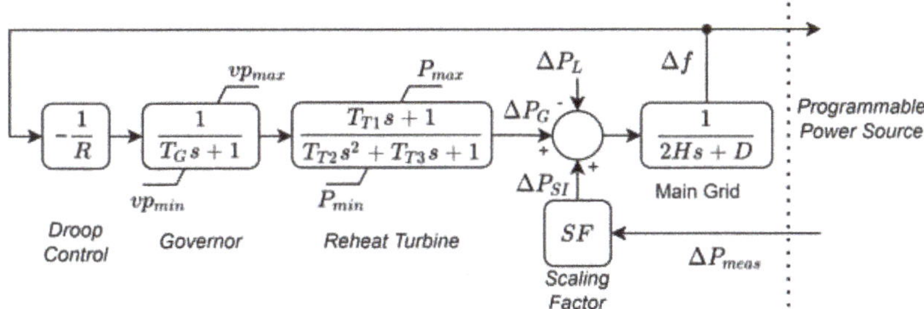

Figure 3. Power system model with synthetic inertia contribution.

Table 1. Power system model coefficients.

R [p.u.]	T_G [s]	T_{T1} [s]	T_{T2} [s]	T_{T3} [s]	H [s]	D	SF [p.u./kW]
0.050	0.20	2.1	2.1	7.3	3.0	0.020	0.0116

Figure 4 shows the transient response of the simulated system to a sudden load step change, when only primary frequency regulation is considered (i.e., the term ΔP_{SI} related to the power contribution of synthetic inertia is set to zero). Since the scope of the following tests is to test inertial control, which happens within the first 500–1000 ms of the transient, secondary frequency regulation is not modeled. At the beginning of the simulation ($t < 0$ s), the system was assumed to be in steady-state conditions, with the synchronous generation perfectly balancing the load. At time $t = 0$ s, an instantaneous upward load step change ($\Delta P_{load} = +0.1$ p.u.) was applied. The primary frequency regulation of the synchronous generation regulated the power output of the quantity ΔP_{gen} so that a new equilibrium point was reached. The primary frequency regulation was modeled according to the transfer functions typical of a thermoelectric plant, whose parameter values are shown in Figure 3.

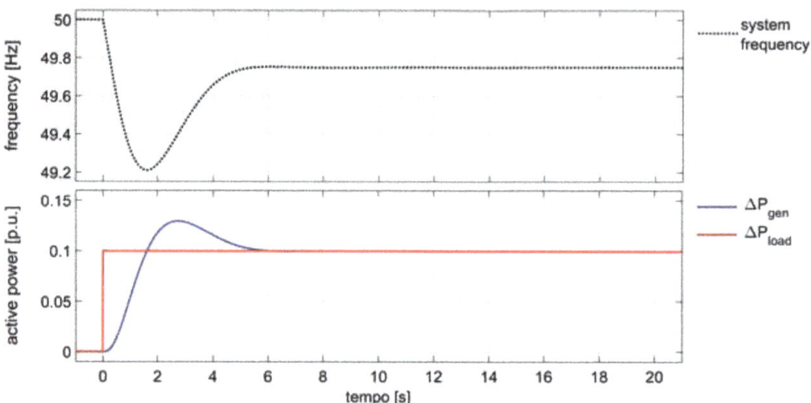

Figure 4. Frequency and active power trajectories without SI control.

As shown in Figure 4, the load step variation gave rise to the frequency transient represented by a dotted black curve. After 1.62 s, the frequency trajectory reached its minimum, or *nadir*, at 49.21 Hz. Then, thanks to primary regulation, system frequency is brought back close to its nominal value, settling to about 49.75 Hz. Without synthetic inertia support, frequency control was only provided by the synchronous generators that, by means of the primary frequency regulation, adjusted active power generation according to the simulated response of the turbines' governor. After about 6 seconds from the contingency, the power response ΔP_{gen}, depicted in Figure 4 with a continuous blue curve, reached the same value of the applied load variation ΔP_{load}. A power overshoot of about 0.030 p.u. was reached during regulation.

3.2. PHIL Tests of Synthetic Inertia by Means of BESS

In this second subsection, the capability of a BESS to modulate its power output and provide frequency support was investigated. The presence of an additional non-null active power contribution ΔP_{SI} in the scheme of Figure 3 was therefore considered. This additional term takes into account the SI contribution provided by BESS in terms of active power balance. The BESS power output was remotely managed, applying a current reference set-point I_{ref} to the BMS. This set-point was calculated by a simulated SI controller on the basis of the RoCoF signal, and then sent via Modbus TCP/IP communication. In these tests, the proposed SI controller is only emulated through the RTS. Frequency and RoCoF measurements are ideal signals obtained by the RTS during the transient simulation.

The active power response of the BESS was measured by the controllable power source and fed back to the RTS in order to close the loop of the PHIL simulation. Since the SI active power control should reproduce the response of a larger number of storage units, ΔP_{SI} was scaled by a factor SF (shown in Figure 3). By means of this scaling factor, the real BESS, whose maximum inertial contribution was set to about 3.5 kW, represents in the real-time simulation a BESS with nominal power equal to 4% of the reference power.

The synthetic inertia control adopted in this test case and simulated through the RTS is represented in Figure 5.

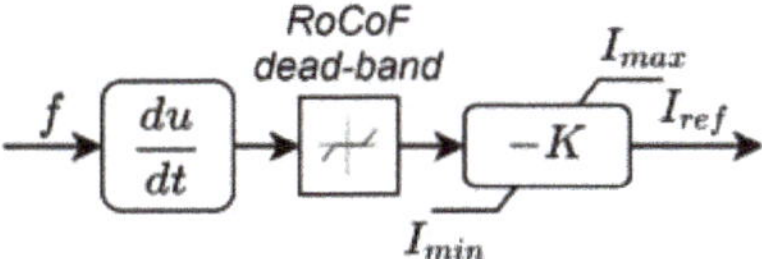

Figure 5. Block diagram of the SI controller simulated with the RTDS.

With the aim to avoid an excessive control activity of BESS around the frequency equilibrium point and endanger the lifespan of the battery, a 10 mHz/s RoCoF dead-band was assumed. Moreover, the saturation block limited the direct current reference I_{ref} to $[-5,+5]$ A, thus about 3.5 kW in charge/discharge mode.

As previously mentioned, in these tests, an ideal SI controller is just emulated in the RTS. The SI control law is obtained according to the scheme Figure 5. The RoCoF signal is calculated during the simulation by deriving the continuous frequency signal. RoCoF is therefore a continuous quantity, known instantaneously at each simulation time step (i.e., 0.125 ms). Due to the absence of delays and measurement errors in the RoCoF signal, as shown in Figure 5, no filters were adopted in the control chain.

In tests, increasing values of gain K in the interval 10–25 A/(Hz/s) were used in order to study the dependence of the system frequency response with respect to the amount of active power provided by the inertial control. Figure 6 shows the frequency response of the simulated system after the introduction of the SI support with various values of gain K. The black dashed line represents the system frequency response without any SI support, which was previously calculated and assumed as base case. Differently, all other curves are the frequency trajectories obtained in the presence of BESS synthetic inertia control. The observed response is coherent with the typical effects of fast SI control actions [13]. In comparison with the base case (no SI control action), higher frequency *nadir* points were reached, even though the settling time was increased.

In general, it can be observed that an increase of gain K allowed to reduce the frequency derivative in the very first instant and reach a higher *nadir*. However, higher magnitude of inertial control corresponds also to larger second overshoot and longer settling time. From Figure 7, it is possible to observe how control gain affects system response. A lower gain reduces the BESS participation in frequency control, letting the synchronous machine more rapidly take care of the power imbalance. The synchronous generator overshoot decreases with SI contribution. In the PHIL tests, this overshoot was reduced from 0.0295 p.u. of the base case to 0.0232 and 0.0230 p.u. for the case with $K = 10$ and $K = 25$, respectively. Having reached the *nadir*, the SI control action changes its sign: for this reason, higher gains cause an increase of the second frequency overshoot and an increase of the time necessary to reach steady state.

Table 2 permits to better compare the results obtained using different gains after the same load-step variation. The settling time was calculated considering a 2% tolerance around the steady state value. The fall time is the time necessary for the frequency to drop from 10% to 90% of the settling value. This fall time was used to average the RoCoF value in the initial characteristic of the frequency transient (see Table 2). To assess the impact of SI on RoCoF, the average RoCoF value is more suitable than the maximum value, since the latter is always experienced in the very first instant after $t = 0$ s, when no actual regulating response of the BESS is possible (in other words, the maximum instantaneous RoCoF is the same in all simulations). The obtained results demonstrated that fall time and average initial RoCoF improve with SI gain increase. However, it can be noticed how these improvements soon saturate with SI gain increase, whereas settling time continues to increase. This is due to the fact that, in the initial moments when RoCoF reaches its maximum, high gains will cause SI control to hit the maximum active power limits of the battery, and therefore no further frequency response improvement can be obtained.

Table 2. Ideal synthetic inertia controller: characteristics of the frequency step response.

Gain K [A/(Hz/s)]	Frequency Reporting Time [s]	Fall Time [s]	Average RoCoF [Hz/s]	Time *nadir* [s]	Frequency *nadir* [Hz]	Overshoot [%]	Settling Time [s]
0	-	0.242	0.824	1.616	49.211	0.050	5.063
10	inst.	0.307	0.650	2.285	49.364	0.057	10.764
15	inst.	0.310	0.645	2.483	49.396	0.068	11.896
20	inst.	0.312	0.640	2.672	49.417	0.076	12.800
25	inst.	0.321	0.623	2.770	49.430	0.080	13.547
10	0.050	0.278	0.720	2.254	49.367	0.056	10.795
10	0.100	0.260	0.769	2.209	49.371	0.053	10.842
10	0.150	0.243	0.824	2.181	49.380	0.049	10.949

This is a known problem of SI control, and, in general, two possible solutions can be adopted. The first solution is to adopt a nonlinear SI control with a gain that is a function of RoCoF. This kind of control allows to exploit all fast frequency regulation resources in the first fall, whereas the SI control action will more rapidly damp out while frequency approaches the *nadir* or the settling value. The disadvantage of this control is that SI response will always be very moderate in frequency events that are not characterized by high-frequency derivatives. Another possible solution is to adapt gain to the specific operating conditions. Gain can be set so that active power control will reach maximum capacity during a credible worst-case event (the one characterized by the highest frequency derivative). In an isolated network, this event might correspond to the sudden loss of the highest load feeder or generating station. Clearly, this kind of approach requires the presence of a control center (a microgrid controller in the case of a small isolated distribution network), plus a communication channel must be established between this control center and the SI controllers installed on the field. Gain K can be easily updated knowing the current battery capability and the expected maximum RoCoF excursion. In both cases, the use of an additional fast frequency control proportional to the frequency deviation (and not to the RoCoF) can help to control transients characterized by small derivatives, but large frequency deviations. Investigations on these control schemes are, however, out of the scope of this paper, since our main aim is to test the fast control capability of the proposed controller. Any kind of SI control rule can be easily programmed in the controller.

Figure 7 allows to assess the speed of the BESS active power response. The first variation in power exchange was measured by the programmable power source after about 80–100 ms from the load step change. Despite this initial delay, the BESS was able to reach its active power peak within 200 ms. The SI controller was just simulated and therefore there were no delays due to frequency and RoCoF measurements. However, some delays were introduced by the Modbus TCP/IP communication and the use of a master node to control the BMS. In our network, these delays are usually in the 50–80 ms range.

Some final PHIL tests were carried out using the simulated ideal controller. These tests, whose results are also synthetically reported in Table 2, were aimed to assess the impact of the frequency reporting rate. This is a relevant issue, since the simulations with a real controller, which will be shown in the next sub-section, are characterized by actual measurement delays. So far, frequency and RoCoF measurements were assumed to be continuous variables, known instantaneously at each time step of the simulation. Figure 8 shows what happens when the reporting rate of frequency (and RoCoF) assumes values closer to the actual time resolution achievable with hardware instrumentation. These tests were performed using gain $K = 10$.

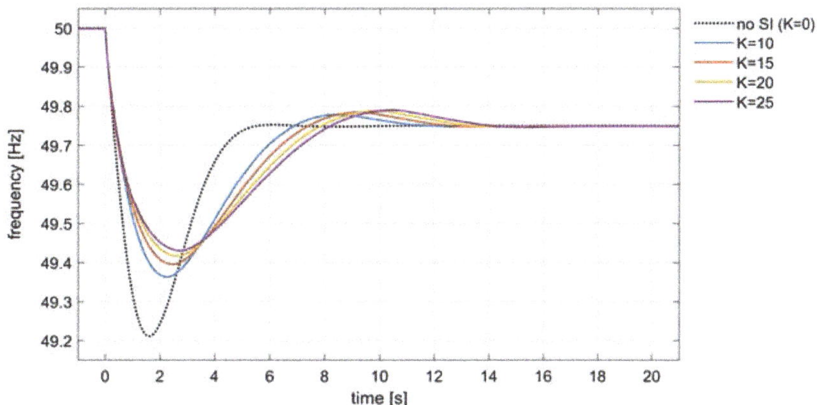

Figure 6. Ideal SI controller: frequency response with different gain K.

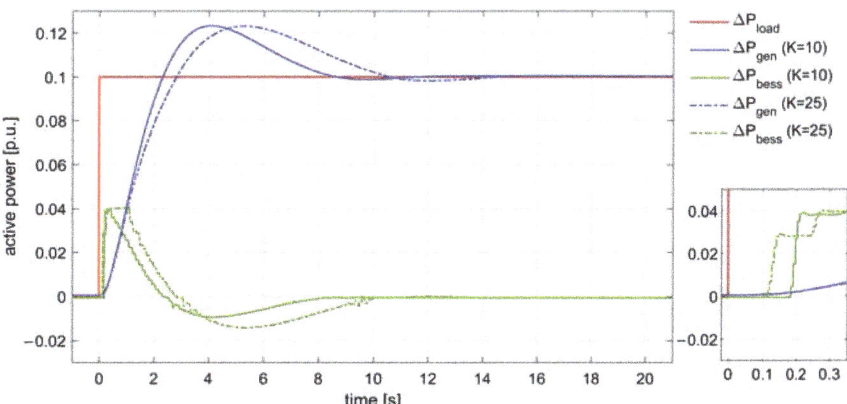

Figure 7. Ideal controller: BESS active power response with different gain K.

If the response of the controlled equipment is slowed by measurement reports, the frequency falls faster. However, up to about 100 ms, the reporting time barely affected the frequency transient. In Figure 9, it is possible to notice how with a 150 ms reporting time, the battery was operated close to its maximum capacity for a longer period, even leading to the highest *nadir*. However, this response is more similar to the one obtained by fast frequency regulation schemes that operate with control laws proportional to the frequency deviation; the response is improved in terms of *nadir*, but the initial RoCoF is barely influenced by the BESS control. In this test case, characterized by low inertia and a very fast frequency transient, reporting times equal to or higher than 150 ms cannot produce any improvement in terms of RoCoF reduction. Table 2 allows us to numerically compare the results of the simulations in terms of RoCoF, *nadir*, and settling times. These simulations will also be used as benchmark for the PHIL tests with the actual controller.

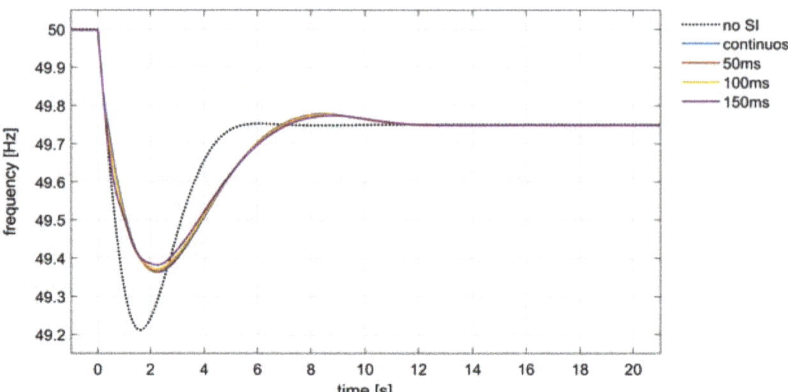

Figure 8. Ideal controller: frequency response with different RoCoF reporting time and $K = 10$.

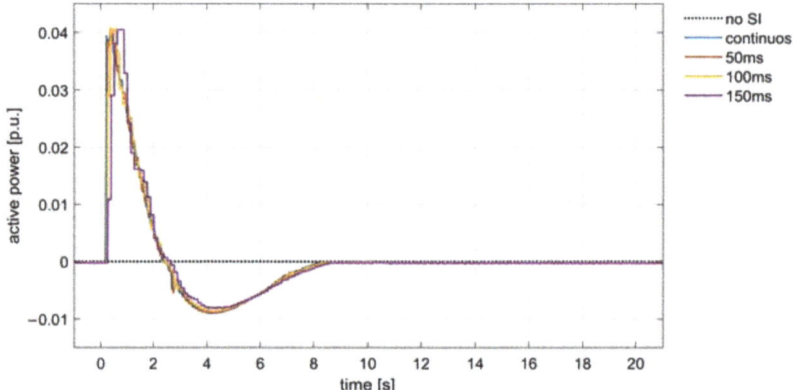

Figure 9. Ideal controller: BESS active power response with different RoCoF reporting time and $K = 10$.

4. Power Hardware-in-the-Loop Tests with the Single-Board SI Controller

This final test results section has been aimed to assess the capability of the proposed SI controller to provide frequency support by means of BESS. This controller is able to measure the grid frequency directly from the voltage signal of the BESS busbar, calculate RoCoF, and implement an SI control law on the BMS. In these tests, the RTS is only employed to simulate the power system electromechanical response and provide the frequency reference to the programmable source.

4.1. Algorithm for Fast Frequency Measurement

The physical SI controller under test is designed to be able to provide frequency support autonomously, without external inputs or measurements. Frequency is measured locally by the same controller using the algorithm discussed in [19]. This algorithm permits to perform fast frequency estimation through the calculation of an autocorrelation integral $A(\tau)$ of a generic waveform with period $\tilde{\tau}$. Based on a discretized version described below, the algorithm proposed in [19] is aimed at finding the period $\tilde{\tau}$ which maximizes the autocorrelation function $A(\tau)$. Please note that, in the following, the square brackets are used to represent discrete functions and values. The method is based on associating a value $[A]$ to a candidate solution $[\tau]$. Each candidate period $[\tau]$ is a discrete value that corresponds to a multiple of the sampling period. Therefore, a continuous value of period (and frequency) can be obtained only by adopting an interpolation rule on consecutive values of $[A]$ and $[\tau]$. In this case, a second-order polynomial approximation is used. The algorithm is organized with the following structure:

1. An initial guess τ_1 is made (for example, the frequency period measured at the previous acquisition).
2. Having considered $[\tau_1] \approx \tau_1$, two more values in its neighborhood are selected, $[\tau_0] = [\tau_1] - \Delta\tau$ and $[\tau_2] = [\tau_1] + \Delta\tau$.
3. $[A_0]$, $[A_1]$, and $[A_2]$ are evaluated in correspondence of values $[\tau_0]$, $[\tau_1]$, and $[\tau_2]$, respectively.
4. A new value τ_1 is calculated as the abscissa of the vertex of the parabola passing through the three points $([\tau_0], [A_0])$, $([\tau_1], [A_1])$, and $([\tau_2], [A_2])$.
5. A narrow neighborhood of $[\tau_1] \approx \tau_1$ is analyzed assuming a $\Delta\tau$ equal to the sampling period and repeating steps 2–4.
6. The measured frequency is $\tilde{f} = \dfrac{1}{\tilde{\tau}}$, with $\tilde{\tau} = \tau_1$.

4.2. Influence of Measurement Errors

Differently from that discussed in the previous subsections, in these tests, the BESS response is affected by both RoCoF sampling period and measurement errors. Due to the methodology used for fast frequency measurement, the reporting time of frequency (and RoCoF) is about 50–60 ms. This time is needed to sample the two entire cycles necessary to evaluate the weighted autocorrelation integral and process the samples to obtain the frequency measurement. Measurement errors are introduced by several factors, including the accuracy of the proposed real-time frequency measurement methodology, the adopted transducer, the ADC sample rate, and voltage quantization. Moreover, the effects of harmonics and other disturbances in the real power circuits must be added. Disturbances and errors sum up, providing an estimated RoCoF signal that can be extremely noisy.

Figure 10 shows the frequency response obtained using the proposed SI controller without applying any filter to the RoCoF signal. Even though a very wide RoCoF dead-band (± 0.100 Hz/s) and a small gain (i.e., $K = 10$) were adopted, the presence of noise on RoCoF introduced excessive errors on BESS control. No improvement was brought to the frequency response in terms of *nadir*. Moreover, measurements errors introduced by the controller resulted in a continuous activation and deactivation of the BESS response, causing unbearable frequency fluctuations around its theoretical steady-state value.

4.3. Tests with Filtered RoCoF Measurements

Since low-cost applications, such as the one proposed with this single-board controller, cannot make use of high-accuracy transducers and signal processing instrumentation, measurement errors must be filtered. In the following tests, a low-pass filter based on exponential smoothing was applied to frequency measurements. The SI control scheme programmed on the single-board SI controller is shown in Figure 11. The filtering action can be modulated by varying the smoothing factor α in the range $[0, 1]$. The maximum filtering action is obtained with $\alpha = 0$, whereas the filter is completely deactivated with $\alpha = 1$. Samples $f_{filt}(k)$ and $f(k)$ are, respectively, the k-th samples of the filtered and not filtered frequency at the time instant t. The RoCoF signal $RoCoF_{filt}(k)$ is calculated from the filtered frequency measurements using a discrete derivative function. Smoothing is needed not only to filter measurements but also to reduce the stress on the controlled component and avoid too many sudden power variations that can impact the cells' lifetime.

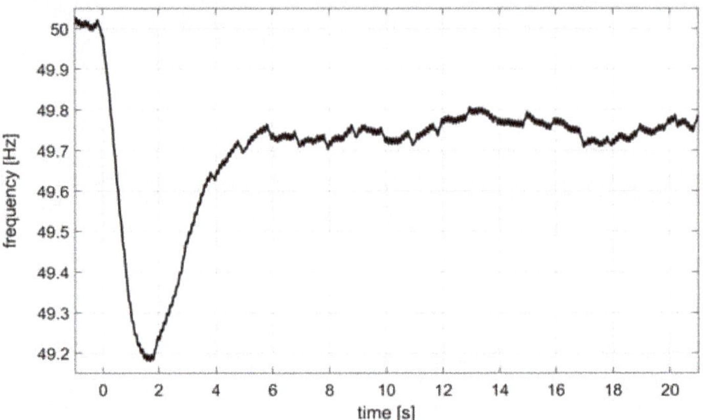

Figure 10. Real controller: frequency response without filters on frequency measurements ($K = 10$).

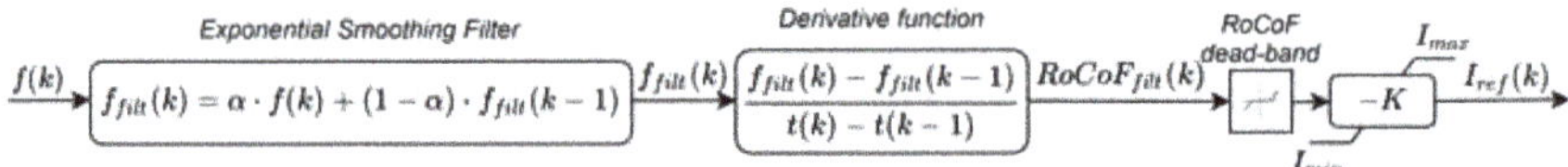

Figure 11. Block diagram of the actual SI controller.

In Figure 12, the frequency response of the simulated power system during a generic frequency event is compared with the filtered and unfiltered frequency values measured by the controller at the microgrid switchboard where the BESS is installed. Figure 13 shows the corresponding values of RoCoF. The filtered signals were obtained with a smoothing factor α equal to 0.1. It can be observed that the filtered signals were slightly delayed, but the beneficial effects of the filter during a frequency measurement disturbance, arising around $t = 7.5$ s, are clearly visible. The RoCoF error is drastically reduced, also allowing to keep a small dead-band on RoCoF (i.e., ± 0.025 Hz/s in all simulations with the real controller).

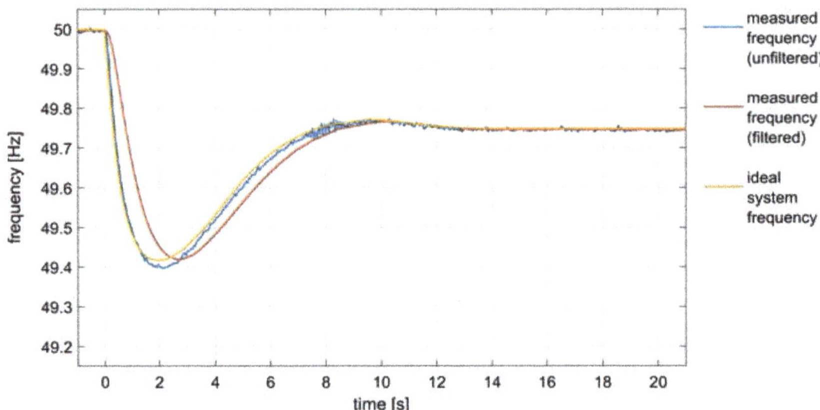

Figure 12. Frequency measured by the controller vs. ideal simulated frequency.

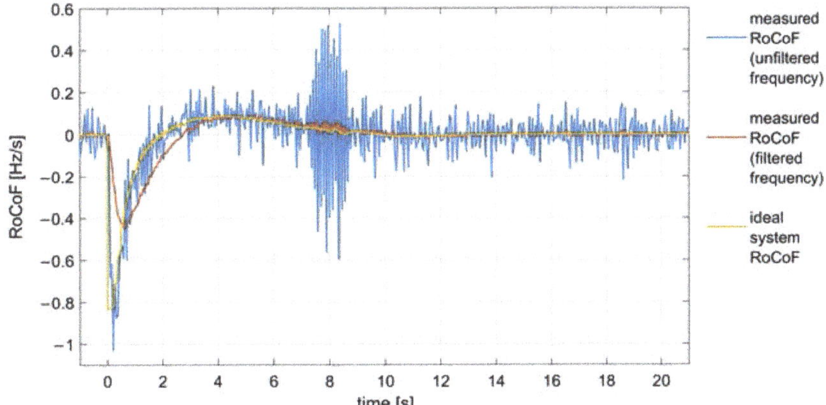

Figure 13. RoCoF measured by the controller vs. ideal simulated RoCoF.

Figure 14 shows the system response obtained using the real SI controller, and adopting different values of gain K. The characteristics of the frequency response are also reported in Table 3. The smoothing factor was set to 0.1 in all tests. With respect to Figure 6, the results obtained with the real controller are slightly worse in terms of initial RoCoF and settling time, but slightly better with regard to *nadir* and second overshoot. It can be observed that higher gains ($K = 20$ and $K = 25$) resulted in bumpier frequency trajectory, due to the amplification of RoCoF errors. Moreover, around the *nadir* point, higher gains created some small oscillations when the RoCoF trajectory entered and exited the dead-band. These small oscillations also introduced some errors in the detection of the *nadir* time and frequency, as in Table 3 for the case $K = 25$.

Figure 15 compares the BESS response in the two extreme cases ($K = 10$ and $K = 25$), showing how a lower gain permitted obtaining a smoother response, with limited stress on the controlled component. According to these tests, higher gain levels produced only marginal improvements, which did not compensate the other drawbacks.

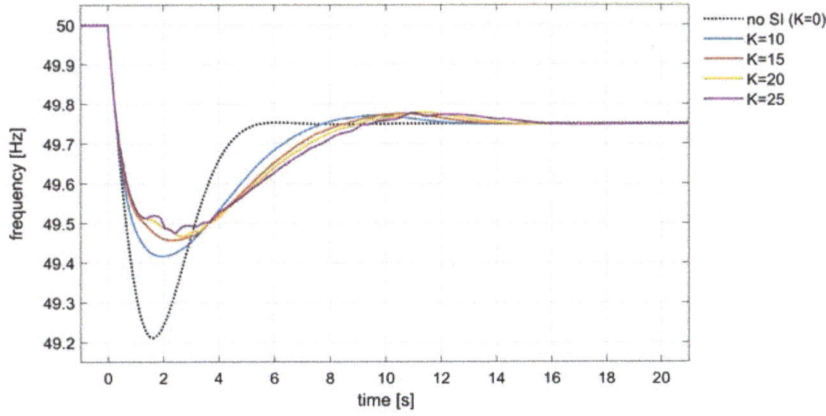

Figure 14. Real controller: frequency response with different gain K.

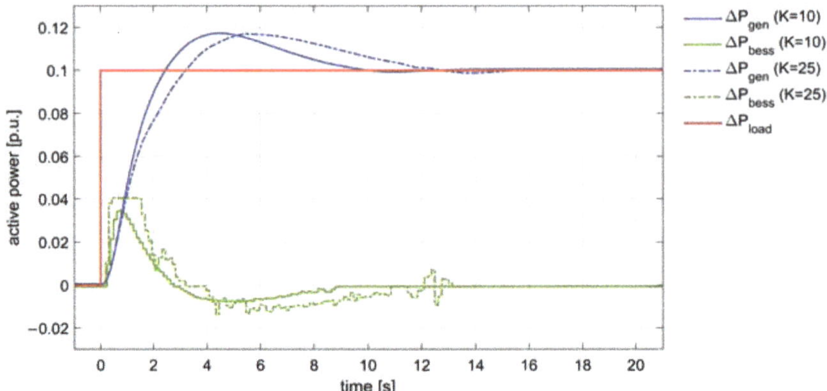

Figure 15. Real controller: BESS active power response with different gain K.

Table 3. Real synthetic inertia controller: characteristics of the frequency step response.

Gain K [A/(Hz/s)]	Frequency Reporting Time [s]	Alpha α	Fall Time [s]	Average RoCoF [Hz/s]	Time *nadir* [s]	Frequency *nadir* [Hz]	Overshoot [%]	Settling Time [s]
0	-	-	0.242	0.824	1.616	49.211	0.050	5.063
10	~0.050	0.10	0.246	0.812	1.948	49.416	0.047	11.792
15	~0.050	0.10	0.249	0.801	2.300	49.457	0.051	13.049
20	~0.050	0.10	0.249	0.801	2.641	49.465	0.055	13.825
25	~0.050	0.10	0.250	0.799	2.419	49.475	0.054	15.155

4.4. Frequency Measurement Reporting Time

Further tests were carried out to investigate the influence of the measurement reporting time with the smoothing factor α. As shown in Figure 16 and Table 4, a higher smoothing factor ($\alpha = 0.25$) allowed reduction of the initial RoCoF, but the frequency response was still characterized by a lower *nadir* and permanent fluctuations due to the effects of RoCoF error. However, since RoCoF estimation is derived from two subsequent frequency measurements, the RoCoF error can be reduced with a slower frequency reporting rate. As observed in [24], since frequency measurement is based on the evaluation of power system voltage, frequency computation can typically be updated every few cycles, 90–120 ms. Since a trade-off between fast enough and accurate frequency measurements is needed, according to ENTSO-E, accurate RoCoF calculations should be based on sliding windows which average results over few consecutive measurements. Robust results can be obtained, for example, in about 0.5 s if a 100 ms time resolution is used. For this reason, some further tests were carried out introducing a small delay in the frequency reporting of the controller, so that more accurate RoCoF estimations can be used by the controller. The effect of increasing the reporting time up to 100 ms is shown in both Figure 16 and Table 4.

Increasing the reporting time to 100 ms allowed obtaining of more stable frequency responses, but it did not prove useful when a stronger filter was applied ($\alpha = 0.10$). A 100 ms reporting time worked well with a higher α ($\alpha = 0.25$), as in the green dotted line in Figure 16, obtaining comparable results to the ones obtained with faster reporting rate and $\alpha = 0.10$ (blue line). These results are comparable with the ones obtained with ideal delayed control (see Figure 8 and Table 2).

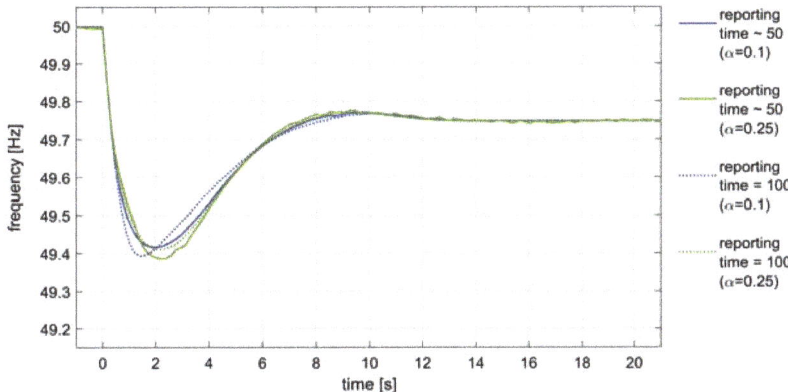

Figure 16. Real controller: frequency response with different reporting time and smoothing factor ($K = 10$).

This result is significant for several reasons. If a slower reporting rate is employed, the controller has a consistent amount of idle time (in this case more than 40 ms) that can be used either in the implementation of more efficient data processing and filters or in the control of more devices. This means that there is enough time to send more control signals and to deal with different protocols and communication media, extending the range of action of the controller from just one BESS to more coordinated DERs. In a microgrid, where different DERs might provide asymmetric frequency control resources (for example, loads vs. RES generation), this is a key improvement since it would allow increasing the extent of the frequency control capacity using just a single controller.

Table 4. Real synthetic inertia controller: comparison of frequency step response with different reporting time.

Gain K [A/(Hz/s)]	Frequency Reporting Time [s]	Alpha α	Fall Time [s]	Average RoCoF [Hz/s]	Time *nadir* [s]	Frequency *nadir* [Hz]	Overshoot [%]	Settling Time [s]
0	-	-	0.242	0.824	1.616	49.211	0.050	5.063
10	~0.050	0.10	0.246	0.812	1.948	49.416	0.047	11.792
10	~0.050	0.25	0.251	0.797	2.285	49.386	0.054	13.526
10	0.100	0.10	0.244	0.817	1.496	49.393	0.047	11.785
10	0.100	0.25	0.246	0.812	2.100	49.410	0.047	11.663

4.5. Discussion and Future Developments

The results obtained by previous PHIL tests validated the capability of the proposed controller to manage BESS, or other distributed flexible resources, in order to provide frequency support to power systems. The data shown in Tables 3 and 4 demonstrated that the inertial response can be sensibly improved in terms of frequency *nadir*. A limitation of the proposed approach is that initial RoCoF can only be minimally reduced by SI control. This is due to the unavoidable delays introduced by the measurement, filtering, and communication processes. An effective SI contribution can be reached only after few hundreds of milliseconds (see Figures 14 and 16). Clearly, the obtained performances cannot be compared with other approaches to virtual inertia, such as, for example, the ones based on virtual synchronous machines, which are able to generate much faster responses. However, one should be reminded that the scope of this paper is to prove the feasibility of enabling, through low-cost technologies, synthetic inertia control of legacy and distributed devices which cannot be programmed to emulate the behavior of a synchronous generator. These devices could be storage systems, such as in our tests, but also other flexible resources which could also have asymmetric active power control capacity (for example, demand response systems, photovoltaics, etc.). Nevertheless, achieved performances are comparable with typical activation times required for wind power or BESS in providing frequency support proportional to the frequency derivative (<0.5 s) [31].

The PHIL tests permitted testing of the response of the proposed control scheme in a realistic scenario, making use of real voltage trajectories affected by noise and including the actual computation and communication delays that can affect the control. Due to the presence of disturbances and uncertainties, such as time delays, and the presence of several blocks that introduce discontinuities (dead-band, saturations, etc.), we have preferred to check the stability of the controller through an extensive set of time-domain simulations and power hardware-in-the-loop tests. These tests were conducted, adopting wide ranges of variation of main parameters such as delays, gains, dead-bands, etc. For the sake of brevity, only the results of few tests have been reported in the paper. However, the system proved to be stable even in the presence of highly distorted measurements, as also demonstrated in Figures 12 and 13. The main aim of this work was to check that the performances obtained with an autonomous controller, able to both measure frequency from voltages and produce a control signal, comply with the time requirements needed for inertial support. Future developments will be aimed to investigate further on the design of the measurement filter, with a more analytic stability analysis.

5. Conclusions

In this work, the capability of a single-board, low-cost controller to be employed at end-user level to provide synthetic inertia with legacy BESS was investigated. The controller presented in this paper is programmed to perform fast frequency measurements using an autocorrelation algorithm to reduce the influence of voltage disturbances that can affect many frequency measurement methodologies. The controller cost is estimated at about USD 100 and is proved to be able to obtain real-time frequency and RoCoF measurements from local voltage samples, even in the case of single-phase systems, and to generate the SI control law to be sent to the remotely controlled BESS within 50–60 ms.

The controller was tested in a power hardware-in-the-loop simulation environment, on a physical LiFePO4 battery, proving to be accurate and fast enough to ensure a stable response and reduce frequency excursions during transients. The controller, however, can be applied to any other component, an inverter, PLC, or actuator, able to receive a set-point via common automation protocols. Therefore, the controllertested with the BESS can be ideally applied to any flexible resource such as switchable and curtailable loads, RES generators, or charging stations. The controller can interact with legacy components whose management system cannot be easily reprogrammed to include SI control schemes.

The PHIL tests shed light on the main issues deriving from the use of fast frequency monitoring and SI control systems. The presence of noise on grid voltage signals is required to filter the acquired frequency. However, delays introduced from these filtering processes barely affected the SI control effectiveness. In addition, test results showed that the frequency reporting time of this controller can be slowed down to 100 ms so that the controller itself has an idle time. This outcome demonstrated how the provision of SI control actions can be intentionally delayed to allocate the remaining time to control additional processes and making possible, at the same time, the implementation of different controls and coordination of multiple resources, without significantly penalizing the effectiveness of the SI support.

Author Contributions: Conceptualization, S.B. and M.L.S.; methodology, S.B.; software, C.I. and C.R.; validation, G.G., C.I. and C.R.; formal analysis, S.B.; investigation, C.I.; resources, C.R.; data curation, G.G.; writing—original draft preparation, S.B., C.I. and C.R.; writing—review and editing, G.G. and M.L.S.; visualization, G.G.; supervision, M.L.S.; project administration, G.G.; funding acquisition, S.B. and M.L.S. All authors have read and agreed to the published version of the manuscript.

Energies **2022**, 15, 3016

Funding: This work was partially financed by Regione Puglia through the Grant NGFKLZ2, within the framework of the Programme Innolabs, FESR-FSE 2014–2020. The work of Dr. Bruno was supported by Regione Puglia through the Grant A1C03120 within the framework of the Programme REFIN—Research for Innovation, FESR-FSE 2014–2020.

Institutional Review Board Statement: Not applicable.

Informed Consent Statement: Not applicable.

Conflicts of Interest: The authors declare no conflict of interest.

References

1. Morch, A.Z.; Siface, D.; Gerard, H.; Kockar, I. Market architecture for TSO-DSO interaction in the context of European regulation. In Proceedings of the 16th International Conference on the European Energy Market (EEM), Ljubljana, Slovenia, 18–20 September 2019; pp. 1–5. [CrossRef]
2. ENTSO-E. *Rate of Change of Frequency (RoCoF) Withstand Capability*; Technical Report; ENTSO-E: Brussels, Belgium, 2018.
3. Winter, W.; Elkington, K.; Bareux, G.; Kostevc, J. Pushing the limits: Europe's new grid: Innovative tools to combat transmission bottlenecks and reduced inertia. *IEEE Power Energy Mag.* **2015**, *13*, 60–74. [CrossRef]
4. ENTSO-E. *Need for Synthetic Inertia (SI) for Frequency Regulation*; Technical Report; ENTSO-E: Brussels, Belgium, 2017.
5. Fang, J.; Zhang, R.; Tang, Y.; Hongchang, L. Inertia Enhancement by Grid-Connected Power Converters with Frequency-Locked-Loops for Frequency Derivative Estimation. In Proceedings of the 2018 IEEE Power Energy Society General Meeting (PESGM), Portland, OR, USA, 5–10 August 2018; pp. 1–5. [CrossRef]
6. Bruno, S.; Giannoccaro, G.; La Scala, M. A Demand Response Implementation in Tertiary Buildings Through Model Predictive Control. *IEEE Trans. Ind. Appl.* **2019**, *55*, 7052–7061. [CrossRef]
7. Subramanian, L.; Debusschere, V.; Gooi, H.B.; Hadjsaid, N. A Distributed Model Predictive Control Framework for Grid-Friendly Distributed Energy Resources. *IEEE Trans. Sustain. Energy* **2021**, *12*, 727–738. [CrossRef]
8. Fang, J.; Li, X.; Tang, Y. Grid-Connected power converters with distributed virtual power system inertia. In Proceedings of the 2017 IEEE Energy Conversion Congress and Exposition, ECCE 2017, Cincinnati, OH, USA, 1–5 October 2017. [CrossRef]
9. Chamorro, H.R.; Sanchez, A.C.; Øverjordet, A.; Jimenez, F.; Gonzalez-Longatt, F.; Sood, V.K. Distributed synthetic inertia control in power systems. In Proceedings of the 2017 International Conference on ENERGY and ENVIRONMENT (CIEM), Bucharest, Romania, 19–20 October 2017; pp. 74–78. [CrossRef]
10. Martínez-Sanz, I.; Chaudhuri, B.; Junyent-Ferré, A.; Trovato, V.; Strbac, G. Distributed vs. concentrated rapid frequency response provision in future great britain system. In Proceedings of the 2016 IEEE Power and Energy Society General Meeting (PESGM), Boston, MA, USA, 17–21 July 2016; pp. 1–5. [CrossRef]
11. The SmartNet Consortium. *Deliverable 1.2—Characterization of Flexibility Resources and Distribution Networks*; Technical Report; The SmartNet Consortium: Torino, Italy, 2017.
12. Bruno, S.; De Carne, G.; Iurlaro, C.; Rodio, C.; Specchio, M. A SOC-feedback Control Scheme for Fast Frequency Support with Hybrid Battery/Supercapacitor Storage System. In Proceedings of the 2021 6th IEEE Workshop on the Electronic Grid (eGRID), New Orleans, LA, USA, 8–10 November 2021; pp. 1–8. [CrossRef]
13. Kuga, R.; Esguerra, M.; Chabot, B.; Avendano Cecena, A. *EPIC 2.05: Inertia Response Emulation for DG Impact Improvement*; Technical Report; 2019. Available online: https://www.pge.com/pge_global/common/pdfs/about-pge/environment/what-we-are-doing/electric-program-investment-charge/PGE-EPIC-Project-2.05.pdf (accessed on 30 March 2022).
14. Fang, J.; Tang, Y.; Li, H.; Li, X. A Battery/Ultracapacitor Hybrid Energy Storage System for Implementing the Power Management of Virtual Synchronous Generators. *IEEE Trans. Power Electron.* **2018**, *33*, 2820–2824. [CrossRef]
15. Conte, F.; Di Vergagni, M.C.; Massucco, S.; Silvestro, F.; Ciapessoni, E.; Cirio, D. Synthetic Inertia and Primary Frequency Regulation Services by Domestic Thermal Loads. In Proceedings of the 2019 IEEE EEEIC/I and CPS Europe 2019, Genova, Italy, 11–14 June 2019. [CrossRef]
16. Trovato, V.; Sanz, I.M.; Chaudhuri, B.; Strbac, G. Advanced Control of Thermostatic Loads for Rapid Frequency Response in Great Britain. *IEEE Trans. Power Syst.* **2017**. [CrossRef]
17. Rezkalla, M.; Zecchino, A.; Pertl, M.; Marinelli, M. Grid frequency support by single-phase electric vehicles employing an innovative virtual inertia controller. In Proceedings of the 2016 51st International Universities Power Engineering Conference (UPEC), Coimbra, Portugal, 6–9 September 2016; pp. 1–6. [CrossRef]
18. Bruno, S.; Giannoccaro, G.; Iurlaro, C.; Scala, M.L.; Rodio, C.; Sbrizzai, R. Fast Frequency Regulation Support by LED Street Lighting Control. In Proceedings of the 2021 IEEE International Conference on Environment and Electrical Engineering and 2021 IEEE Industrial and Commercial Power Systems Europe (EEEIC/I CPS Europe), Bari, Italy, 7–10 September 2021; pp. 1–6. [CrossRef]
19. Bruno, S.; Giannoccaro, G.; Iurlaro, C.; Scala, M.L.; Rodio, C. A Low-cost Controller to Enable Synthetic Inertia Response of Distributed Energy Resources. In Proceedings of the 2020 IEEE International Conference on Environment and Electrical Engineering and 2020 IEEE Industrial and Commercial Power Systems Europe (EEEIC/I&CPS Europe), Madrid, Spain, 9–12 June 2020; pp. 1–6. [CrossRef]

20. Rodio, C.; Giannoccaro, G.; Bruno, S.; Bronzini, M.; La Scala, M. Optimal Dispatch of Distributed Resources in a TSO-DSO Coordination Framework. In Proceedings of the 12th AEIT International Annual Conference, AEIT 2020, Catania, Italy, 23–25 September 2020; pp. 1–6. [CrossRef]
21. Orrù, L.; Petrocchi, L.; Siviero, A.; Silletti, F.; Lisciandrello, G.; Albimonti, G.; Ronzio, D.; Losa, I.; Lazzaro, M. H2020 OSMOSE Project: The Italian demonstrator. Testing flexibilities resources in a coordinated approach. In Proceedings of the 2021 IEEE PES Innovative Smart Grid Technologies Europe (ISGT Europe), Espoo, Finland, 18–21 October 2021; pp. 1–5. [CrossRef]
22. Daly, P.; Qazi, H.W.; Flynn, D. RoCoF-Constrained Scheduling Incorporating Non-Synchronous Residential Demand Response. *IEEE Trans. Power Syst.* **2019**, *34*, 3372–3383. [CrossRef]
23. Tosato, P.; Macii, D.; Brunelli, D. Implementation of phasor measurement units on low-cost embedded platforms: A feasibility study. In Proceedings of the 2017 IEEE International Instrumentation and Measurement Technology Conference, Turin, Italy, 22–25 May 2017. [CrossRef]
24. ENTSO-E. *Frequency Measurement Requirements and Usage*; Technical Report Version 7; ENTSO-E: Brussels, Belgium, 2018.
25. Romano, P.; Paolone, M. Enhanced Interpolated-DFT for Synchrophasor Estimation in FPGAs: Theory, Implementation, and Validation of a PMU Prototype. *IEEE Trans. Instrum. Meas.* **2014**, *63*, 2824–2836. [CrossRef]
26. Zhang, Z.; Fu, P.; Gao, G.; Jiang, L.; Wang, L. A Rogowski Digital Integrator With Comb Filter Signal Processing System. *IEEE Trans. Plasma Sci.* **2018**, *46*, 1338–1343. [CrossRef]
27. Haque, M.E.; Sakib Khan, M.N.; Islam Sheikh, M.R. Smoothing control of wind farm output fluctuations by proposed Low Pass Filter, and Moving Averages. In Proceedings of the 2015 International Conference on Electrical Electronic Engineering (ICEEE), Rajshahi, Bangladesh, 4–6 November 2015; pp. 121–124. [CrossRef]
28. Bruno, S.; Giannoccaro, G.; Scala, M.L.; Lopopolo, G. First activities and power-hardware-in-the-loop tests at the public research laboratory LabZERO. In Proceedings of the 2018 110th AEIT International Annual Conference, AEIT 2018, Bari, Italy, 3–5 October 2018; pp. 1–6. [CrossRef]
29. De Carne, G.; Bruno, S.; Liserre, M.; La Scala, M. Distributed Online Load Sensitivity Identification by Smart Transformer and Industrial Metering. *IEEE Trans. Ind. Appl.* **2019**, *55*, 7328–7337. [CrossRef]
30. Donnini, G.; Carlini, E.; Giannuzzi, G.; Zaottini, R.; Pisani, C.; Chiodo, E.; Lauria, D.; Mottola, F. On the Estimation of Power System Inertia accounting for Renewable Generation Penetration. In Proceedings of the 2020 AEIT International Annual Conference (AEIT), Catania, Italy, 23–25 September 2020; pp. 1–6. [CrossRef]
31. ENTSO-E. *Future System Inertia 2*; Technical Report; ENTSO-E: Brussels, Belgium, 2020.

Article

Lessons Learnt from Modelling and Simulating the Bottom-Up Power System Restoration Processes

Roberto Benato, Sebastian Dambone Sessa and Francesco Sanniti *

Industrial Engineering Department, University of Padova, 35131 Padova, Italy; roberto.benato@unipd.it (R.B.); sebastian.dambonesessa@unipd.it (S.D.S.)
* Correspondence: francesco.sanniti@phd.unipd.it

Abstract: This paper aims to present the gained experience in modeling and simulating bottom-up power system restoration processes. In a system with low inertia, such as a restoration path, the Common Information Models for the regulation systems appear to no longer be suitable for the estimation of the frequency behavior, and thus a detailed model must be considered. On the other hand, due to the predominantly inductive behavior of the HV transmission network, the assumption of decoupling the power-frequency behavior to study the restoration stability seems to be licit. All these issues are discussed and justified in the paper by means of the use of different software packages and of the comparison with on-field recordings.

Keywords: power system restoration; primary frequency control; frequency nadir estimation; low inertia systems

Citation: Benato, R.; Dambone Sessa, S.; Sanniti, F. Lessons Learnt from Modelling and Simulating the Bottom-Up Power System Restoration Processes. *Energies* **2022**, *15*, 4145. https://doi.org/10.3390/en15114145

Academic Editor: David Schoenwald

Received: 30 March 2022
Accepted: 1 June 2022
Published: 5 June 2022

Publisher's Note: MDPI stays neutral with regard to jurisdictional claims in published maps and institutional affiliations.

1. Introduction

1.1. Motivations

In recent years, the studies on power system restoration are growing in importance since the massive replacement of the conventional generation set by converter-based resources is leading the bulk power system to be more sensitive to any transient contingency [1].

Therefore, the frequency deviations are becoming more and more prominent, and the eventuality of a system blackout is no longer a remote scenario. Besides the preventive and curative countermeasures, it is necessary to conceive strategies and restoration plans adapted to the new system paradigm: this ought to guarantee an electricity service recovery as quickly and safely as possible in case of wide outage. The modelling and simulation of the restoration plans, in addition to advanced transmission line modellings [2], make it possible to check, validate, and quickly modify the plan of the restoration processes.. From the authors' gained practice on modelling and simulation of bottom-up power system restoration processes, two questions arise:

- Is it licit to totally decouple the voltage and the frequency behavior for the dynamic simulation of a restoration process? How much does this assumption impact on the estimation of the frequency behaviour?
- Is it licit, as a first approximation, to simulate the restoration process by adopting the Common Information Models (CIMs) in the regulation sets? How much does this choice impact on the estimation of the frequency behaviour?

In order to answer these questions, this paper evaluates the impacts of these assumptions, by comparing the results of the implemented models with the recordings of a real restoration process.

1.2. Literature Review

Research on power system restoration have traditionally focused on system operation issues [3], for which the main objective is the optimal allocation of resources to restore and to maximize the portion of recovered load and minimize the outage time [4,5]. This is generally formulated as a constrained optimization problem [6]. The dynamic stability analysis of the restoration process has been widely investigated in the most recent decades of the last century [7–10] and has also been studied through mock drills [11–13].

In the last years, the research on power system restoration dynamic stability is growing in relevance [14–19] since the frequency deviations are becoming more and more prominent, and so a correct estimation of transient behaviour is crucial.

In particular, in [14], a simple yet effective model for the simulation of restoration processes has been implemented by the authors. In that model, the decoupling between the voltage and frequency behaviours is the starting assumption. This assumption allows the authors to very significantly simplify the model by considering the restoration dynamic only based on the swing equation. In the present paper, a final validation of the model proposed in [14] is provided by comparing the decoupled model and the complete electromechanical model of a restoration path. Although this approximation is well-known in literature [20], this paper aims to quantitatively evaluate the impacts of this assumption by exploiting the recordings of a real restoration test.

With regard to the CIM modelling, the reference is the standard IEC 61970-302 [21], which is related to power systems dynamics: it currently has the primary use of facilitating the exchange of power system network data between organizations. In line with the ever-increasing standardization of power system devices, the authors think it is worth testing the dynamic performance of these models.

CIM models have already proven to be suitable for the estimation of the transient behaviour in highly interconnected power systems [22,23], but they do not provide any guarantee on their performance in low inertia systems, of which the restoration backbone is an example.

1.3. Contribution

The main contribution of this paper is summarized in Figure 1. The goal of this study is to quantitively evaluate the influence of different modelling approaches to represent the key elements involved in the restoration process by comparing the results of the simulations with the recordings of a real restoration process.

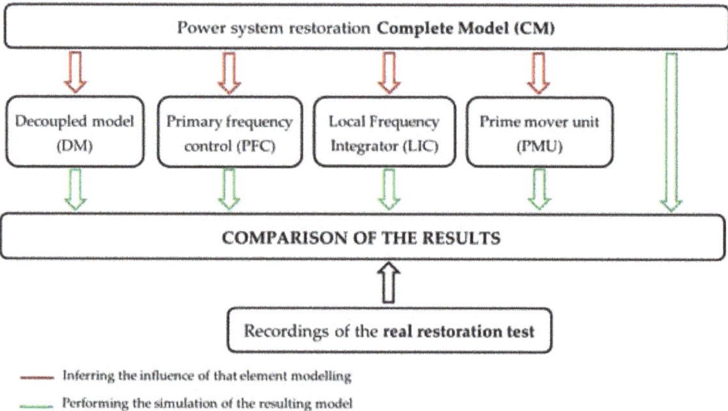

Figure 1. Visual scheme of the studies performed in this paper.

In Figure 1, the Complete Model (CM) must be intended as the detailed model of the real restoration path, including the actual configuration of the Primary Frequency Control (PFC), the Local Integrator Control (LIC), and the Prime Mover Unit (PMU) installed on the pilot generators. On the other hand, the Decoupled Model (DM) represents the same model but with the assumption of power-frequency decoupling. The red arrows indicates that an assumption is made on the CM. More specifically, the influence of decoupling the frequency behaviour is studied as a first step. Then, the detailed models of PFC and the PMU are replaced one by one by their equivalent CIMs, leaving all the other elements of the CM unaltered. With regards to the LIC model, since there is no CIM equivalent model to represent it, its influence is studied by simulating the restoration with and without that element.

All the results are compared in terms of frequency deviations. The reason why the paper mainly focuses on frequency behavior is that this is the most critical dynamic in a bottom-up restoration process. In particular, it is important to be able to correctly estimate the frequency nadir, i.e., the minimum value the frequency reaches during the transient period. This index is important in a restoration framework since large frequency unbalances are expected after each load supply.

1.4. Organization

This paper is organized as follows. Section 2 describes the complete model of the restoration, the assumptions made for the decoupled model, and the simplified CIM models adopted for the purpose of comparison. Section 3 presents the results of the comparison between the measurement recording, the complete model, the decoupled model, and the adoption of the CIM models for the frequency regulation sets.

2. Power System Restoration Model

In this section, all the modelling features that influence the frequency behaviour during a restoration process are described. First, the focus is on the power system mathematical model and on assumptions needed to assume the decoupling between the frequency and the voltage behaviour. Thus, the models of the key elements that mostly affect the power-frequency behaviour are described: the PFC, the LIC, and PMU. In particular, PFC, LIC, and PMU are modelled by considering their actual set up, i.e., the real configuration of the regulators used during the real restoration is represented, without introducing simplifying hypotheses. Hence, the detailed models of PFC and PMU are compared with their equivalent CIM.

2.1. Power System Decoupling

With the aim of analyzing the rotor angle and the voltage stability of power systems, they are typically modelled by a set of non-linear Differential Algebraic Equations (DAEs), as follows:

$$\begin{aligned} \dot{x} &= f(x, y) \\ 0 &= g(x, y), \end{aligned} \tag{1}$$

where f and g are the differential and the algebraic equations, $x, x \in \mathbb{R}^n$ and $y, y \in \mathbb{R}^m$ are the state and algebraic variables, respectively. This model is widely adopted to represent the electromechanical behaviour of the system for transient stability analysis, i.e., for the time scale from 0.01 s to 10 s [20]. Now, let us consider a single source–a two-bus system with a lossless line and a load on bus 2, as depicted in Figure 2.

Assumed as a reference the phasor $v_2 \angle \vartheta_2$, such a system could represent the bulk power system of a bottom-up restoration. The lossless line is an acceptable assumption considering the low r/x ratio of the transmission lines. The set of equations in (1), if represented with the power-injection model [20], has two state equations:

$$\begin{aligned} \dot{\delta} &= \omega_n \Delta\omega \\ \dot{\omega} &= (p_m - p_e - D\Delta\omega)/T_a \end{aligned} \tag{2}$$

and two algebraic equations, related to the power-injections at bus (if one neglects the generator internal algebraic equations), as follows:

$$\begin{aligned} p_1 &= \frac{v_1 v_2}{x_L} \sin(\vartheta_1 - \vartheta_2) \\ q_1 &= \frac{v_1^2}{x_L} - \frac{v_1 v_2}{x_L} \cos(\vartheta_1 - \vartheta_2) \end{aligned} \tag{3}$$

where δ is the rotor angle, ω_n is the nominal angular speed in rad/s, $\Delta\omega_n$ is the angular speed deviation from the reference $\Delta\omega_{ref}$, p_m, and p_e are the mechanical and the electric power, respectively, D is the damping of the generator, T_a is the starting time of the generator ($T_a = 2H$ where H is the well-known inertia constant), p_1 and q_1 are the active and reactive power injected at bus 1, $v_1\angle\vartheta_1$ and $v_2\angle\vartheta_2$ are the voltage phasors at bus 1 and 2, respectively, and x_L is the line reactance.

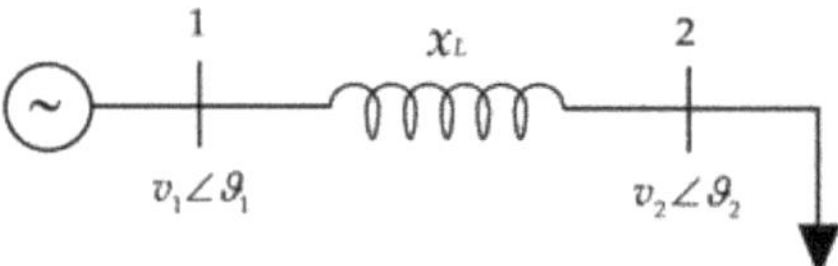

Figure 2. One machine, two-bus, one load system.

The differentiation of p_1 and q_1 in (3) leads to:

$$dp_1 = \frac{v_2}{x_L}\sin(\vartheta_1 - \vartheta_2)dv_1 + \frac{v_1 v_2}{x_L}\cos(\vartheta_1 - \vartheta_2)d\vartheta_1$$
$$dq_1 = \frac{2v_1 - v_2\cos(\vartheta_1 - \vartheta_2)}{x_L}dv_1 + \frac{v_1 v_2}{x_L}\sin(\vartheta_1 - \vartheta_2)d\vartheta_1. \tag{4}$$

Assuming as a first approximation that the angle deviation $\Delta\vartheta = \vartheta_1 - \vartheta_2$ is small, then $\sin\Delta\vartheta \approx \Delta\vartheta$ and $\cos\Delta\vartheta \approx 1$. Equation (4) yields:

$$dp_1 \approx \frac{v_1 v_2}{x_L}d\vartheta_1$$
$$dq_1 \approx \frac{2v_1 - v_2}{x_L}dv_1 \tag{5}$$

where the well-known concept that the angle deviation participates the most to the active power balance and the voltage deviations most affect the reactive power balance becomes clear. Therefore, since the commonly accepted definition of the frequency is the derivative of the voltage angle $\dot{\vartheta}$, the frequency of the system can be well-estimated by considering only the active power unbalances. In practice, this means that the model of the system is reduced to the second equation of (2), i.e., the swing equation of the system. The frequency, in the case of a single machine, is equivalent to the rotor angular velocity ω, and in the case of multiple machines, to the frequency of the Center of Inertia (COI) ω_{COI}, as:

$$\omega_{COI} = \frac{\sum_{i=1}^{N} T_{ai} S_i \omega_i}{\sum_{i=1}^{N} T_{ai} S_i} \tag{6}$$

where T_{ai}, S_i, ω_i are the starting time, the apparent power and the angular velocity of the i-th machine, respectively, and N is the number of machines involved in the restoration process.

Thus, the swing equation becomes:

$$\sum_{i=1}^{N} T_{ai}\dot{\omega}_{COI} = \sum_{i=1}^{N} p_{mi} - p_e - \sum_{i=1}^{N} D_i \Delta\omega_{COI} \tag{7}$$

where p_{mi} and D_i are the mechanical power and the damping coefficient of the i-th machine and p_e is the active power absorbed by the load, as the transmission lines are *transparent* for the active power flow. In the remainder of this paper, the notation CM indicates the implementation of the whole set of equations in (1), and DM indicates the implementation of (7).

Note that p_{mi} in (7) is:

$$p_{mi} = (p_{PFCi} + p_{LICi}) \cdot p_{PMUi} \tag{8}$$

where p_{PFC} is the contribution of the PFC, p_{LIC} is the contribution of the LIC, and p_{PMU} involves the dynamic of the PMU. In the reminder of this section further details on the modelling of this contributions are given.

2.2. Primary Frequency Control

The detailed model of the primary frequency control is depicted in Figure 3.

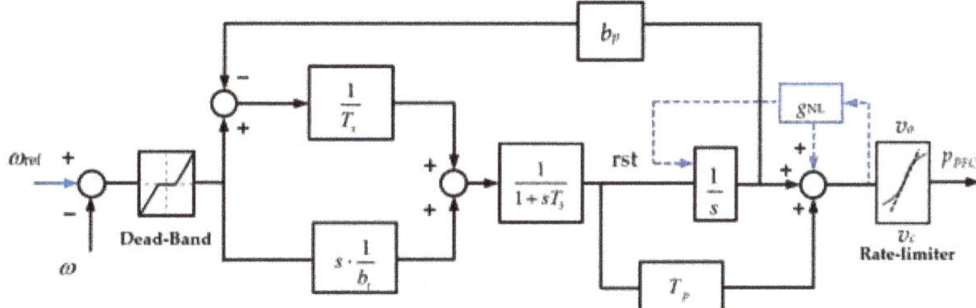

Figure 3. Control scheme of the detail model of the Hydro primary frequency regulation.

The model in Figure 3 represents the real regulation scheme implemented in the power plant object of the restoration path analysed in this paper.

Blue lines in Figure 3 refer to the control logic involved in the black-start phase during the dynamic transition from $\omega = 0$ to $\omega = \omega_n$. The starting phases of the generator start-up are the following:

- ω_{ref} is initialized to 0 and is then driven by a piece-wise ramp signal to reach the no-load speed value;
- when the generator reaches the no-load speed value, the actual value of p_{PFC} is memorized in the g_{NL} block and it is added to the total regulator output. At the same time, the integrator of the PFC is reset. This operation is required to memorize the value of the no-load gate opening of the turbine to avoid an unexpected response by the PFC during the normal operation;
- The generator parallel breaker is closed and the reference value ω_{ref} is set to its nominal value.

The equivalent transfer function of the PFC around the nominal speed is:

$$G_{PFC}(s) = \frac{1}{b_p} \frac{\left(1 + s\frac{T_x}{b_t}\right)\left(1 + sT_p\right)}{\left(1 + s\frac{T_x}{b_p}\right)\left(1 + sT_3\right)}. \tag{9}$$

The equivalent model of this regulator in the CIM library is the GovHydroIEEE2 (from now on called HG2) model [21] depicted in Figure 4. Note that all equivalent models found in literature do not involve the generator start up logics and instead only describe the model around its nominal speed.

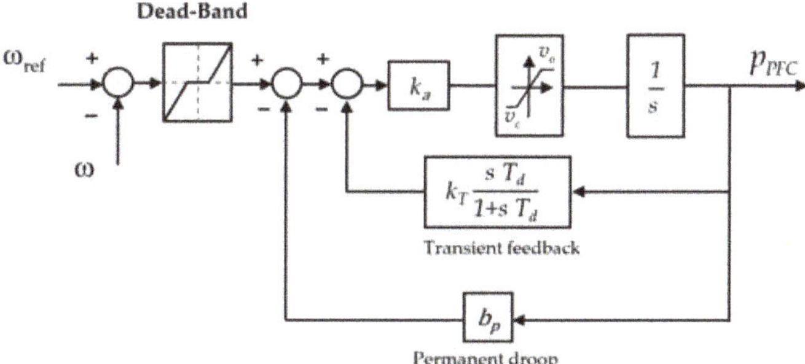

Figure 4. Control scheme of the HG2 model.

The conversion between the parameter set of the PFC and of the HG2 is as follows:

$$\begin{cases} T_d = T_x / b_p \\ k_T = b_t - b_p \end{cases} \tag{10}$$

2.3. Local Integration Control

The LIC is generally enabled for emergency condition frequency deviations and is always installed on the black start units. It ensures the perfect tracking of the frequency also for isolated generators without Automatic Generation Control (AGC) [24]. Compared to the AGC, however, the LIC is characterized by a higher gain to speed up the frequency restoration. The control scheme of the LIC is depicted in Figure 5.

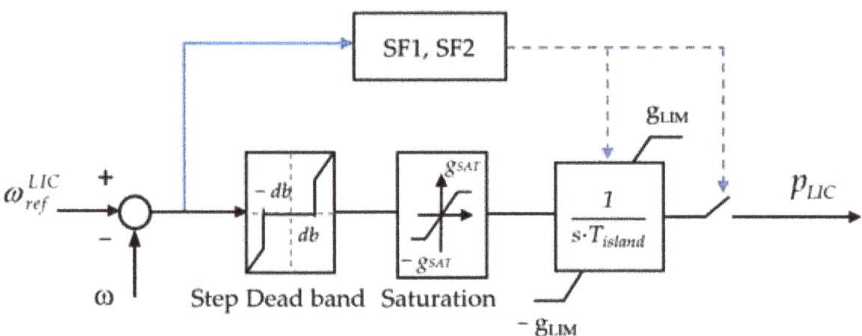

Figure 5. Control scheme of the LIC.

The logic function that controls the LIC action is highlighted by blue in Figure 5. The SF1 constant is the frequency threshold which determines the activation of the LIC control, whereas SF2 determines the condition for the disconnection of the LIC, i.e., when the system frequency stays beyond the SF2 threshold for at least T_{SF2} seconds. The disconnection action also resets the LIC integrator. The saturation block limits the frequency deviation to a maximum value g_{SAT} to avoid overshot behaviour in the LIC response in case of grid frequency peaks. As already mentioned, since there is no CIM representation for the LIC, no comparison is possible and so the LIC influence is studied in Section 3 by simulating the restoration with and without this element.

2.4. Prime Mover Model

The prime mover unit of the hydroelectric power plant considered in this study consists of a Francis turbine supplied by a forebay through a penstock pipe.

The adopted model involves the turbine model as well as the model of the hydro penstock. For the purpose of this study, the most detailed model of the penstock is adopted, i.e., the model that considers the elasticity of the penstock. Starting from the prime mover model of [25] (p. 416), the function $F(s)$ is derived as follows:

$$F(s) = \frac{dH}{dU} = -\left(\phi_p + Z_p \tanh(T_{ep}s)\right) \tag{11}$$

where dH is the deviation in the head of the water column, dU is the deviation on the water flow, ϕ_p is the penstock friction, Z_p is the equivalent penstock impedance, and T_{ep} is the penstock elastic time. Now, by replacing the hyperbolic tangent with its exponential series, it yields:

$$dH = dU \cdot Z_p(e^{-2T_{ep}s} - 1 - \phi_p) - dH \cdot e^{-2T_{ep}s} \tag{12}$$

This relation is implemented in the global prime mover model by exploiting the block algebra, as shown in Figure 6.

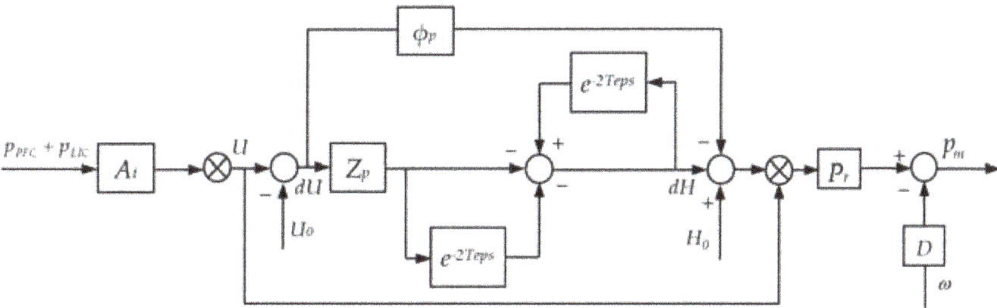

Figure 6. Detailed model of the prime mover.

With regards to the CIM equivalent representation of the PMU, there are many different equivalent representations in literature depending on the turbine type. Nevertheless, for the purpose of this study, the authors adopt the commonly accepted model of the prime mover for transient stability studies, which links the mechanical power to the gate opening through a transfer function with one zero and one pole, as follows [25]:

$$p_m = (p_{PFC} + p_{LIC})\frac{1 - T_w s}{1 + 0.5 T_w s} \tag{13}$$

where T_w is the water starting time. The relationship between T_w and the quantities that describe the detailed model are represented by (14):

$$T_w = Z_p T_{ep} \tag{14}$$

3. Case Study

The results presented in this section are based on a real bottom-up power system restoration test performed in the Italian Transmission Grid in 2019. The recordings of the restoration test are compared to the simulation results of the models presented in the previous section. Moreover, an extensive comparison between the detailed models and the standard models is carried out, even by changing the parameter settings.

The restoration path is made up by:

- The pilot power plant, which is a hydroelectric power plant, where three parallel synchronous generators are installed;
- The target power plant, which is a thermoelectric power plant;
- Portions of the HV transmission grid;
- Portions of the distribution grid;

The restoration process starts with the black-start of the three hydro generators, then a portion of the HV transmission grid is supplied and at the same time some loads are restored by supplying the portions of the distribution grid. The final step is the parallel between the restoration path and the target power plant.

This work focuses on the study of the frequency behaviour during the start-up of the pilot generators and during the first two load restoration steps. A simplified scheme of the simulated restoration path is shown in Figure 7. In Figure 7, the black lines represent the medium voltage level (15 kV and 20 kV), the green lines represent the transmission voltage level (220 kV), and the blue lines represent the subtransmission voltage level (132 kV).

In Figure 7, G1, G2, G3 are the three hydro generators; L1 and L2 are the equivalent load representing a portion of a distribution grid; S_{G1}, S_{G2}, S_{G3}, S_{L1}, S_{L2} are the brakers that are closed during the restoration process to build the restoration path. With the aim of evaluating the performance of the model in terms of frequency deviations, the PQ power exchange of L1 and L2 is imposed externally by a measurement file taken from the measurement recordings.

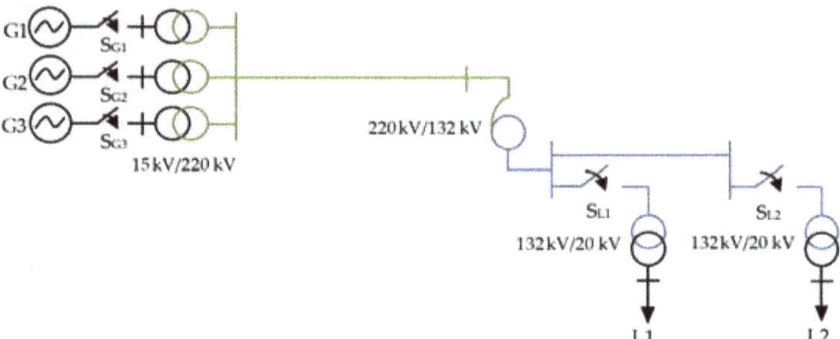

Figure 7. Simplified single-line scheme of the restoration path.

The time domain simulations of the complete model are performed by the software package DIgSILENT PowerFactory; the simulations related to the decoupled model are instead performed in the Matlab-Simulink environment.

3.1. Measurements vs. Complete Model

Figure 8 shows the global results of the comparison between the real recording of the restoration test (Meas.) and the simulation results performed with the complete model. All the five switching closures of Figure 7 are simulated. The (a) graph of Figure 8 shows that the actual grid frequency is well-reproduced by CM, especially for the load step events S_{L1} and S_{L2}, which are the most critical operations for the restoration process. The (c) graph of Figure 8 shows the gate opening signal of G1. Even in this case, one can appreciate how the model correctly estimates the real behaviour. The (b) graph of Figure 8 is reported just for figuring the total active power absorbed by the restoration path. The simulated power is equivalent to the measured one, as expected, since the active flow is an input of the model, as discussed above.

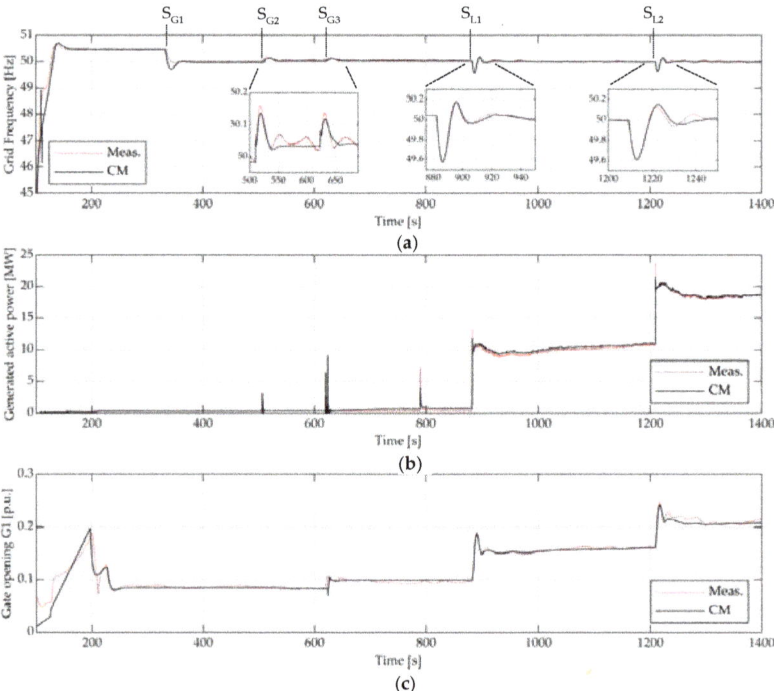

Figure 8. Comparison between simulation results and measurement recordings of the restoration test for (**a**) the grid frequency, (**b**) the generated active power and (**c**) the gate opening of G1.

3.2. Influence of the Power Frequency Decupling

In this section, the influence of the frequency–voltage decoupling assumption is discussed by comparing the simulation results of the decoupled model (DM) described in Section 2.1 with the results of the complete model (CM) and with the recording (Meas.). The focus, from now on, is on the results of the two load step events S_{L1} and S_{L2}, since they originate a transient response on which the models of interest have a higher influence. Figure 9 shows the transient behaviour of (a) the grid frequency, (b) the gate opening of the supply system of GRA, and (c) the output signal of the LIC of G1. The results show that the DM gives a perfect estimation of the frequency behaviour except for the frequency overshoot, for which the maximum error respect to the real behaviour is 0.04 Hz, see Figure 9c. This important result allows us to consider DM as a licit approximation of the CM to study the frequency behaviour during a restoration process. Nevertheless, in the remainder of this section, the simulations are performed by adopting the CM approach. This choice is made in order to separate the error (even if very small) introduced by adopting the DM from the influence of the single components modelling approach. In this way, it is possible to better allocate the cause of potential displacements in the simulation results.

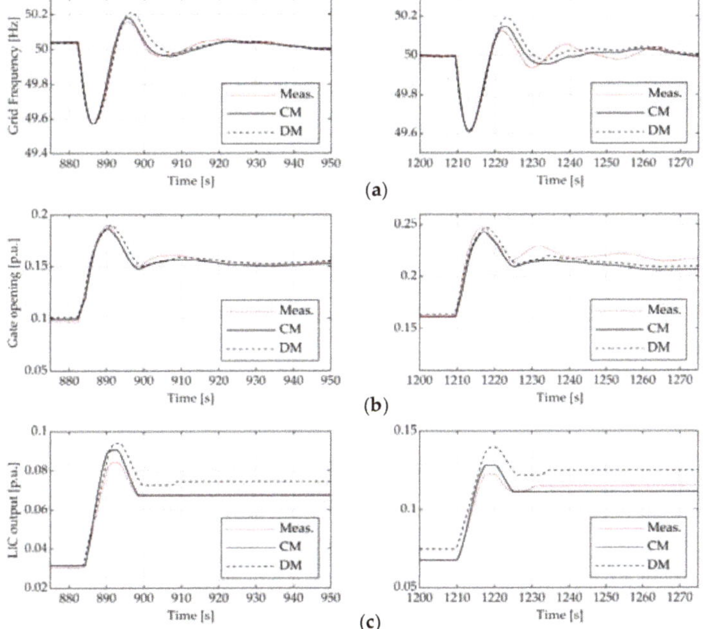

Figure 9. Comparison between simulation results and measurement recordings of the S_{L1} and S_{L2} events for (**a**) the grid frequency, (**b**) the gate opening of G1, and (**c**) the output signal of the LIC of G1 for different mathematical models.

3.3. Primary Frequency Regulation Influence

In this section, the influence of different modelling of the PFC described in Section 2.2 is discussed. In particular, the complete model (CM) is compared with the HG2 model with the parameters derived from (10) (HG2), and the HG2 model with the standard parameters given in [21] (HG2 std.). The parameters of the PFC set in the three models are given in Table 1. The result of the comparison is depicted in Figure 10 for (a) the grid frequency and for (b) the gate opening of the supply system of G1.

Table 1. Parameters related to different models of the PFC.

Model	b_p [-]	b_t [-]	$T3$ [s]	T_p [s]	v_o [pu/s]	v_c [pu/s]	T_x [s]	T_d [s]	k_a [-]	k_t [-]
CM	0.04	0.4	0	0	0.07	0.15	2.3	-	-	-
HG2	0.04	0.4	-	-	0.07	0.15	2.3	5.75	100	0.36
HG2 std	0.05	0.55	-	-	0.1	0.1	6	12	2	0.5

The results show that the HG2 model returns a good estimation of the frequency behaviour, with a relative error on the frequency undershoot of 6.2% instead of the 25% given by the HG2 std. model. The quantitative behaviour of HG2 std. is also delayed with respect to the real behaviour for 4 s.

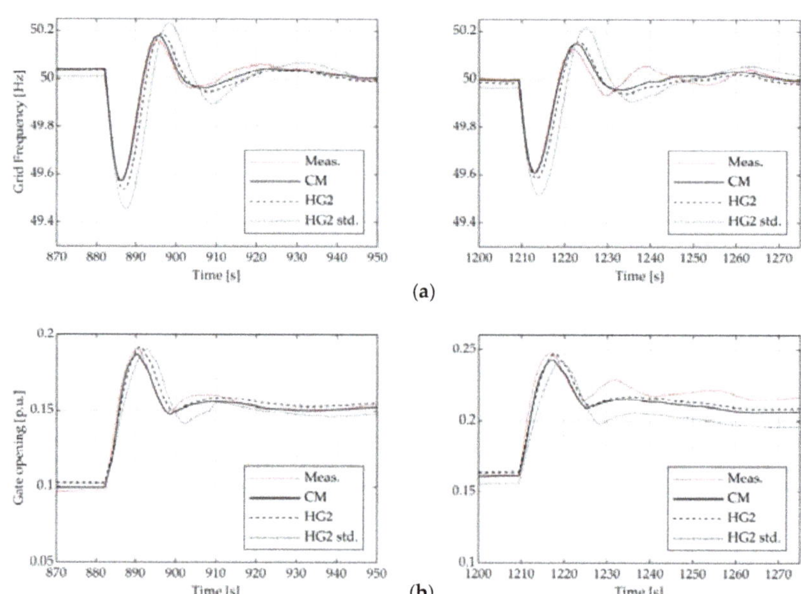

Figure 10. Comparison between simulation results and measurement recordings of the S_{L1} and S_{L2} events for (**a**) the grid frequency and (**b**) the gate opening of G1 for different model of the PFC.

3.4. Prime Mover Unit Influence

In this section, the influence of the different modelling of the PMU described in Section 2.3 is discussed. In particular, the complete model (CM) is compared with the complete model without the prime mover model (No PMU), the complete model with the standard PMU model and with the parameters derived from (13) (PMU), and the complete model with the standard PMU model and the standard parameters given in [21] (PMU std.). The parameters of the PMU set in the three models are given in Table 2. The results of the comparison are depicted in Figure 11 for (a) the grid frequency and for (b) the gate opening of the supply system of G1.

Table 2. Parameters related to different models of the PMU.

Model	A_t [-]	Z_p [-]	ϕ_p [-]	T_{ep} [s]	D [pu]	P_r [pu]	T_w [s]
CM	1.12	1.52	0.001	0.23	0.08	0.77	-
PMU	-	0.4	-	0.23	-	-	0.35
PMU std	-	-	-	-	-	-	2

The results show how the model of the prime mover does not significantly affect the frequency behaviour, since the relative error of the frequency undershoot between Meas and No PMU is 6%. Moreover, the influence of using the standard prime mover model with the real parameters is extremely low; indeed, the curves of No PMU and PMU are overlapped. On the contrary, the standard prime mover model with the standard parameter leads to the rise of permanent wide oscillations in the frequency and in the gate opening behaviour, resulting in a completely wrong estimation of the real behaviour.

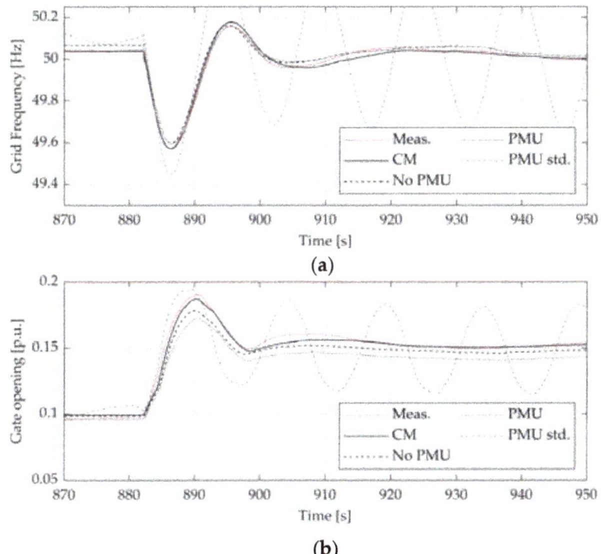

Figure 11. Comparison between simulation results and measurement recordings of the S_{L1} event for (**a**) the grid frequency and (**b**) the gate opening of G1 for different model of the PMU.

3.5. Local Integrator Control Influence

In this section, the influence of the presence of the LIC on the overall behaviour of a restoration process is discussed. In this case, the comparison with the equivalent standard model is not performed since there is no equivalent CIM model for the LIC.

In particular, the complete model (CM), which includes the LIC model, is compared with the complete model without the LIC (No LIC). The parameters of the LIC are given in Table 3. The results of the comparison are depicted in Figure 11 for (a) the grid frequency and for (b) the gate opening of the supply system of G1.

Table 3. Parameters of the LIC model.

Model	T_{island} [s]	$SF1$ [Hz]	$SF2$ [Hz]	db [Hz]	g_{SAT} [Hz]	g_{LIM} [pu]	T_{SF2} [s]
LIC	120	0.3	0.03	0.075	0.3	1	300

It is worth remembering that a standard AGC normally works in a higher time scale with respect to the PFC. On the contrary, the LIC action is faster, as it is possible to appreciate from Figure 12, since the presence of the LIC completely alters the PFC response and significantly contributes to the frequency restoration and to the nadir containment.

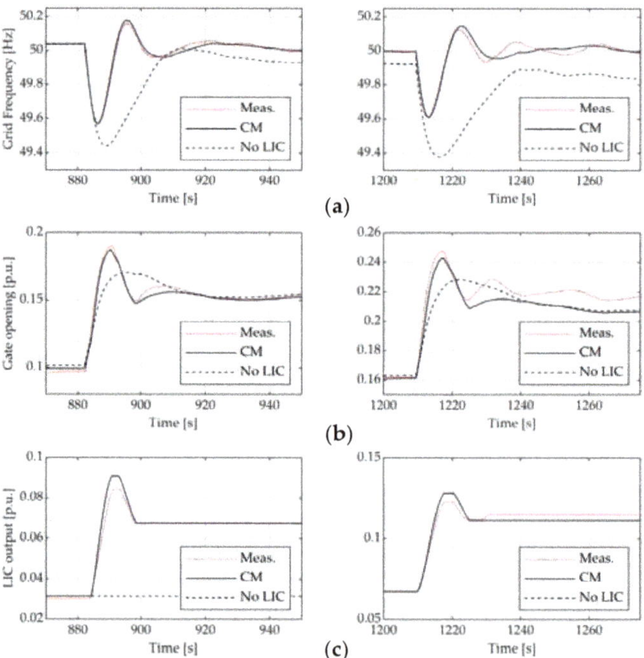

Figure 12. Comparison between simulation results and measurement recordings of the S_{L1} and S_{L2} events for (**a**) the grid frequency, (**b**) the gate opening of G1 and (**c**) the output signal of the LIC of G1, with and without the LIC model.

4. Conclusions

In this paper, a detailed model simulating a power system restoration process is presented and tested on a real restoration test. The proposed model correctly estimates the frequency behaviour during the load restoration and is proven to be effective, even for out-of-nominal conditions. (e.g., during the start-up of the generator).

Moreover, the results highlight how the assumption of the voltage-frequency decoupling does not significantly affect the frequency behaviour. This confirms the effectiveness of the simplified model presented in [13] as a powerful tool for the estimation of the frequency deviations during a restoration process. It could give the TSO useful decision support to determine the feasibility of a given restoration path.

The results of the comparison between the detailed model of the PFC and the PMU with their equivalent CIM models highlight, in some cases, a displacement on the estimation of the frequency Nadir. This happens when the CIM models are adopted by applying the IEC proposed parameters. Although the CIM models with standard parameters are more than suitable to estimate the transient behavior for a highly interconnected bulk power system, the paper demonstrates that, in the case of an islanded transmission system, the CIM models only work correctly if real parameters are adopted. In fact, for the analyzed case study, it was possible to combine the real parameters with the CIM structure. However, in general, it is very difficult to use real parameters by maintaining the IEC proposed CIM structure.

Hence, the knowledge of the actual regulation sets is of great relevance to correctly estimate the transient behaviors. In fact, a simplified representation of these components could lead to a wrong estimation of the frequency behaviour., e.g., when considering the standard parameters of the regulations sets.

Eventually, the results show how the influence of the LIC during a restoration process is prominent and how its action is overlapped with the action of the PFC, differently from what one might expect from an ACG in normal operation.

Author Contributions: Conceptualization, R.B. and F.S.; methodology, F.S. and S.D.S.; software, F.S.; validation, F.S.; formal analysis, F.S. and S.D.S.; investigation, F.S.; resources, F.S. and R.B.; data curation, F.S.; writing—original draft preparation, F.S.; writing—review and editing, F.S., R.B. and S.D.S.; visualization, S.D.S.; supervision, R.B.; project administration, R.B.; funding acquisition, F.S. All authors have read and agreed to the published version of the manuscript.

Funding: Italian Ministry of Education, Universities and Research: 35th PhD Cycle.

Acknowledgments: The authors wish to thank the Italian Transmission System Operator for providing the measurement recording of the real restoration test.

Conflicts of Interest: The authors declare no conflict of interest.

Acronyms

AGC	Automatic Generation Control
CIM	Common Information Model
CM	Complete Model
COI	Center Of Inertia
DAE	Differential Algebraic Equation
DM	Decoupled Model
HG2	GovHydroIEEE2
HV	High Voltage
LIC	Local Integrator Control
PFC	Primary Frequency Control
PMU	Prime Mover Unit
TSO	Transmission System Operator

References

1. Milano, F.; Dörfler, F.; Hug, G.; Hill, D.J.; Verbič, G. Foundations and Challenges of Low-Inertia Systems (Invited Paper). In Proceedings of the 2018 Power Systems Computation Conference (PSCC), Dublin, Ireland, 11–15 June 2018; pp. 1–25.
2. Benato, R.; Dambone Sessa, S.; Gardan, G.; Palone, F.; Sanniti, F. Experimental Harmonic Validation of 3D Multiconductor Cell Analysis: Measurements on the 100 km Long Sicily-Malta 220 kV Three-Core Armoured Cable. *IEEE Trans. Power Deliv.* **2022**, *37*, 573–581. [CrossRef]
3. Liu, Y.; Fan, R.; Terzija, V. Power System Restoration: A Literature Review From 2006 to 2016. *J. Mod. Power Syst. Clean Energy* **2016**, *4*, 332–341. [CrossRef]
4. Patsakis, G.; Rajan, D.; Aravena, I.; Rios, J.; Oren, S. Optimal Black Start Allocation for Power System Restoration. *IEEE Trans. Power Syst.* **2018**, *33*, 6766–6776. [CrossRef]
5. Liu, S.; Podmore, R.; Hou, Y. System Restoration Navigator: A Decision Support Tool for System Restoration. In Proceedings of the 2012 IEEE Power and Energy Society General Meeting, PES 2012, San Diego, CA, USA, 22–26 July 2012; IEEE Computer Society: San Diego, CA, USA, 2012.
6. Sun, W.; Liu, C.-C.; Zhang, L. Optimal Generator Start-up Strategy for Bulk Power System Restoration. *IEEE Trans. Power Syst.* **2011**, *26*, 1357–1366. [CrossRef]
7. Special Consideration in Power System Restoration. The Second Working Group Report. *IEEE Trans. Power Syst.* **1994**, *9*, 15–21. [CrossRef]
8. Adibi, M.M.; Borkoski, J.N.; Kafka, R.J.; Volkmann, T.L. Frequency Response of Prime Movers during Restoration. *IEEE Trans. Power Syst.* **1999**, *14*, 751–756. [CrossRef]
9. Adibi, M.M.; Borkoski, J.N.; Kafka, R.J. Analytical Tool Requirements for Power System Restoration. In *Power System Restoration: Methodologies and Implementation Strategies*; Wiley-IEEE Press: Hobocken, NJ, USA, 2000; pp. 86–95.
10. Delfino, B.; Invernizzi, M.; Morini, A. Knowledge-Based Restoration Guidelines. *IEEE Comput. Appl. Power* **1992**, *5*, 54–59. [CrossRef]
11. Mariani, E.; Mastroianni, F.; Romano, V. Field Experiences in Reenergization of Electrical Networks from Thermal and Hydro Units. In *Power System Restoration: Methodologies and Implementation Strategies*; Wiley-IEEE Press: Hobocken, NJ, USA, 2000; pp. 354–360.
12. Sforna, M.; Bertanza, V.C. Restoration Testing and Training in Italian ISO. *IEEE Trans. Power Syst.* **2002**, *17*, 1258–1264. [CrossRef]

13. Delfino, B.; Denegri, G.B.; Invernizzi, M.; Morini, A.; Cima Bonini, E.; Marconato, R.; Scarpellini, P. Black-Start and Restoration of a Part of the Italian HV Network: Modelling and Simulation of a Field Test. *IEEE Trans. Power Syst.* **1996**, *11*, 1371–1379. [CrossRef]
14. Benato, R.; Bruno, G.; Sessa, S.D.; Giannuzzi, G.M.; Ortolano, L.; Pedrazzoli, G.; Poli, M.; Sanniti, F.; Zaottini, R. A Novel Modeling for Assessing Frequency Behavior During a Hydro-to-Thermal Plant Black Start Restoration Test. *IEEE Access* **2019**, *7*, 47317–47328. [CrossRef]
15. Benato, R.; Giannuzzi, G.M.; Poli, M.; Sanniti, F.; Zaottini, R. A Fully Decoupled Model for the Study of Power-Frequency Behaviour during a Power System Restoration Test. In Proceedings of the 2020 AEIT International Annual Conference (AEIT), Bari, Italy, 23–25 September 2020; pp. 1–6.
16. Benato, R.; Giannuzzi, G.M.; Masiero, S.; Pisani, C.; Poli, M.; Sanniti, F.; Talomo, S.; Zaottini, R. Analysis of Low Frequency Oscillations Observed During a Power System Restoration Test. In Proceedings of the 2021 AEIT International Annual Conference (AEIT), Milan, Italy, 4–8 October 2021; pp. 1–6.
17. Benato, R.; Giannuzzi, G.M.; Poli, M.; Sanniti, F.; Zaottini, R. The Italian Procedural Approach for Assessing Frequency and Voltage Behavior during a Bottom-up Restoration Strategy. In Proceedings of the 2020 IEEE International Conference on Environment and Electrical Engineering and 2020 IEEE Industrial and Commercial Power Systems Europe (EEEIC/I CPS Europe), Madrid, Spain, 9–12 June 2020; pp. 1–6.
18. Benato, R.; Bruno, G.; Pedrazzoli, G.; Giannuzzi, G.M.; Ortolano, L.; Zaottini, R.; Poli, M.; Sanniti, F. An Experimental Validation of Frequency Model for Studying a Real Black Start Restoration. In Proceedings of the 2019 AEIT International Annual Conference (AEIT), Florence, Italy, 18 September 2019; pp. 1–6.
19. Hachmann, C.; Lammert, G.; Lafferte, D.; Braun, M. Power System Restoration and Operation of Island Grids with Frequency Dependent Active Power Control of Distributed Generation. In Proceedings of the 5th Conference on Sustainable Energy Supply and Energy Storage Systems, NEIS 2017, Hamburg, Germany, 21–22 September 2017; pp. 45–50.
20. Milano, F. *Power System Modelling and Scripting*, 2010 ed.; Springer: Berlin/Heidelberg, Germany, 2010.
21. *IEC 61970-302*; Common Information Model (CIM) Dynamics, Version 1. IEC: Geneva, Switzerland, 2018.
22. Gomez, F.J.; Vanfretti, L.; Olsen, S.H. CIM-Compliant Power System Dynamic Model-to-Model Transformation and Modelica Simulation. *IEEE Trans. Ind. Inform.* **2018**, *14*, 3989–3996. [CrossRef]
23. Semerow, A.; Hohn, S.; Luther, M.; Sattinger, W.; Abildgaard, H.; Garcia, A.D.; Giannuzzi, G. Dynamic Study Model for the Interconnected Power System of Continental Europe in Different Simulation Tools. In Proceedings of the IEEE Eindhoven PowerTech, PowerTech 2015, Eindhoven, The Netherlands, 29 June–2 July 2015; Institute of Electrical and Electronics Engineers Inc.: Eindhoven, The Netherlands, 2015.
24. Ross, H.B.; Giri, J.; Kindel, B. An AGC Implementation for System Islanding and Restoration Conditions. In *Power System Restoration: Methodologies and Implementation Strategies*; Wiley-IEEE Press: Hobocken, NJ, USA, 2000; pp. 75–85.
25. Kundur, P. *Power System Stability and Control*; McGraw Hill: New Delhi, India, 2008.

Article

Power System Stability Analysis of the Sicilian Network in the 2050 OSMOSE Project Scenario †

James Amankwah Adu [1], Alberto Berizzi [2,*], Francesco Conte [3], Fabio D'Agostino [3], Valentin Ilea [2], Fabio Napolitano [1], Tadeo Pontecorvo [1] and Andrea Vicario [2]

[1] Department of Electrical, Electronic and Information Engineering, University of Bologna, 40126 Bologna, Italy; jamesamankwah.adu@unibo.it (J.A.A.); fabio.napolitano@unibo.it (F.N.); tadeo.pontecorvo@unibo.it (T.P.)

[2] Department of Energy, Politecnico di Milano, 20156 Milan, Italy; valentin.ilea@polimi.it (V.I.); andrea.vicario@polimi.it (A.V.)

[3] Department of Electrical, Electronics and Telecommunication Engineering and Naval Architecture, University of Genoa, 16145 Genoa, Italy; fr.conte@unige.it (F.C.); fabio.dagostino@unige.it (F.D.)

* Correspondence: alberto.berizzi@polimi.it

† This paper is an extended version of our paper published in 2021 AEIT International Annual Conference (AEIT), Milano, Italy, 4–8 October 2021; pp. 1–6.

Abstract: This paper summarizes the results of a power system stability analysis realized for the EU project OSMOSE. The case study is the electrical network of Sicily, one of the two main islands of Italy, in a scenario forecasted for 2050, with a large penetration of renewable generation. The objective is to establish if angle and voltage stabilities can be guaranteed despite the loss of the inertia and the regulation services provided today by traditional thermal power plants. To replace these resources, new flexibility services, potentially provided by renewable energy power plants, battery energy storage systems, and flexible loads, are taken into account. A highly detailed dynamical model of the electrical grid, provided by the same transmission system operator who manages the system, is modified to fit with the 2050 scenario and integrated with the models of the mentioned flexibility services. Thanks to this dynamic model, an extensive simulation analysis on large and small perturbation angle stability and voltage stability is carried out. Results show that stability can be guaranteed, but the use of a suitable combination of the new flexibility services is mandatory.

Keywords: large perturbation angle stability; small perturbation angle stability; voltage stability; synthetic inertia; demand response; reactive compensation

Citation: Adu, J.A.; Berizzi, A.; Conte, F.; D'Agostino, F.; Ilea, V.; Napolitano, F.; Pontecorvo, T.; Vicario, A. Power System Stability Analysis of the Sicilian Network in the 2050 OSMOSE Project Scenario. *Energies* **2022**, *15*, 3517. https://doi.org/10.3390/en15103517

Academic Editor: Saeed Golestan

Received: 30 March 2022
Accepted: 5 May 2022
Published: 11 May 2022

Publisher's Note: MDPI stays neutral with regard to jurisdictional claims in published maps and institutional affiliations.

1. Introduction

The energy transition process will lead the penetration of Renewable Energy Sources (RESs) in the European power system to drastically increase. In a few decades, traditional generation plants will be progressively decommissioned, and their fundamental contribution to the power system flexibility will be missed. Flexibility is understood as the power system's ability to cope with variability and uncertainty in demand and renewable generation over different timescales to guarantee a stable and efficient operation [1]. Therefore, the challenge is to keep the current level of power system flexibility by exploiting a new mix of resources, i.e., the same RESs, but also loads by Demand Side Response (DSR) and energy storage systems.

Academic and industrial research studies have been (and are) extensively active in proposing methods and technologies able to transform RESs and loads into sources of flexibility by modifying their control procedures and/or combining them with Battery Energy Storage Systems (BESSs). More specifically: in [2–4] the authors propose control strategies to enable wind generators to provide Synthetic Inertia (SI); in [5] wind generators are made capable to provide primary droop frequency regulation; in [6,7] converter-interfaced generators are designed to provide SI; in [8–10] aggregates of thermal loads are controlled to provide primary and secondary frequency regulation services; in [11] a control strategy for a building air cooling system that provides secondary frequency regulation is experimentally validated; in [12] multiple BESSs are controlled to provide frequency regulation; in [13] BESSs are coupled with wind generators to provide frequency regulation services; in [14] a BESS is coupled with a large scale photovoltaic (PV) generator to provide primary frequency regulation; in [15] a decentralized control strategy is proposed to allow wind generators to provide primary voltage regulation; in [16] primary voltage regulation is realized by multiple BESSs via broadcast control at subtransmission level, while [17] generalizes this technique for any type of RES, applies it at distribution network level, upgrades it with an additional coordination level and compares it against centralized optimal reactive power control; in [18,19] voltage regulation is realized through DSR considering residential loads; in [20] voltage regulation performance form Distributed Energy Resources (DERs) are validated by Hardware-in-the-Loop simulations; in [21,22] voltage regulation is realized by PV generators; in [23] a chance-constrained optimization method is proposed to provide voltage regulation by DERs; in [24] voltage regulation is provided by distributed BESSs via reinforcement learning.

To summarize, the result is a huge collection of possible solutions by which the new mix of resources can provide several classes of flexibility, such as SI, support to frequency, voltage regulation, and balancing services.

In this context, the European project OSMOSE [25–27] has the objective of defining the optimal mix of flexibility resources among all the available options in order to avoid stand-alone solutions that might be less efficient in terms of overall efficiency. The project, involving nine European countries, three Transmission System Operators (TSOs), many companies, universities, and research institutions, covers several aspects to evaluate flexibility needs and assess flexibility services.

In this paper, we focus on Stability Aspects within the scope of the work performed by EnSiEL [28] under Task 1.4.3. More specifically, the Work Package 1 (WP1) of the project started with Task 1.1, led by the Technical University of Berlin (TUB), by proposing long-term future scenarios for the years 2030 and 2050 [25], which differ in load levels, installed capacities, investment possibilities, and the quantity of flexibility options. Next, Task 1.2 (carried out by RTE, the French TSO) assessed and validated these scenarios [27,29] using a static reserve adequacy analysis. Finally, in Task 1.4.3, EnSiEL evaluated the impact of innovative flexibility sources on the power system stability, testing these scenarios for 2030 and 2050 in a realistic model of the electrical network of Sicily. Sicily, one of the two main islands of Italy, is synchronously connected to the national power system. The grid model was provided by Terna, the Italian TSO.

Since results for 2030 were briefly presented in [30], this paper focuses on the 2050 scenario. In particular, we evaluate the response of the power system to some typical perturbations (e.g., loss of a large generator, slow increase in loads, and contingencies of branches) by simulations carried out on the DIgSILENT PowerFactory 2019 platform [31].

According to the 2050 scenario, the flexibility sources considered in this analysis are:

- SI provided by wind generators;
- FFR provided by BESSs;
- SI and FFR provided by controlled loads (DSR);
- Voltage regulation support provided by RESs (PV and wind generators).

The following topics are investigated:

1. *Large-perturbation angle stability (electromechanical stability)*: To analyze the response of the power system to the variations in frequency and voltage on a time scale from tens of milliseconds up to tens of seconds. In particular, the electromechanical stability is assessed, taking into account the dynamics of generators and loads and the triggering of protection schemes. The objective is to state if synchronous machines can keep synchronism after a severe transient disturbance [32].

2. *Small-perturbation angle stability*: To evaluate the power system stability when small variations in loads and generation occur, as continuously happens in a real scenario, not necessarily related to a transient disturbance.
3. *Voltage security*: To assess the capability of a power system to keep an acceptable steady voltage at all busses, either under normal conditions or after disturbances.

The results prove that in the 2050 scenario, with an extremely high penetration of RESs, the new flexibility sources are essential to guarantee both angle and voltage stability. Literature provides many works with power system stability analyses taking into account the lowering of the system inertia [33] and the high penetration of inverter-based generation [34,35]. However, in most cases, these studies focus on one or two stability issues and simulation tests are carried out on relatively small (less than one hundred busses) benchmark network models. Differently, this paper proposes an original and useful analysis which:

- Is based on validated realistic scenarios for the year 2050;
- Takes into a complete set of flexibility options, as listed before (SI provided by wind generators; FFR provided by BESSs; SI and FFR provided by controlled loads (DSR); voltage regulation support provided by RESs);
- Uses a detailed network model composed by more than 600 busses, provided by the TSO that owns and manages the modeled system.

The reminder of the paper is organized as follows. Section 2 describes the network model and the simulation scenarios. Section 3 presents the results of the large perturbation stability analysis. Section 4 reports the results of the small-perturbation stability analysis. Section 5 presents the results of the voltage stability analysis. Finally, Section VI provides the general conclusions of the study.

2. Network Model and Scenarios

The grid model employed for this analysis is the Sicilian Island, a portion of the Italian transmission network (Figure 1); basic data were provided by Terna.

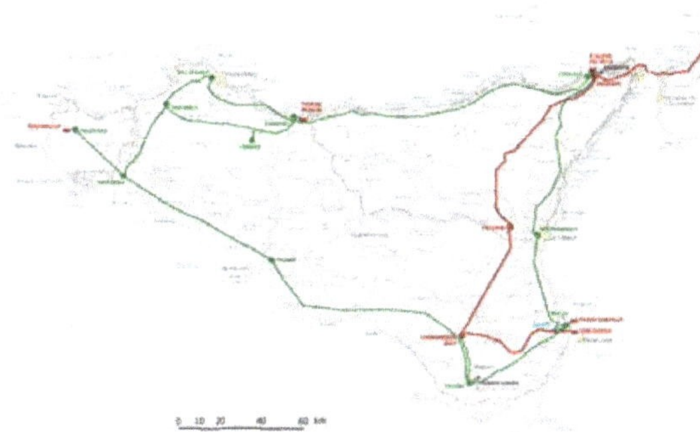

Figure 1. The studied grid [36].

The Sicilian Island corresponds to the European market zone 56IT used by Task 1.1 and 1.2 of the OSMOSE project [25,29]. The provided grid model is detailed and presents:

1. More than 600 busbars at 400, 230, 150, and 132 kV;
2. 441 lines;
3. 516 substations;
4. 30 large (nominal power higher than 10 MW) static generators, representing the wind and solar plants connected to the HV grid;
5. 379 loads, mimic both equivalents and HV loads;
6. 72 synchronous machines representing both thermal and hydro plants with their associated controllers.

The grid presents only few lines with a voltage higher or equal to 230 kV (in green in Figure 1); considering the geographic dimension and the high amount of generation capacity installed, it is poorly meshed. The 400 kV system (in red in Figure 1) essentially consists of a single backbone starting from the Sorgente substation in the north-east and ending in the Syracuse petrochemical nucleus in the south-eastern part of the region. It goes through the powerful interconnection substations of Sorgente, Paternò, and Chiaramonte Gulfi up to the ISAB plants near Priolo Gargallo. Since 2016, Sicily has been connected to the Italian system through a new AC interconnection (two parallel undersea cables) at 400 kV. Both the AC interconnections start from Rizziconi substation on the mainland and get to the Sorgente substation on the Island.

Regarding the installed conventional generators, Table 1 reports Sicily's major thermal power plants in 2020.

Table 1. Sicily's major thermal power plants in 2020.

Plant	Owner	Type	Rating (MW)
Anapo (SR)	Enel	Storage Hydro	500
Augusta (SR)	Enel	Fossil Fuel	210
Priolo Gargallo, Nuce Nord (SR)	ERG	CCGT *	480
Priolo Gargallo, ISAB energy (SR)	ERG	IGCC *	528
Priolo Gargallo Archimede (SR)	Enel	CCGT	750
Trapani (TP)	E.On	OCGT *	169
Gela (CL)	Eni	Fossil Fuel	260
San Filippo del Mela (ME)	A2A Energie Future	Fossil Fuel	1280
Milazzo (ME)	Termica Milazzo	CCGT	365
Termini Imerese (PA)	Enel	CCGT	1340

* CCGT—Combined Cycle Gas Turbine, IGCC—Integrated Gasification Combined Cycle, OCGT—Open Cycle Gas Turbine.

As mentioned, all conventional generators provide droop primary frequency regulation (by the governor), primary voltage regulation (by the Automatic Voltage Regulator), and Power System Stabilizers (PSS). In this paper, focusing on frequency regulation, we indicate the primary active power reserve as Frequency Containment Reserve (FCR) according to the nomenclature introduced in [37].

In the Rizziconi substation, end of the link in the continent, a frequency relay is installed; it is able to disconnect the Island from the rest of the system in particular critical conditions. Usually, Sicily exports to Italy active power, keeping the Sicilian power system in operation in case of the trip of the interconnection and avoiding load shedding on the Island. Active power transit is always monitored, and if the exported power is higher than a given amount, specific devices can disconnect some generators in Sicily.

Moreover, the Rizziconi–Sorgente Islanding relay trips when severe underfrequency events occur in the continental power system; the tripping relay operates according to these rules:

- The frequency overtakes the threshold of 49.7 Hz and the frequency derivative is lower than -0.2 Hz/s;
- The frequency is lower than 49.5 Hz.

These rules also explain why Sicily mainly exports power; indeed, in case of an underfrequency event, if the Island was importing, a disconnection could worsen the frequency decay. Sicilian four large pumping units of Anapo plant are always disconnected as a first solution if operating in pumping mode.

In the case of underfrequency events, moreover, a load shedding scheme is in operation, where its settings are shortly described in Table 2: according to different thresholds, a prescribed load shedding step is activated, as described.

Table 2. Load shedding settings [38].

Threshold	Starting Frequency (Hz)	Frequency Derivative Threshold (Hz/s)	Pure Frequency Threshold (Hz)	Percentage of Shed Load
1	49.3	−0.3	49.0	9%
2	49.2	−0.6	48.9	8%
3	49.1	−0.9	48.8	7%
4	49.1	−1.2	48.7	7%

2.1. Scenarios

2.1.1. Data Scenario Application

The scope of this work is to simulate the dynamical stability of the Sicilian grid with a different mix of loads and generation deployment (coming from Task 1.1 and Task 1.2) and evaluate possible security issues and possible suggestions for countermeasures. The Sicilian grid has been updated with the new values of capacities and loads identified for 2050. Each scenario identified in Task 1.1 provides data for the 99 European zones defined by e-Highway 2050. In detail, for each zone, the total installed generation capacity, differentiated by primary sources (biomass, geothermal, wind, solar, gas, and hydro), is given, as well as the total load installed.

Task 1.1 provided the installed capacity, differentiated by technology, for each European market zone, either connected to the HV or MV grids. This new capacity was allocated to the HV and MV grids, according to the 2020 shares available on the GAUDÌ portal [39], the Terna's website with the technical characteristics of all the power plants. Table 3 shows the RES deployment in Sicily (2020) and the percentage connected to either MV or HV level; it is clear that the PV plants are almost always connected to the distribution grid, and the wind and hydro plants to the transmission grid. Such percentages were used to share the RES data provided by Task 1.1, while some conventional thermal plants were simply switched off to mimic their decommissioning as RESs take over. The Power to Gas (P2G) technology was considered as a repowering of some of the old thermal plants and located accordingly, biogas (bioenergy) and waste-to-energy technologies were modelled assuming they are brown field.

Table 3. Current RES installed in Sicily (beginning of 2020) [39].

Technology	Total Installed (MW)	MV Connected (%)	HV Connected (%)
Photovoltaic	1422	97.00	3.00
Wind	1887	6.00	94.00
Hydro	274	4.00	96.00

In the following, the major changes applied to the 2020 grid to meet the 2050 data of Task 1.2, showed in Table 4, are described.

1. Power rating of the cables with the mainland were increased to 2300 MW to meet the Available Transfer Capacity (ATC) data provided.
2. An equivalent synchronous machine was installed close to the continental terminal of the link Sicily-continental Italy to mimic the dynamic behavior of the rest of the Italian system. Its rated power equals the sum of the active power of the synchronous machines of the other Italian market zones at the time frame considered, divided by a power factor (0.8), and starting time constant equal to 10 s. This approach is conservative in the sense that the equivalent machine represents only a portion of the generating units of continental Italy able to provide primary frequency control. This machine is equipped with a governor, an automatic voltage regulator, and a power system stabilizer properly tuned.
3. 15 Full-Converter Wind Turbines (FCWTs) and 20 Doubly-Fed Induction Generators wind turbines (DFIG) were equipped with a controller able to provide SI (detailed in Section 2.3).

4. Already existing HV PV plants ere modelled according to [40], including its LV/MV transformer. The same PV model was installed at the MV busbars of the 285 primary substations to mimic the contribution of the dispersed generation connected to the distribution grids. All PV models are equipped with the overfrequency protections, set according to the Italian standards [38,41].

5. A BESS model able to provide FFR (detailed in Section 2.4) was considered. BESSs are installed either near each new wind plant or at the sites of the few synchronous generators not decommissioned. The total amount of storage power was installed according to Task 1.1. These BESS were sized equal to the 20% of the capacity of the already installed generation plant.

6. Controlled loads able to provide both SI and FFR (detailed in Section 2.5) were installed near 16 already existing loads considered large enough (at least 13 MW in the Terna base model).

7. The 230 kV ring circuit of the Island was doubled, even if no indications regarding the zonal grid reinforcements are given by the other Tasks. This solution was carried out to accommodate the demand expected for 2050.

8. Capacities of all generators were upgraded according to the data provided by Task 1.1 (Table 4). Table 5 shows the ratings and the location of the remaining synchronous machines and their identification names.

Table 4. Installed capacity (MW) provided by Task 1.1 for 2050 for Sicily [25].

Zone	Battery	PV	Hydro	Wind	Waste	Gas	P2G
56IT	1572	6075	313	6360	566	162	1947

Table 5. Ratings and location of the synchronous power plants in 2050.

Location	Identification Name	Rating (MVA)
Contrasto	CNTP	24
Dittaino	DITP	23
Priolo Gargallo	EGNP	576
Città Giardino	EGSP	90
Augusta	ESSP	80
Priolo Gargallo	ISBP	344
Paternò	PATP	9
Priolo Gargallo	PRGP	658
Milazzo	TEMP	185
Termini Imerese	TIMP	946
Troina	TROP	14

2.1.2. Typical Generation and Demand Conditions—Dispatching Profiles

In this analysis, only the scenario "Current Goal Achievement" (GCA) is considered. Capacities of the GCA scenario are implemented for 2050 related to the generators, while the following most typical and critical generation/demand conditions were tested, with reference to Sicily:

- Very low load/very low rotating generation in operation;
- High load/low rotating generation in operation;
- Maximum export/import of areas;
- Operational conditions with a weak network (lines out of service);
- Islanding conditions.

These snapshots were selected to represent the weaker grid conditions. They were identified by analyzing the load demand and the balance between the traditional generation and the renewable one.

From Task 1.2, the level of generation, differentiated by the primary sources, the active power demand, and the import (or export) in MW for each hour of one entire year are available. The profiles provided were carefully analyzed, and, to achieve the aforementioned critical conditions, the most appropriated hours, considered to better resemble the desired situations, were picked up. Reactive power was not taken into account by Task 1.2. To keep a realistic load behavior, typical values of power factor, based on the characteristics of the Italian power system, were assigned to the loads of the Sicilian zone. Finally, the active load demand was adjusted using a suitable scaling factor to increase, or decrease, the total demand and achieve the desired values.

The following dispatching profiles (DPs), derived from the data provided by RTE, were picked up and implemented for the 2050 Sicilian grid:

- *High Export*: represented by 24th of May of MC 4 at 2:00 a.m., characterized by quite high wind production, no photovoltaic, and low load demand.
- *High Import*: represented by 11th of January of MC 1 at 1:00 a.m., characterized by almost no renewable production and medium/high demand.
- *High Load*: represented by 27th of May of MC 1 at 4:00 p.m., characterized by high load demand, high wind, and photovoltaic production.
- *Island*: represented by the 23rd of June of MC 5 at 3:00 p.m.; with high photovoltaic and wind production and the link with the mainland out of service.
- *Low Load*: represented by the 4th of June of MC2 at 2:00 a.m., characterized by low load and almost zero wind production.
- *Lines out of service*: it is the *Low Load* profile with the Favara/Chiaramonte and Caracoli/Sorgente 230 kV lines out of service.

A detailed description of these six DPs is reported in Table 6. Table 7 reports the operating points of the 16 controlled loads in any of the dispatching profiles introduced.

Table 6. Hourly dispatching profiles (MW) [27,29].

Dispatching Profile	High Export	High Import	High Load	Island	Low Load	Lines Out of Service
Time	24 May 2050 02:00	11 January 2050 01:00	27 May 2050 16:00	23 June 2030 15:00	4 June 2050 02:00	4 June 2050 02:00
MC Year	4	1	1	5	2	2
Zonal export (MW)	725	−1188	0	0	−578	−578
Loads						
Total Load (MW)	1837	3218	4099	4294	1514	1514
Traditional load (MW)	1837	3218	2318	2500	1514	1514
DSR Electrical Vehicle (MW)	0	0	0	0	0	0
Pumping (MW)	0	0	0	0	0	0
Electrolyser (MW)	0	0	1781	1794	0	0
Battery Storage (MW)	0	0	0	0	0	0
DSR Heat Pump (MW)	0	0	0	0	0	0
Generation						
Total Generation (MW)	2562	2030	4099	4294	936	936
ROR (MW)	30	11	30	25	25	25
WIND (MW)	1480	253	2059	1124	195	195
SOLAR (MW)	0	0	1211	2429	0	0
NUCLEAR (MW)	0	0	0	0	0	0
COAL (MW)	0	0	0	0	0	0
GAS (MW)	0	0	0	0	0	0
BATTERY (MW)	786	0	0	0	0	0

Table 6. *Cont.*

Dispatching Profile	High Export	High Import	High Load	Island	Low Load	Lines Out of Service
			Generation			
PSP (MW)	0	0	0	0	0	0
P2G (MW)	0	1500	533	450	450	450
CHP (MW)	0	0	0	0	0	0
BIOENERGY (MW)	266	266	266	266	266	266
H. STOR (MW)	0	0	0	0	0	0
SPIL. ENRG (MW)	0	0	0	0	0	0
UNSP. ENRG (MW)	0	0	0	0	0	0
			Continental Italy Equivalent Generator			
NOMINAL POWER (MVA)	10,411	1493	23,053	9607	13,533	13,533

Table 7. Loads associated with the DSR models.

	Active Power (MW)					
Load	High Export	High Import	High Load	Island	Low Load	Lines Out of Service
Load 1	42.8	42.8	59.92	59.92	21.4	21.4
Load 2	13.8	27.6	38.64	38.64	13.8	13.8
Load 3	14.5	29	40.6	40.6	14.5	14.5
Load 4	13.9	27.8	38.92	38.92	13.9	13.9
Load 5	13.3	26.6	37.24	37.24	13.3	13.3
Load 6	15	30	42	42	15	15
Load 7	126.6	189.9	177.24	177.24	0	0
Load 8	13.5	27	37.8	37.8	13.5	13.5
Load 9	16.9	33.8	47.32	47.32	16.9	16.9
Load 10	15	30	42	42	15	15
Load 11	13.4	26.8	37.52	37.52	13.4	13.4
Load 12	52.4	52.4	73.36	73.36	26.2	26.2
Load 13	13.8	27.6	38.64	38.64	13.8	13.8
Load 14	13.6	27.2	38.08	38.08	13.6	13.6
Load 15	14.8	29.6	41.44	41.44	14.8	14.8
Load 16	13.3	26.6	37.24	37.24	13.3	13.3
TOTALS	**406.6**	**654.7**	**827.96**	**827.96**	**232.4**	**232.4**

In the following, we provide a detailed description of models of wind generators, BESSs, and flexible loads, which are all able to provide frequency and/or voltage regulation services. This description is of particular interest since the main question of the present study is if the 2050 European power system will be stable and how much flexibility services from RES, BESSs, and DSR would be necessary.

2.2. Wind Generators Models

2.2.1. Full-Converter with Synthetic Inertia

SI corresponds to the controlled action of an inverter-based generating unit to mimic the kinetic energy exchanged by a synchronous generator with the power system. Different mechanisms are available in the literature to relate the frequency variation with the converter injection; here, the emulation is obtained by acting on the control of a back-to-back converter [42]. Basically, this strategy emulates the inertial response supplying a temporary contribution during the first instants of a frequency transient, exploiting the energy content of the capacitor of the DC-link [6,7], suitable oversized, thanks to the use of supercapacitors, or "ultra-capacitors".

This control model was implemented (referred as the *current-controlled* model) and tested in the wind turbine model shown in Figure 2a. This SI control could be applied to whatever type of plant connected with a fully rated converter, introducing a sufficient energy buffer.

Figure 2b depicts the adopted control diagram (a typical synchronous-frame structure is considered), where:

-   Two linear Proportional-Integral (PI) regulators (suitably equipped with saturation) are introduced. These two independent controllers are associated with the DC bus stabilization and reactive injection. They provide the direct i_d^{ref} and quadrature i_q^{ref} references for the internal current loop.
-   An additional signal p_{in}, introduced on the direct-axis reference and proportional (by the inertia coefficient K_{in}) to the frequency approximate derivative, enables the inertial control. Its negative sign is introduced to increase active power injection in case of underfrequency events, exploiting the discharge of the DC capacitor. Indeed, the DC voltage can temporarily vary in response to the system frequency disturbance and allow the inertial response of the converter.

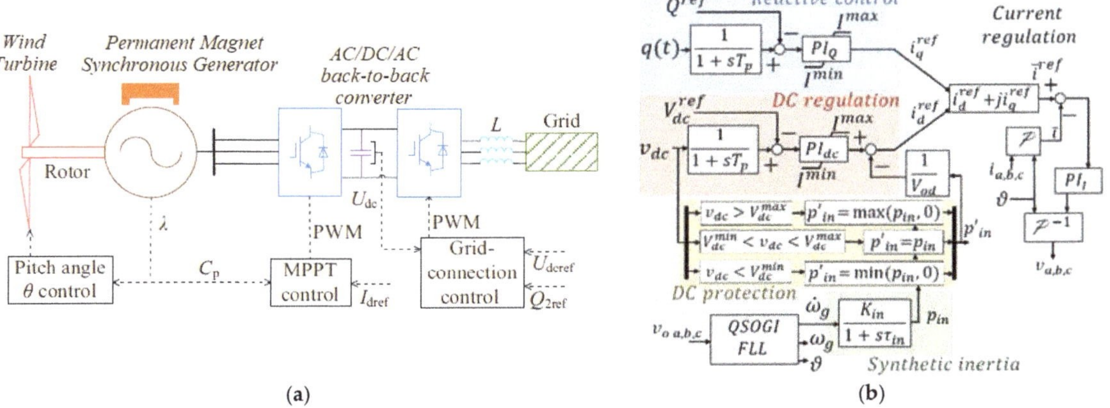

(a)

(b)

Figure 2. Full-converted wind turbine model [43] (**a**), and current controlled model for SI [6,7] (**b**).

In order to extract the angular frequency derivative $\dot{\omega}_g$, the QSOGI-FLL (Quadrature Second-Order Generalized Integrator—Frequency Locked Loop) algorithm is adopted [44]. The QSOGI-FLL system is also exploited to determine the synchronization angle for the Park transform (the capital P in Figure 2b). An additional filtering effect has also been introduced, combining the FLL dynamics with a first-order low-pass (with a time-constant τ_{in}):

$$p_{in} = \frac{K_{in}}{1 + s\,\tau_{in}}\, \dot{\omega}_g \tag{1}$$

Assuming a variability range $\Delta\dot{V}_{dc} = 0.1$ p.u., the sizing of the capacitor is given by:

$$C_{DC} = \frac{10\%\ A_b\ \Delta t}{\frac{1}{2}\left(V_{dc}^{nom2} - V_{dc}^{min2}\right)} \tag{2}$$

where:

-   A_b is the nominal power of the equivalent wind generator exploited;
-   Δt is the maximum time-length of the frequency transient supported;
-   V_{dc}^{nom} is the nominal DC voltage;
-   V_{dc}^{min} is the minimum allowed DC voltage.

The "DC protection" was introduced to decouple the design of the inertia coefficient K_{in} from the grid dynamics. This block guarantees the maximum exploitability of the available energy reserve and fulfills the technical/security constraints associated with the converters. Indeed, SI control is active only when the discharge process is compatible with the DC voltage ratings, in particular:

-   When DC voltage is lower than the minimum allowed value (V_{dc}^{min}), discharge is inhibited, and only charging operations are allowed;
-   Discharging signals can be sent to the control whenever the voltage is close to its maximum and charging operations are inhibited.

This logic can be identified in Figure 2b.

2.2.2. DFIG with Synthetic Inertia

For the DFIG wind models turbines considered, as shown in Figure 3, a different architecture to realize the DC-link SI known as the voltage-controlled model was adopted.

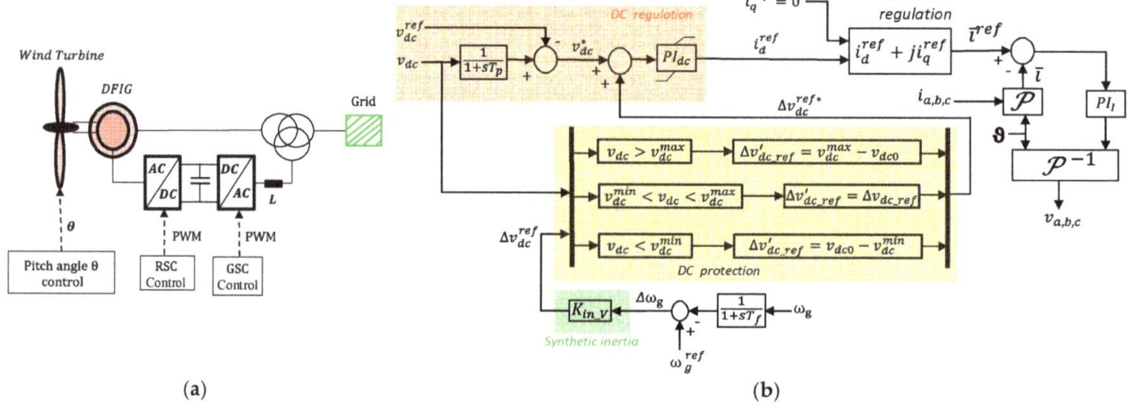

Figure 3. DFIG model [45] (**a**), and voltage controlled model for SI [4,46,47] (**b**).

Figure 3b depicts the control scheme, which considers a proportional relation between the DC-link voltage reference and the measured network frequency deviation. A frequency controller with a constant gain K_{in_V} takes the frequency deviation signal as the input and the variation of the DC voltage reference as output.

The dynamics of the DC-link voltage controller is made very fast to allow the stored electrostatic energy to be released/absorbed in response to the DC-link voltage deviation and allow SI provision. The scheme links frequency and DC-link voltage in such a way that: (i) the change in network frequency will lead to a change in DC-link voltage and (ii) the derivative of the resulting change in DC-link voltage causes the absorption or release of power from DC-link capacitors [4]. By setting the q-axis reference rotor current to zero, the reactive power flow from the Grid Side Converter (GSC) is regulated to zero to reduce the GSC rating.

The Voltage Control Mode DC protection scheme, although a bit different from the Current Control Mode SI DC protection scheme (they both have similar objectives of keeping the DC voltage within v_{dc}^{max} and v_{dc}^{min}) takes the output of the SI scheme and measured DC voltage as the input and limits the inertial response outputs to values compatible with the DC voltage ratings.

2.3. Battery Energy Storage Systems Model

BESSs were implemented using the library model available in DIgSILENT PowerFactory [31]. As shown in Figure 4, the battery is represented as an equivalent DC voltage generator whose voltage U_{DC} depends on the battery State of Charge SoC. Current I_{DC} exchanged by the battery is driven by a Pulse-Width Modulation (PWM) AC/DC converter. I_{DC} depends on the direct and quadrature AC currents I_d and I_q, which are imposed by the PWM converter, according to the reference signals I_d^{ref} and I_q^{ref}, in turn defined by the "PQ controller". The latter implements PI regulators to realize the exchange of active and reactive powers, P^{ref} and Q^{ref}, required by the user.

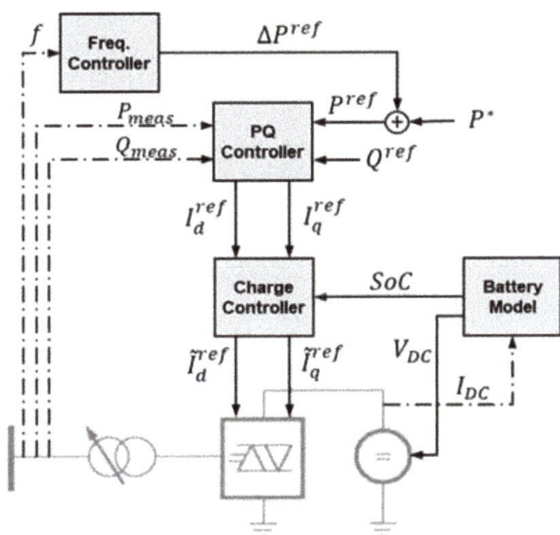

Figure 4. BESS model available from DIgSILENT PowerFactory [31].

As shown in Figure 4, P^{ref} is the sum of the set-point P^* and the variation ΔP^{ref} defined by the "Frequency Controller" to realize the FFR service. The scheme in Figure 5 shows that the FFR is realized by a simple droop controller with a dead-band set to 0.01 Hz and the droop coefficient b.

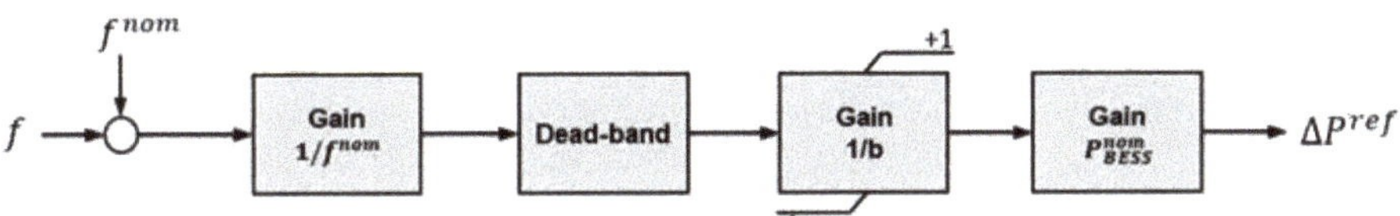

Figure 5. Frequency Controller model.

The "Charge Controller" block limits the current import, when the battery is fully charged ($SoC = 1$), and the current export, when the battery is fully discharged ($SoC = 0$). The battery SoC is determined by a simple Coulomb counting model:

$$SoC(t) = -\frac{1}{C_b \cdot 3600} \int_0^t I_{DC}(\tau)d\tau + SoC_0, \tag{3}$$

where C_b is the battery capacity in Ah and SoC_0 is the initial condition. Two important assumptions worth to be mentioned are that capacity C_b is constant, and voltage U_{DC} is linearly dependent on the SoC. These approximations are acceptable in the context of the present study since dynamical simulations are executed in a time scale from tens of milliseconds up to tens of seconds.

Within the DIgSILENT model, capacity C_b is determined as $C_b = N_c^{par} \cdot C_c$, where N_c^{par} is the number of parallel cells and C_c is the capacity of the single cell. The nominal DC voltage is determined as $U_{DC}^{nom} = N_c^{ser} \cdot u_{max}$, where N_c^{ser} is the number of series cells and u_{max} is the voltage of full cells. Moreover, the DIgSILENT model allows setting the BESS rated power P_{BESS}^{nom}. All implemented BESSs adopt the parameters reported in Table 8, whereas N_c^{par} is defined to obtain a desired capacity C_{BESS} in Wh such that $P_{BESS}^{nom} = C_{BESS}/1\,\text{h}$.

Table 8. BESS common parameters [1].

Parameter	Symbol	Value
Single cell capacity	C_c	50 Ah
Voltage of full cells	u_{max}	15 V
Voltage of empty cells	u_{min}	12 V
Number of series cells	N_c^{ser}	60
Battery nominal voltage	U_{DC}^{nom}	0.9 kV
Converter AC side nominal voltage	U_{AC}^{nom}	0.4 kV
Cells internal resistance	Z_i	0.001 Ω

[1] For details of parameters not introduced in the text, the reader is referred to [31].

2.4. DSR

By suitably modulating their power consumption, electrical loads equipped with a form of energy storage (e.g., thermal energy in thermal loads) can provide frequency regulation services to the power system. Scientific literature has proposed many solutions in order to allow commercial buildings [11], industrial compounds and/or aggregates of domestic loads [9,10] to provide frequency regulation services. In general, even if with different levels of robustness, depending on the proposed control architecture and on the typology of loads, in all these works, loads are made able to follow a reference signal designed to provide SI and FFR.

Ideally, the reference signal is

$$\Delta P_L^{ref}(t) = K_{FFR}\Delta f(t) + K_{SI}Df(t), \tag{4}$$

where: $\Delta f(t)$ is the frequency deviation at time t, Df is the frequency derivative at time t (RoCoF); K_{FFR} is the FFR coefficient, and K_{SI} is the SI coefficient. The values of these two last coefficients depend on the type of loads and on the adopted control strategy. K_{FFR} is an equivalent droop coefficient, whereas K_{SI} is the provided amount of SI.

In any case, the reference signal ΔP_L^{ref} cannot be perfectly followed, mainly due to the fact that the input values of the controller are not the real-time values of frequency deviation Δf and frequency derivative Df, but the measured values Δf^m and Df^m, collected with a certain sampling time and then filtered with a low-pass filter, as shown in Figure 6 (according to the ENTSO-E "Demand Connection Code" [48]).

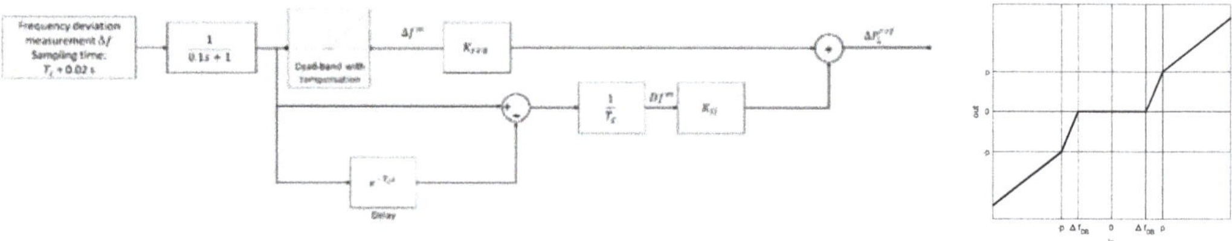

Figure 6. Measurement chain for frequency and RoCoF (**left**). Detail of the dead-band with compensation (**right**).

Therefore, (4) is substituted by:

$$\Delta P_L^{ref}(t) = K_{FFR}\Delta f^m(t) + K_{SI}Df^m(t), \tag{5}$$

where $\Delta f^m(t)$ and $D f^m(t)$ are the measured values of the two mentioned quantities. The dead-band is defined by two parameters: Δf_{DB}, which determines the band amplitude and the break point p. In all simulations, $\Delta f_{DB} = 0.02$ Hz and $p = 0.03$ Hz.

The value of K_{FFR} should be defined in function of the FCR, in [37] is defined as the *active power reserves available to contain system frequency after the occurrence of an imbalance*. According to [37], each TSO must secure that the combined reaction of FCR respects with the following requirements:

A.   FCR is not artificially delayed and begins as soon as possible after a frequency deviation;

B.   If the frequency deviation is equal to or larger than 200 mHz, at least 50% of the full FCR capacity must be delivered at the latest after 15 s; and 100% of the full FCR capacity must be delivered at the latest after 30 s. Moreover, the activation of the full FCR capacity must rise at least linearly from 15 to 30 s;

C.   If the frequency deviation is smaller than 200 mHz, the activated FCR capacity must be at least proportional with the same time behavior referred in points (A) and (B).

2.5. Methodology

The methodology adopted in this study, detailed in the following sections, is described by the flowchart depicted in Figure 7.

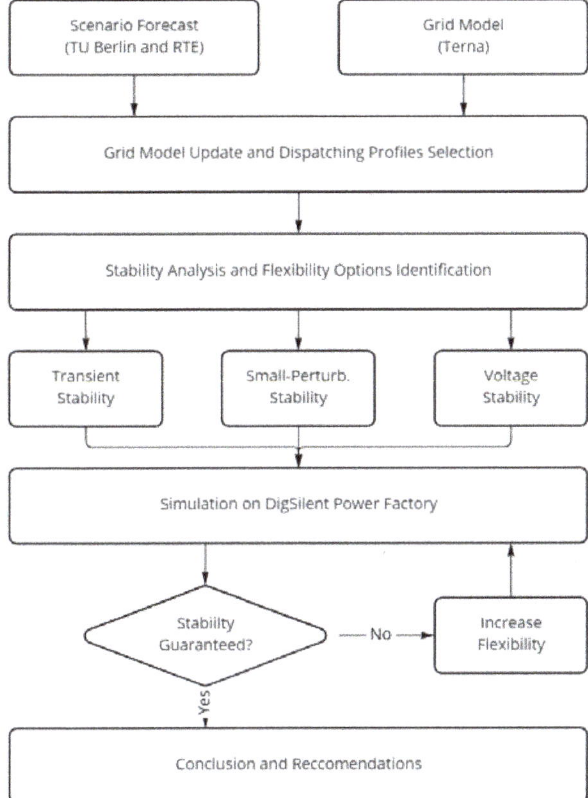

Figure 7. Stability analysis: methodology flowchart.

As shown in the figure, the grid model is updated and the DPs are selected taking as inputs the 2050 scenario forecast provided by Task 1.1 and 1.2 and the 2020 Italian grid model provided by Terna. For each DP a transient stability analysis, a small-perturbation stability analysis and a voltage stability analysis are performed in DIgSILENT PowerFactory. If the DP does not comply with one of the stability conditions, additional flexibility tools are implemented (e.g., SI, DSR) or their contribution is increased to the extent until stability is guaranteed. Finally, the identified solutions able to provide stability even in the most critical operating conditions are examined in the conclusions and recommendations section.

3. Large Perturbation Stability Analysis

Large perturbation stability analysis is carried out by dynamical simulations on a time scale from tens of milliseconds up to tens of seconds. In particular, electromechanical stability is assessed, considering the dynamics of generators and loads and the triggering of protection schemes.

The six DPs of Table 6 are considered, and for each of them, a set of events is simulated. Specifically, the following events are simulated:

- The outage of one or two of the three connections with the mainland (except for the Island DP);
- A 500 MW load-step occurring in continental Italy (load-step in continental Italy is not considered in the Island DP);
- The outage of groups of large generating units.

3.1. Frequency Stability Definitions

According to the Italian grid code [49]:

- In Normal Operating Conditions (NOC), the frequency must remain within the range 49.9–50.1 Hz;
- For the special case of Sicily disconnected from the mainland, the NOC range is 49.5–50.5 Hz;
- In Emergency Operating Conditions (EOC), the frequency must be kept in the range 47.5–51.5 Hz.

Notice that if frequency exceeds the EOC range of 47.5 Hz or 51.5 Hz, a system blackout can hardly be avoided [50]. Therefore, for our simulations, stability is guaranteed if the numerical integration converges, and frequency is kept within 47.5–51.5 Hz. Moreover, a simulation is labelled as follows:

1. *Strongly stable*, if stability is guaranteed and frequency remains within the NOC limits;
2. *Stable*, if stability is guaranteed, but NOC are violated;
3. *Unstable*, if stability is not guaranteed.

3.2. Network Configurations

The objective of the present study is to analyze if the 2050 power system of Sicily can assure frequency stability. According to the results provided in the OSMOSE Task 1.1 [25], both BESSs operating FFR and flexible loads providing DSR will be available. Therefore, simulations are carried out with the *base configuration*, without BESS FFR and DSR, and with further network configurations with BESS FFR and/or DSR.

3.2.1. Configurations with BESS FFR

As described in Section 2.2, BESS FFR is implemented as a droop controller, essentially defined by the droop coefficient b, which states the amount of FCR provided by the BESS. Specifically, the i-th BESS with nominal power $P_{BESS,i}^{nom}$ will provide, at steady-state, the following power variation:

$$\Delta P_{BESS,i} = \frac{100}{b} \frac{\Delta f}{f^{nom}} P_{BESS,i}^{nom}. \tag{6}$$

In this analysis, the same droop coefficient is assumed for all BESSs in service with values within the interval 1–5%, and the relevant network configuration is indicated with BESS-b. It is worth noticing that 4–5% are typical droop coefficient values adopted for traditional (hydro or thermal) generators. In this study, lower droop values are considered since BESSs could potentially be installed to exclusively provide FFR. Therefore, the droop can be lowered more than in the case of generators. In particular, since the simulated BESSs are associated with wind and traditional generators, and their nominal powers are assumed equal to the 20% of the relevant generators, a droop coefficient equal to 1% means that the BESS emulates the 5% droop frequency response of the associated generator. Obviously, a general hypothesis is that all BESSs are provided with an energy reserve sufficient to realize the regulation service, i.e., their SoC at the event occurrence is such that power variation in (5) can be kept up to a prescribed maximal time interval (a typical value is 15 min).

By adopting $\Delta f_{max} = 0.2$ Hz as the frequency deviation at which the full FCR should be released (according to [37]), the FCR provided by the i-th BESS is:

$$FCR_{BESS,i} = \min\left(\frac{100}{b} \cdot \frac{0.2}{f_{nom}} \cdot P_{BESS,i}^{nom} , \; P_{BESS,i}^{nom} - \left| P_{BESS,i}^{0} \right| \right), \qquad (7)$$

where $P_{BESS,i}^{0}$ is the i-th BESS working point. In the following, FCR_{BESS}^{tot} will indicate the total FCR provided by all BESSs providing FFR, i.e., $FCR_{BESS}^{tot} = \sum_{i} FCR_{BESS,i}$.

3.2.2. Configurations with DSR

DSR is operated by the 16 loads listed in Table 7. To simplify the sizing of the DSR service, described in detail in Section 2.3, parameters FCR_{FFR} and FCR_{SI} are set equal to the same tuning value FCR_{DSR}. For the i-th load, $FCR_{DSR,i}$ is defined as a percentage p (%) of the load working point $P_{L,i}$:

$$FCR_{DSR,i} = \frac{p}{100} \cdot P_{L,i}. \qquad (8)$$

Given p, the relevant network configuration will be indicated as DSR-p. Moreover, FCR_{DSR}^{tot} will indicate the total FCR and SI provided by all the 16 loads, i.e., $FCR_{DSR}^{tot} = \sum_{i} FCR_{DSR,i}$.

3.2.3. Simulations Procedure

For each DP and for each event, the following procedure is followed. The *base configuration* is firstly tested. Then, if strong stability cannot be guaranteed, network configurations with BESS FFR are simulated by progressively decreasing the droop coefficient from 5% to 1%. If this is not sufficient to obtain strong stability, further network configurations with both BESS FFR and DSR are tested by progressively increasing the DSR percentage parameter p up to 35%.

3.3. Results

Table 9 provides an overview of the obtained results. For each simulation, the table reports if strong and/or simple stability was guaranteed with the *base configuration*. In the cases where one of these two conditions is not satisfied, Table 9 reports the amount of FCR provided with the "best" BESSs-DSR configuration ($FCR_{BESS}^{tot} + FCR_{DSR}^{tot}$) and if, in this case, the two conditions were successfully satisfied. As best BESSs-DSR configuration, we intend the one with the lowest level of contribution from BESSs and DSR (highest droop coefficient and lowest p) allowing frequency regulation performances to be improved.

Table 9. Large perturbation stability analysis: simulation results.

Event	Base Configuration		Best BESSs—DSR Configuration			
	Stable?	Strongly Stable?	Configuration	FCR	Stable?	Strongly Stable?
High Export						
Load-step in continental Italy (500 MW)	YES	NO	BESS-1 DSR-25	655 MW	YES	YES
Outage of 1 link with continental Italy	YES	YES	Not required	-	-	-
Outage of 2 links with continental Italy	YES	YES	Not required	-	-	-
Generation outage (220 MW)	YES	YES	Not required	-	-	-
High Import						
Load-step in continental Italy (500 MW)	YES	NO	BESS-5 DSR-15	203 MW	YES	YES
Outage of 1 link with continental Italy	YES	YES	Not required	-	-	-
Outage of 2 links with continental Italy	NO	NO	BESS-1	522 MW	YES	YES
Generation outage (321 MW)	YES	YES	Not required	-	-	-
Generation outage (569 MW)	YES	NO	BESS-3 DSR-25	338 MW	YES	YES
Generation outage (817 MW)	YES	NO	BESS-1 DSR-35	751 MW	YES	NO
High Load						
Load-step in continental Italy (500 MW)	YES	YES	Not required	-	-	-
Outage of 1 link with continental Italy	YES	YES	Not required	-	-	-
Outage of 2 links with continental Italy	YES	YES	Not required	-	-	-
Generation outage (337 MW)	YES	YES	Not required	-	-	-
Island						
Generation outage (337 MW)	YES	NO	BESS-3	174 MW	YES	YES
Low Load						
Load-step in continental Italy (500 MW)	YES	NO	BESS-1 DSR-30	362 MW	YES	YES
Outage of 1 link with continental Italy	YES	YES	Not required	-	-	-
Outage of 2 links with continental Italy	YES	YES	Not required	-	-	-
Generation outage (337 MW)	YES	NO	BESS-1	292 MW	YES	YES
Lines out of service						
Load-step in continental Italy (500 MW)	YES	NO	BESS-1 DSR-30	362 MW	YES	NO
Outage of 1 link with continental Italy	YES	YES	Not required	-	-	-
Outage of 2 links with continental Italy	YES	YES	Not required	-	-	-
Generation outage (337 MW)	YES	NO	BESS-3 DSR-30	362 MW	YES	YES

As we can observe in Table 9, with the *base configuration*, frequency stability is guaranteed in all the DPs and for all events, except for the outage of two connections with continental Italy in the *High Import* DP. In this case, the loss of the two connections causes the outage of the third remaining one. This outage happens because the high level of imported power, equal to 1188 MW, leads the current on the third cable to overcome the rated value. In this condition, we assume that an overcurrent protection is installed, causing the trip of the cable and the islanding of Sicily. It is worth remarking that such a situation is unrealistic and would never occur. However, we study this case as an extreme situation to assess the potential of the flexibility available.

Figure 8 shows the frequency profile and its time derivative obtained in this specific scenario. In this figure, results obtained with the *base configuration* are not reported since they are unstable. In Figure 8a, we observe that in configuration BESS-5, with a total additive FCR of 105 MW, frequency reaches a minimum of 48.9 Hz, before starting to increase toward a steady-state value of about 49.1 Hz. This behavior is obtained thanks to the activation of two load shedding steps. Indeed, at about 2.8 s, frequency overtakes the threshold of 49.3 Hz and, as shown in the zoom in Figure 8, the frequency derivative is lower than -0.3 Hz/s. According to Table 2, this causes the shedding of 9% of loads (equal to 290 MW). Then, always according to Table 2, when the frequency reaches the threshold of 48.9 Hz a further shedding of the 8% of loads (equal to 257 MW) is triggered.

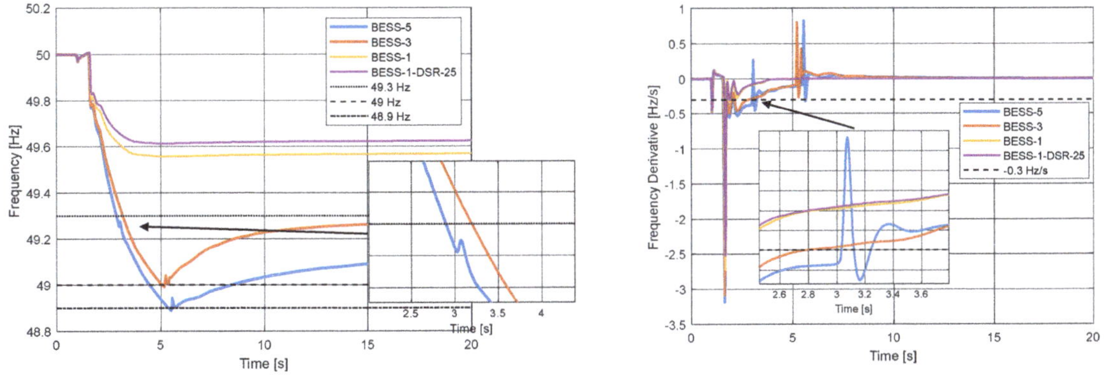

(a)

(b)

Figure 8. *High Import* DP, outage of two of the three connections with continental Italy: (**a**) frequency profiles; (**b**) frequency derivative profiles.

Increasing the BESS contribution, in configuration BESS-3, with a total additive FCR of 174 MW, the shedding at 49.3 Hz is avoided since the overtaking of this threshold is delayed at about 3.2 s (as we can observe in the zoom in Figure 8a), when frequency derivative is higher than the threshold of −0.3 Hz seconds (as we can observe in the zoom in Figure 8b). However, frequency reaches the threshold of 49 Hz, causing the shedding of the 9% of loads (equal to 290 MW), according again to Table 2. Finally, the frequency reaches a steady-state value close to 49.3 Hz.

Therefore, with the support of BESSs, in the two configurations BESS-5 and BESS-3, frequency stability is guaranteed but: (i) load shedding is activated and (ii) the NOC threshold of 49.5 Hz is violated. Augmenting the BESS contribution again, with a lower droop coefficient, in the configuration BESS-1, with a total additive FCR of 522 MW, load shedding is avoided, and frequency is kept higher than the 49.5 Hz threshold. A similar result, with a lower frequency deviation, is obtained with the BESS-1 DSR-25 configuration, where BESSs and DSR provide a total additive power reserve of 684 MW.

Always observing results in Table 9, we notice that in some cases, with the *base configuration*, NOC are violated, even if stability is guaranteed. This never happens in the *High Load* DP, where the number of generating units providing FCR is sufficient to keep NOC without requiring the contribution of BESSs and DSR. Differently, in the other DPs with Sicily connected to the Italian peninsula, the 500 MW load-step occurring in continental Italy causes the violation of NOC with the *base configuration*. In these cases, the support of BESSs and DSR to frequency regulation is always sufficient to keep frequency in the range 49.9–50.1 Hz.

Further cases where the contribution of BESSs and DSR is required to keep NOC are the outages of large traditional generating units in the *High Import, Island, Low Load,* and *Lines out of service* DPs. In most cases, a suitable mix of BESSs and DSR FCR is sufficient to guarantee NOC. In only two cases, this result is not possible. The first one is the outage of a large group of generating units in the *High Import* DP, which causes the loss of 817 MW of generated power. As shown in Figure 9a, with configuration BESS-1 DSR-35, corresponding to an additive FCR of 751 MW, the violation is only 0.02 Hz for about 4 s, and the steady-state frequency value is equal to 49.9 Hz. It is clear that augmenting the DSR participation percentage again from 35% will allow not violating the threshold. For this simulation, Figure 10a shows the active power exchanged by the BESSs associated with the in-service traditional generators, FCWTs and DFIGs. Here, BESSs are exporting power when the simulation starts. Then, when the perturbation occurs, they increase their power export to support the frequency regulation. As expected, the provided power variation is higher as lower is the droop coefficient. Figure 10b reports the variation of the active power absorbed by loads operating DSR. We can observe that, as frequency decreases, the power demand is reduced to support frequency regulation.

The second case where NOC cannot be guaranteed is the load-step in continental Italy in the *Lines out of service* DP. As shown in Figure 9b, an additive FCR of 362 MW provided by the mix BESS-1 DSR-30 is not sufficient to keep frequency higher than 49.9 Hz. In this case, we overtake the limit of $p = 35\%$ by simulating configuration BESS-1 DSR-50, corresponding to an additive FCR of 406 MW. These results are again not sufficient to keep the NOC, even if the violation of the 49.9 Hz threshold is only temporary (about 7 s) and limited to about 0.02 Hz.

It is finally worth remarking that in all simulations, all voltages levels always remain within the Italian grid code limits, and no congestion issues occur.

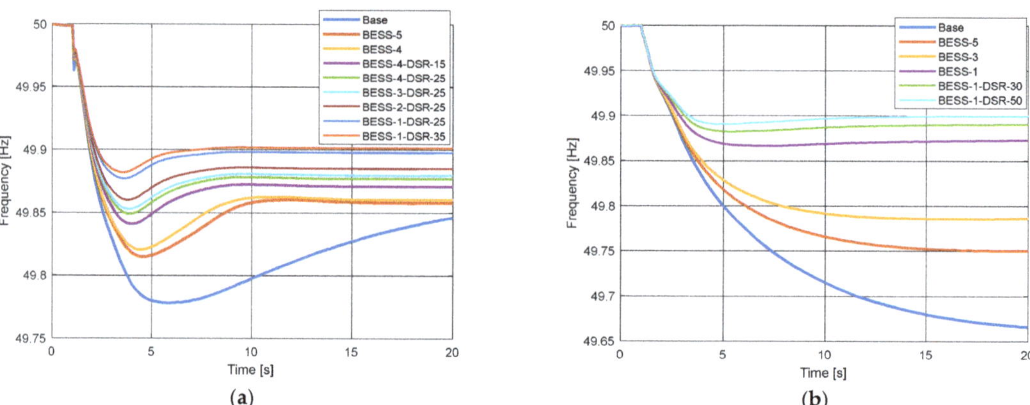

Figure 9. Frequency profiles for: (**a**) *High Import* DP, generation outage of 817 MW: (**b**) *Lines out of service* DP, 500 MW load-step in continental Italy.

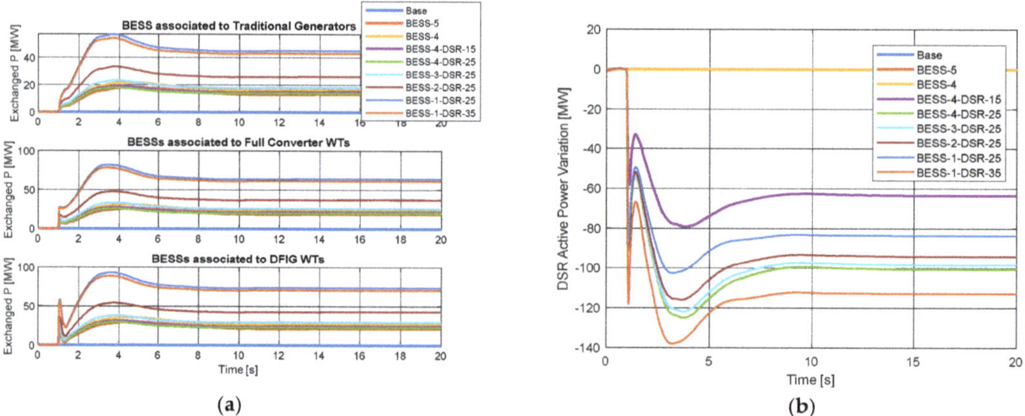

Figure 10. *High Import* DP, generation outage of 817 MW: (**a**) power exchanged by BESSs (positive means export, negative means import); (**b**) total power consumption variation provided by DSR.

4. Small Signal Stability Analysis

The small perturbation stability analysis allows the evaluation of the dynamic of the generators in response to a small variation in loads and generation. Even if small signal angle instability can be related to either the lack of synchronizing or damping torque, this instability mostly concerns insufficient damping of the system's oscillations due to large groups of closely coupled machines connected by weak tie lines [32].

4.1. Small Signal Stability Definitions

Modal analysis was performed for each of the 6 DPs presented in Table 6 to assess the small signal stability analysis of the forecasted network. A dynamic system described by state matrix A is said to be stable to small angle perturbations if, for every mode, the associated eigenvalue has a negative real part.

Damping refers to pairs of complex eigenvalues $\lambda = \sigma \pm j\omega$, which introduce oscillatory modes with a frequency equal to $f = \omega/2\pi$, and it is a measure of the rate of decay of the amplitude of the oscillations. The damping ratio is hence introduced:

$$\zeta = \frac{-\sigma}{\sqrt{\sigma^2 + \omega^2}} \tag{9}$$

which corresponds to the opposite of the real part of the eigenvalue divided by its module [51]. In the practice of electric power systems, a system is said to be stable and oscillations properly damped if the damping ratio is greater than 5%. In the following, modes with a damping ratio above 10% are not reported, and those below 5% are considered unstable; furthermore, electromechanical modes are the primary focus.

For our analysis, the *oscillation vector*, composed of the magnitude of the participation vector and the angle of the observability [52] can be used to easily identify the electromechanical modes. Indeed, the oscillation vectors allow identifying the state variables mainly involved in that mode and, by looking at the angle differences, understanding how the oscillations are manifested in the state variables, e.g., if in phase or counter-phase. In formulas, the oscillation vector of the state variable x_i with respect to mode k is:

$$ov_{ki} = |pf_{ki}| \angle w_k(i) \tag{10}$$

where pf_{ki} is the participation factor of the state variable x_i in the kth mode and $w_k(i)$ is the ith element of observability vector k (i.e., the right eigenvector associated with the mode k). This feature will be used to identify the properties specific to each relevant electromechanical mode and the behavior of the involved generators. Electromechanical modes can be generally classified into three categories depending on the location of the generators involved. In order of increasing frequency, inter-area modes involve generators belonging to distinctly different areas of the grid (e.g., Sicily and the peninsula), inter-plant or local modes are modes involving units of the same area, but different plants, intra-plant modes are modes involving units of the same plant [32].

4.2. Network Configurations

In the following sections, the small signal stability of the network in its *base configuration*, without any support provided by RES, is evaluated. Then the contribution of the SI provided by FCWT is considered, and, lastly, the gain of the PSS of the relevant generators are tuned to increase the overall stability of the grid, in particular, the damping ratio of inter-area mode called M1. Finally, the effect of the two combined actions is presented. The frequency support provided by the BESSs and DFIGs has also been investigated. For the sake of brevity, they are not presented as they provide just a minor contribution.

In Figure 11 the 12 substations' locations and denomination of the conventional plants reported in Section 2, Table 5, are shown. Each plant can have one or more units. In the following discussion, we will be referring to these units with the denomination of the plant (Identification names in Table 5) followed by a sequential number if needed. Of the 24 synchronous generators in the network, just those with a rated power equal to 70 MVA or greater, i.e., 15 of them, are equipped with a PSS. The latter provides a further input signal to the exciter system to provide additional damping to the electromechanical oscillations of the power system through exciter control.

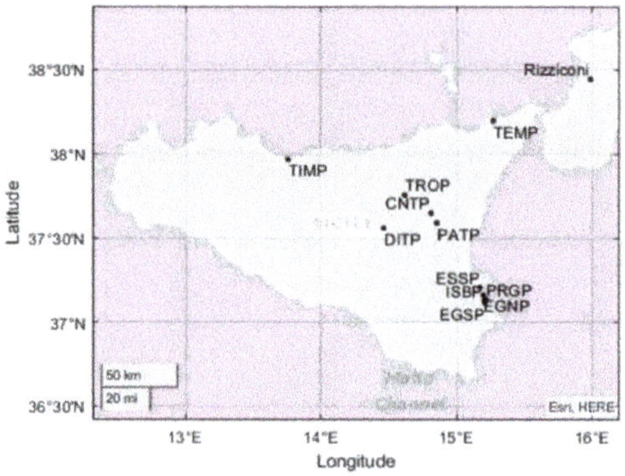

Figure 11. Locations and denominations of the traditional power plants in the Sicilian network.

4.2.1. Base Configuration

The *High Export*, *High Load*, and *Island* DPs do not present any mode with a damping ratio below 10% and are thus considered stable. The following focuses on those DPs that do present some modes with a damping ratio below 10% that is, the *High Import*, *Low Load*, and *Lines out of Service* DPs.

In the *High Import* DP, the modal analysis revealed twenty oscillatory modes with a damping ratio below 10% and all of them with a damping ratio above 5%, thus stable. Fifteen of these (modes D1 to D15) concern the DFIG units and are all very similar. For this reason, just mode D1 is discussed. The results are reported in Table 10.

Table 10. Electromechanical modes below the threshold of 10% for the *High Import* DP.

Mode	Eigenvalues	Frequency (Hz)	Damping (%)	Involved Plants
M1	-0.27 ± 4.81j	0.77	5.66	TIMP, PRGP, SLACK
D1	-0.67 ± 11.29j	1.80	5.90	DFIGs in ANPP
M2	-0.59 ± 8.18j	1.30	7.24	EGSP, PRGP
M3	-0.88 ± 9.98j	1.59	8.83	EGSP
M4	-0.89 ± 9.97j	1.59	8.90	EGSP
M5	-1.04 ± 10.44j	1.66	9.91	DITP, ESSP, PRGP

As anticipated, inspecting the oscillation vectors of the modes allows us to identify the generators involved and whether the rotors swing in phase or counter-phase.

Mode M1 has a frequency of 0.77 Hz and a damping factor of 5.66%. In Figure 12a, the oscillation vector is shown. The generators mainly affected by this oscillatory mode are generators TIMP1 and TIMP2 of the thermoelectric plant in Termini Imerese (TIMP). The mode is an inter-area mode, as the frequency suggests, in which the slack generator in Rizziconi oscillates against TIMP1, TIMP2 and generator PRGP1 in Priolo Gargallo (PRGP). The participation factors of PRGP1 and the slack are significantly lower than those of the TIMP units.

Mode M2 has a frequency of 1.30 Hz and a damping factor of 7.24%. In Figure 12b, the oscillation vector is shown. This is a local mode in which the three generators of the Erg Plant, EGSP1, EGSP2, and EGSP3 oscillate against the units in Priolo Gargallo, PRGP1, and PRGP2.

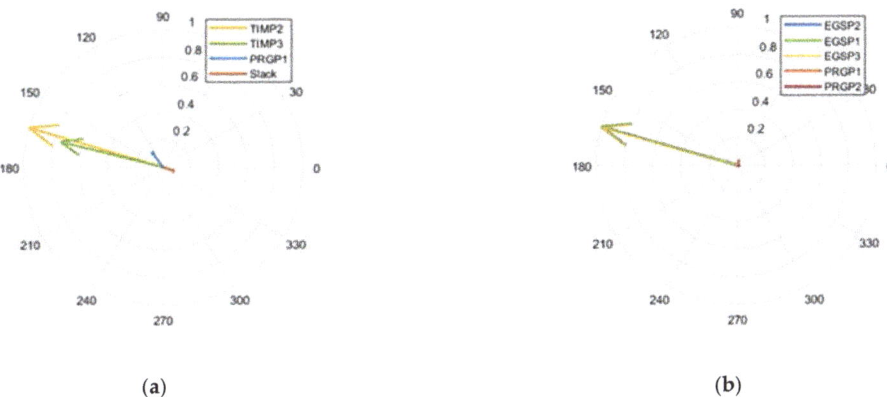

(a)   **(b)**

Figure 12. *High Import* DP: (**a**) oscillation vector of mode M1; (**b**) oscillation vector of mode M2.

Modes M3 and M4 are both intra-plant modes concerning the units of EGSP with the same frequency, 1.59 Hz, and a damping factor of 8.83% and 8.90%, respectively. Their oscillation vectors are shown in Figure 13a,b, respectively. In mode M3, EGSP3 oscillates against EGSP1 and EGSP2, while in mode M4, EGSP1 oscillates against EGSP2 and EGSP3.

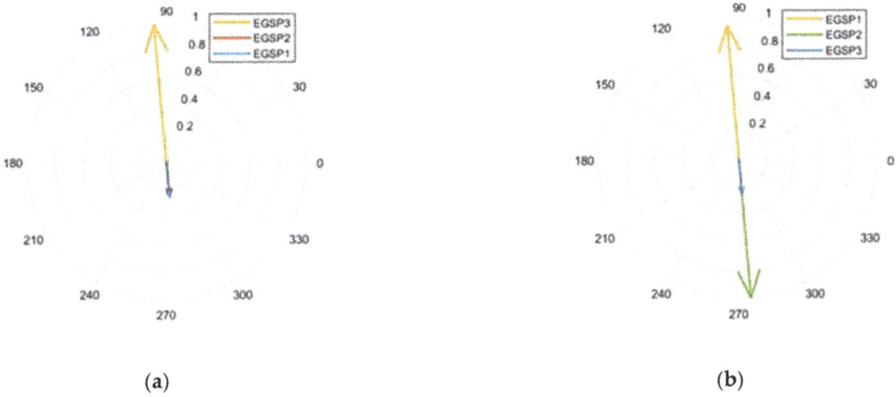

(a)   **(b)**

Figure 13. *High Import* DP: (**a**) oscillation vector of mode M3; (**b**) oscillation vector of mode M4.

Mode M5 has a frequency of 1.66 Hz and a damping factor of 9.91%. The oscillation vector is shown in Figure 14a. This mode is a local mode in which the 22.8 MVA generator in Dittaino (DITP) and the 58.1 MVA generator in Augusta (ESSP2), both not equipped with PSS, oscillate against the larger generator located in Priolo Gargallo, PRGP1, with a rated power of 370 MVA, a large constant of inertia of 7.5 s and equipped with PSS. For this reason, the magnitude of the oscillation vector of PRGP1 effectively appears like a dot in the origin.

Mode D1, with a frequency of 1.80 Hz and a damping factor of 5.90%, involves two DFIG units located in the substation of Anapo (ANPP). In total, there are 15 modes like this one, and they involve the 15 DFIGs in the network. Since all these modes behave very similarly, for the sake of conciseness, just the first one is shown in Figure 14b. This mode is a local mode in which the rotor speed of the DFIG unit ANPP2 oscillates against the DFIG unit ANPP3.

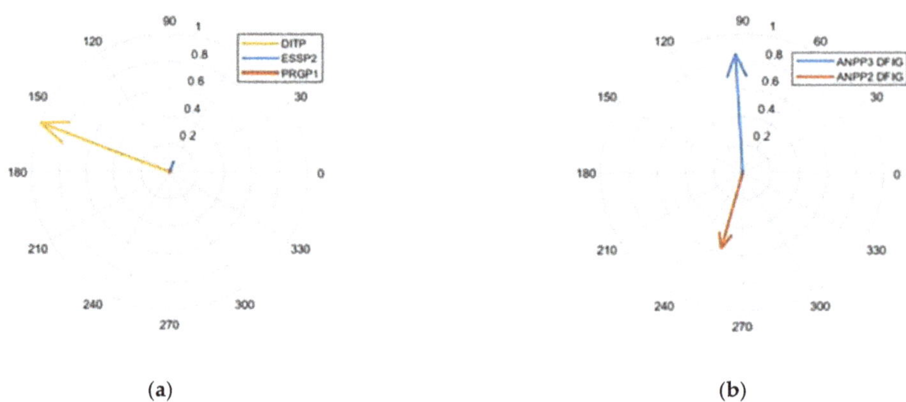

Figure 14. *High Import* DP: (**a**) oscillation vector of mode M5; (**b**) oscillation vector of mode D1.

To analyze the characteristics of these 15 modes, four different wind turbine power setpoints, i.e., $P = [0, 0.1, 0.2, 0.3]$ MW, are considered. The effects on mode D1 are presented in Table 11. While the frequency of D1 does not appear to be linked with the active power setpoint, the damping ratio of D1 is somehow correlated with the power produced. Its damping is always above the stability threshold of 5% and increases above the conservative threshold of 10% for an active power set point greater than 0.1 MW, that is, around 7% of the turbine rated power.

Table 11. Effect of the DFIG active power production on mode D1. Check marks indicate those modes with a damping factor greater than 10%.

DFIG Active Power Setpoint (MW)	Frequency (Hz)	Damping (%)
0	1.80	6.1
0.1	1.80	5.9
0.2	✓	✓
0.3	✓	✓

In the *Low Load* DP, the modal analysis revealed three electromechanical modes with a damping ratio below 10%, the inter-area mode M1, with a value of 3.88%, thus unstable. The results are reported in Table 12. As shown, the inter-area mode can be properly damped, either with the contribution of SI provided by RES (contribution of both FCWTs and BESSs), by tuning the gain of the relevant PSS or by a combination of both.

Table 12. Electromechanical modes below the threshold of 10% for the *Low Load* DP.

Mode	Eigenvalues	Frequency (Hz)	Damping (%)	Involved Plants
M1	$-0.19 \pm 4.95j$	0.79	3.88	TIMP, PRGP, SLACK
M6	$-0.80 \pm 10.86j$	1.73	7.38	CNTP, PRGP
M7	$-0.85 \pm 11.09j$	1.77	7.67	PATP, CNTP

Mode M6 has a frequency of 1.73 Hz and a damping factor of 7.38%, and in Figure 15a the oscillation vector is shown. In this mode, the relatively small generator in Contrasto (CNTP, 24 MVA) oscillates against the larger unit in Priolo Gargallo (PRGP1, 370 MVA), which, being also equipped with a PSS, is practically unaffected by the oscillations. For this reason, its oscillation vector in the graph appears like a dot in the origin.

Mode M7 has a frequency of 1.77 Hz and a damping factor of 7.67%, in Figure 15b, the oscillation vector is shown. This mode involves the 9 MVA unit in Patternò (PATP) oscillating against the 24 MVA unit in Contrasto (CNTP), both not equipped with PSS.

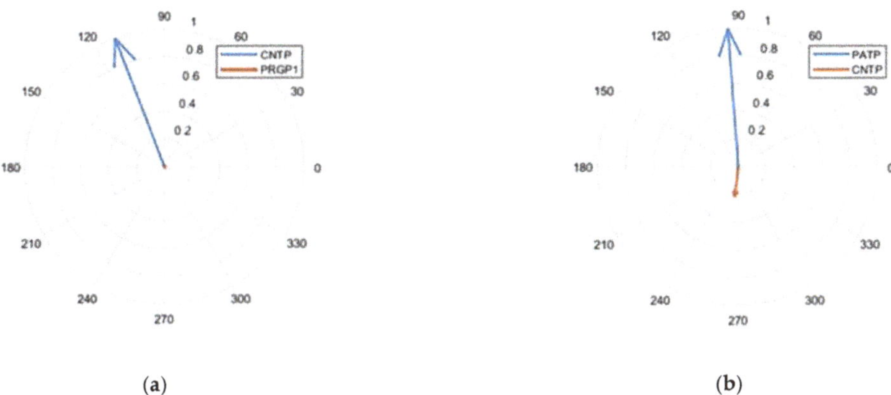

Figure 15. *Low Load* DP: (**a**) oscillation vector of mode M6; (**b**) oscillation vector of mode M7.

The *Lines out of Service* DP being a variation of the *Low Load* one, in which two important 230 kV lines (lines Favara-Chiaramonte and Caracoli-Sorgente) are out of service, presents the same three modes. The results are reported in Table 13. Mode M1 is particularly affected because the line Caracoli-Sorgente effectively connects the plant in Termini Imerese (TIMP) to the interconnection with continental Italy, right before the substation of Rizziconi.

Table 13. Electromechanical modes below the threshold of 10% for the *Lines out of Service* DP.

Mode	Eigenvalues	Frequency (Hz)	Damping (%)	Involved Plants
M1	$-0.05 \pm 4.73j$	0.75	1.08	TIMP, PRGP, SLACK
M6	$-0.86 \pm 10.82j$	1.72	7.95	CNTP, PRGP
M7	$-0.92 \pm 11.05j$	1.76	8.26	PATP, CNTP

The damping ratio of mode M1 decreases from 5.66% in the *High Import* DP, to 3.88% in the *Low Load* DP, and finally to a value of 1.08% in the *Lines out of Service* DP. In the following, a combination of SI support schemes and PSS tuning are adopted to properly damp the critical inter-area mode of the network.

4.2.2. Configuration with FCWT SI

In the base configuration analysis, it appears clear that the main issues of the small perturbation stability are the swings caused by the inter-area mode M1 between the plant in Termini Imerese and the slack on the Italian peninsula. The damping factor of this mode is 5.66% in the *High Import* DP, below 3.88% in the *Low Load* DP, and 1.08% in the *Lines out of Service* DP.

This section aims at determining the contribution of DC-link-based SI provided by the FCWT units. The initial setting of the SI gain K_{in} (see Figure 2b) is 0; then, its value is gradually increased to 50 by steps of 5. The modal analysis is then repeated for each value of K_{in} and for each DP; for the sake of conciseness, the results are reported just for a selected number of values, i.e., $K_{in} = [0, 10, 30, 50]$.

For the *High Import* DP, the results are shown in Table 14. Increasing the gain K_{in} provides a significant improvement to the damping ratio of mode M1. M2 and M5 experience a minor decrease in damping with increasing values of K_{in}. D1, M3, and M4 are not affected by K_{in}.

Table 14. Effect of increasing SI contribution from FCWTs on the modes of the *High Import* DP.

K_{in} (p.u.)	M1		D1		M2		M3		M4		M5	
	f (Hz)	ζ (%)	f (Hz)	ζ (%)	f (Hz)	ζ (%)	f (Hz)	ζ (%)	f (Hz)	ζ (%)	f (Hz)	ζ (%)
0	0.77	5.66	1.80	5.9	1.3	7.24	1.59	8.83	1.59	8.90	1.66	9.91
10	0.77	5.67	1.80	5.9	1.3	7.23	1.59	8.83	1.59	8.90	1.66	9.91
30	0.77	7.39	1.80	5.9	1.3	7.07	1.59	8.83	1.59	8.90	1.66	9.69
50	0.78	7.35	1.80	5.9	1.3	7.03	1.59	8.83	1.59	8.90	1.67	9.66

For the *Low Load* DP, the results are presented in Table 15. Increasing values of K_{in} provide a significant improvement to the damping ratio of mode M1, making it properly damped, a slightly detrimental effect on mode M6 and no effect whatsoever on M7.

Table 15. Effect of increasing SI contribution from FCWTs on the modes in the *Low Load* DP.

K_{in} (p.u.)	M1		M6		M7	
	f (Hz)	ζ (%)	f (Hz)	ζ (%)	f (Hz)	ζ (%)
0	0.79	3.88	1.73	7.38	1.77	7.67
10	0.79	4.37	1.73	7.34	1.77	7.67
30	0.79	5.18	1.73	7.30	1.77	7.67
50	0.79	5.37	1.73	7.28	1.77	7.67

For the *Lines out of Service* DP, the results are reported in Table 16. Increasing values of K_{in} provide a significant improvement to the damping ratio of mode M1, making it properly damped, until the value of K_{in} reaches 30, then the damping ratio of M1 becomes worst. The effect on the other two modes is the same as in the *Low Load* one.

Table 16. Effect of increasing SI contribution from FCWTs on the modes in the *Lines out of Service* DP.

K_{in} (p.u.)	M1		M6		M7	
	f (Hz)	ζ (%)	f (Hz)	ζ (%)	f (Hz)	ζ (%)
0	0.75	1.08	1.72	7.95	1.76	8.26
10	0.75	3.00	1.72	7.92	1.76	8.27
30	0.77	5.17	1.73	7.88	1.76	8.27
50	0.78	4.85	1.73	7.87	1.76	8.27

In conclusion, regarding the *Low Load* and the *Lines out of service* DPs, the SI provided by FCWT can increase the damping ratio of inter-area mode M1 to a value above the 5% threshold. However, it can do so just by a small margin. For a K_{in} value of 30, in the *Low Load* DP the damping ratio increases to a value of 5.18% and to a value of 5.17% in the *Lines out of service* DP. For this reason, in order to guarantee a more comfortable margin, other measures must be implemented. As shown in the following, tuning the PSS of the plants mainly involved in mode M1 allows to increase this margin to a more comfortable level.

4.2.3. Combined Configuration

A separate analysis has shown how values of the gain of the PSS of the TIMP plant ranging from 0.5 to 1 p.u. can provide satisfactory damping to inter-area mode M1 without compromising the stability of the local modes. For the *High Import* DP, the damping ratio gets above 10% and between 5 and 10% for the *Low Load* and *Lines out of Service* DP.

In this subsection, the combined influence of the FCWT frequency support schemes and the PSS tuning is investigated. For the FCWT units, the SI gain value of $K_{in} = 30$ p.u. is selected; for the PSS in TIMP, the value of the gain equal to 1 p.u is adopted.

The results reported in the following tables show that a combined effort from RES and PSS allows very good damping improvement for all the modes, even above the conservative threshold of 10%. In particular, this is true for the critical inter-area mode M1. For the *High Import* DP, the results are presented in Table 17.

Table 17. Electromechanical comparison between the base case and the combined contribution of RESs frequency support and PSS tuning to the small signal stability for the *High Import* DP. Check marks indicate those modes with a damping factor greater than 10%.

Case	M1		D1		M2		M3		M4		M5	
	f (Hz)	ζ (%)	f (Hz)	ζ (%)	f (Hz)	ζ (%)	f (Hz)	ζ (%)	f (Hz)	ζ (%)	f (Hz)	ζ (%)
Base	0.77	5.66	1.80	5.9	1.3	7.24	1.59	8.83	1.59	8.90	1.66	9.91
Comb.	✓	✓	1.80	5.9	✓	✓	✓	✓	✓	✓	✓	✓

For the *Low Load* DP, the results are presented in Table 18. It can be seen how, with a combined effort, the critical inter-area mode is properly stabilized and damped above the 10% threshold. On the other hand, modes M6 and M7, due to the size and the location of the plants involved, are not significantly affected; nonetheless maintain a damping ratio above 5%.

Table 18. Comparison between the base case and the combined contribution of RESs frequency support and PSS tuning to the small signal stability for the *Low Load* DP. Check marks indicate those modes with a damping factor greater than 10%.

Case	M1		M6		M7	
	f (Hz)	ζ (%)	f (Hz)	ζ (%)	f (Hz)	ζ (%)
Base	0.79	3.88	1.73	7.38	1.77	7.67
Combined	✓	✓	1.73	7.30	1.77	7.68

For the *Line out of services* DP, the results are presented in Table 19. This DP is a variation of the *Low Load* one, and the conclusions are very similar.

Table 19. Comparison between the base case and the combined contribution of RESs frequency support and PSS tuning to the small signal stability for the *Lines out of Service* DP. Check marks indicate those modes with a damping factor greater than 10%.

Case	M1		M6		M7	
	f (Hz)	ζ (%)	f (Hz)	ζ (%)	f (Hz)	ζ (%)
Base	0.75	1.08	1.72	7.95	1.76	8.26
Combined	✓	✓	1.73	7.88	1.76	8.27

The results show how the Sicilian grid forecasted for 2050 presents good damping of oscillatory modes and displays a variety of tools to tackle frequency instabilities. For every identified DP a modal analysis was performed. The *High Export, High Load,* and *Island* DPs do not present any electromechanical modes with a damping ratio below 10%. The *High Import, Low Load,* and *Lines out of Service* DP present some local or intra-plant modes that are stable and one inter-area mode with the peninsula, which is either barely stable or unstable. This inter-area mode, between Sicily and the Italian peninsula, is the main concern for the small perturbation stability. It was shown how it can be properly damped thanks to the contribution of SI provided by FCWT and tuning the PSS of the plant in Termini Imerese.

In conclusion, the amount of RES forecasted for the year 2050 does not compromise the small signal stability of the Sicilian grid; on the contrary, it provides a variety of tools to enhance it.

5. Voltage Stability Analysis

This last section presents the results of the voltage stability analysis. The voltage stability analysis consists of a steady-state analysis (i.e., slow dynamics) to identify the maximum loading conditions keeping acceptable voltages at all busses, i.e., between the 0.9 and the 1.1 of the per-unit rated voltage.

First, the voltage stability analysis was carried out considering the reactive support given only by the synchronous generators in service. Then, the RES power plants were equipped with the specifications adopted by Terna in the Italian grid code [53–55] for the reactive provision, and the second set of tests was carried out. Finally, a sensitivity analysis of the parameters of the RES reactive controllers was performed to improve the voltage levels in particularly weak grid conditions.

5.1. Procedure

The PV curve calculation tool of DIgSILENT was used and adapted for this task to carry out the voltage stability analysis. Basically, we found the critical points of voltage instability by increasing the power demand of loads (zero and negative loads have not been considered for this evaluation) until the load flow calculation no longer converges. In particular, the HV voltage magnitudes were monitored to fulfil the 0.9–1.1 p.u. limits.

Reactive support of the RES was also considered and the current capability limits given by Terna in [53,54] were used as reference (Figure 16a, red solid line). The maximum/minimum reactive support must be equal to ±35% of the currently active power available. Regarding the controller logic, the injection of reactive power, depending on the voltage values, was assumed to be in accordance with Figure 16b, where $|\Delta V_{max}| = |\Delta V_{min}| = 0.05 \cdot V_{nom}$:

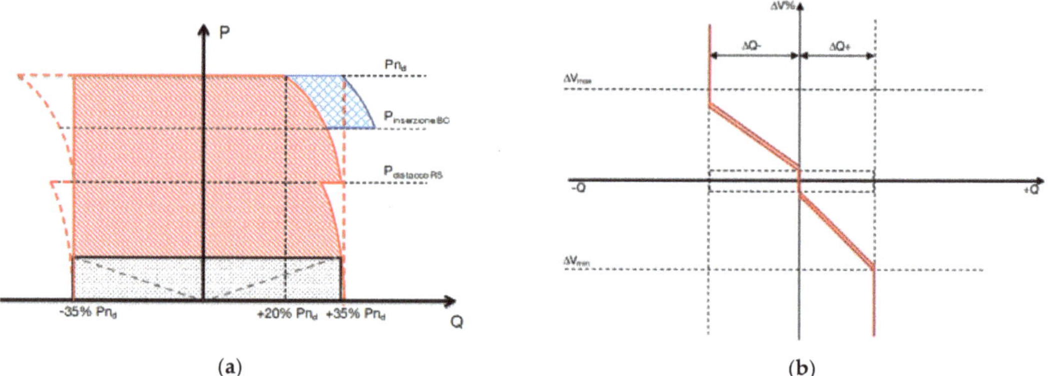

(a) (b)

Figure 16. Capability curve implemented for wind and PV plants (**a**) [53,54], Q/V curve given by Terna in [54] (**b**).

According to the annexes [53] and [54], the reactive droop was set equal to 14%:

$$droop = \frac{\Delta V}{\Delta Q} = \frac{|0.05|}{|0.35|} = 14\% \tag{11}$$

In its base case, the Sicilian grid presents a quite high loadability margin, equal to 65%. However, for our studies, a loadability equal to/higher than 40% is also considered acceptable.

5.2. Results

The *base configuration* results are first analyzed, where only the synchronous machines provide reactive support. These preliminary results are not realistic, but they can be used as a first insight and benchmark for the subsequent assessments. Then, a second study was performed considering the reactive contribution given by the RES and BESS [55] as well. The analysis was stopped as soon as the voltage profile of one single HV busbar goes below the lower 0.9 limit. The 2050 DPs present a quite low percentage of synchronous machines: hence, a higher voltage instability is expected.

The results are summarized in Table 20 (the critical DPs are highlighted in bold), along with the loadability margin in MW and percentage. It can be easily appreciated that the *High Import* and *Island* DPs present the lowest loadability margins, less than 15% and, along with loadability of the *High Load* one, they are much lower than the adopted threshold of 40%: additional resources must be here employed to increase voltage stability margins since the synchronous machines only are not enough to guarantee a satisfactory stability level. The remaining DPs already present enough loadability.

Table 20. Loadability margins for the selected DPs.

Dispatching Profile	Initial Load (MW)	Final Load (MW)	Loadability (MW)	Loadability (%)
High Export	1835	3717	1882	102.5
High Import	**3220**	**3596**	**376**	**11.7**
High Load	4088	5498	1410	34.5
Low Load	1510	3346	1836	121.5
Island	**4289**	**4857**	**568**	**13.2**
Lines out of service	1510	3225	1715	113.5

5.2.1. High Import and Island Dispatching Profiles

The voltage profiles of the 400 kV busbars of the *High Import* DP are shown in Figure 17a, starting from the initial demand of 3220 MW; the maximum demand achievable is equal to 3596 MW. The voltage profiles of the 400 kV busbars of the *Island* DP are shown in Figure 17b, starting from the initial demand of 4289 MW; the maximum demand achievable is equal to 4857 MW.

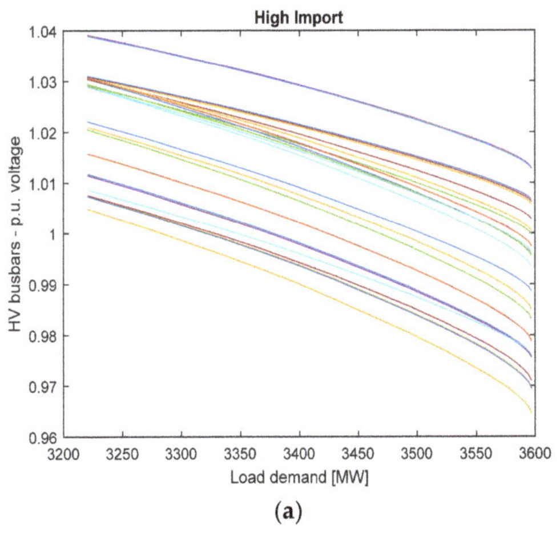

(a)

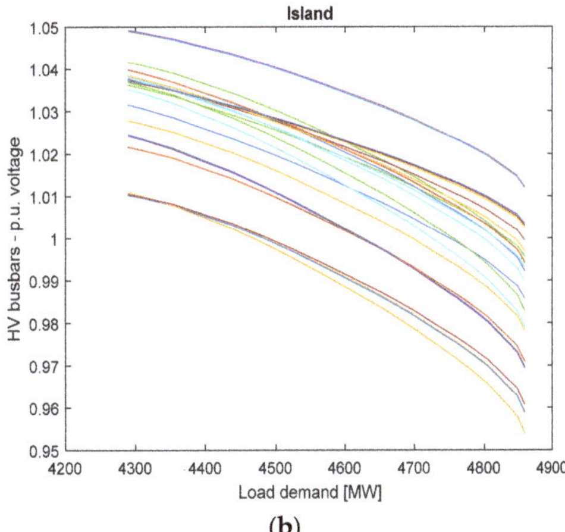

(b)

Figure 17. HV voltage profile of the *High Import* DP (**a**), HV voltage profile of the *Island* DP (**b**).

5.2.2. RES Contribution

The reactive contribution of the RES was considered in the voltage stability analysis. All the HV connected wind plants, the photovoltaic plants connected to the HV and MV network, representing the MV dispersed generation, and the BESS were modified to provide reactive power support. All the inverters were set up with the current Italian standards as described. Since a low level of loadability was detected, analysis was carried out to maximize the reactive support with no restrictions for the inverters: they can provide their full capability and hence overreach the existing limits given by Terna in [53–55].

The results are shown in Table 21. This solution, with "full inverters capability", offers quite satisfactory results compared to the base case, as in all the DPs the margin is increased. Despite this, in the *High Import* and *Island* ones, the loadability is still too low and hence cannot be considered acceptable, even if it is almost doubled compared to the *base configuration*. The *High load* profile reached an acceptable margin instead.

Table 21. Loadability margins for the selected DPs with the full RES reactive contribution.

Dispatching Profile	Initial Load (MW)	Loadability (MW)	Loadability (%)	Loadability Base Case (%)
High Export	4592	2757	150.2	102.5
High Import	**3993**	**773**	**24.0**	**11.7**
High Load	5702	1614	40.0	34.5
Low Load	4887	3376	223.5	121.5
Island	**5791**	**1502**	**35.0**	**13.2**
Lines out of service	4615	3105	205.5	113.5

5.2.3. Critical Dispatching Profiles

Analyzing the results, in both the *High Import* and *Island* DPs, the "critical busses" are two HV nodes close to Priolo, where two large thermal plants are now connected, but will be considered decommissioned in 2050: Erg Nuce Nord and Priolo Gargallo. For the *High Import* profile, the critical node is the PRGP busbar, and for the *Island* one, the EGNP busbar, both connected to the HV 150 kV substation of Melilli, which is the interconnection with the 230 and 400 kV lines.

According to this analysis, the synchronous machines of the Priolo power plants (four machines of the Erg Nuce Nord and two for the Priolo Gargallo (PRGP)) were assumed as still in operation as synchronous condensers only to guarantee suitable reactive support. This solution is cheap, reasonable, and already acceptable since Terna is managing to install synchronous condensers in a few old thermal plants in the south of Italy to provide inertia and voltage control to stabilize the grid.

Assuming the full reactive capability of the RES converters and thanks to the Priolo compensators, the loadability can be increased for the *Island* DP up to 40% (Figure 18a), and for the *High Import* DP up to 60% (Figure 18b):

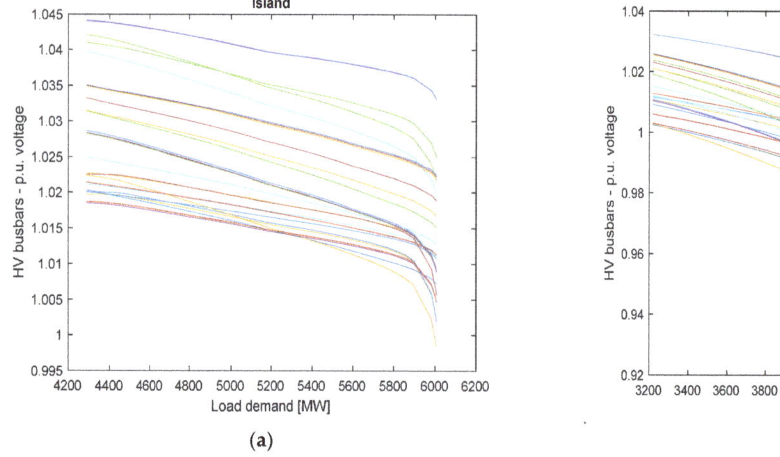
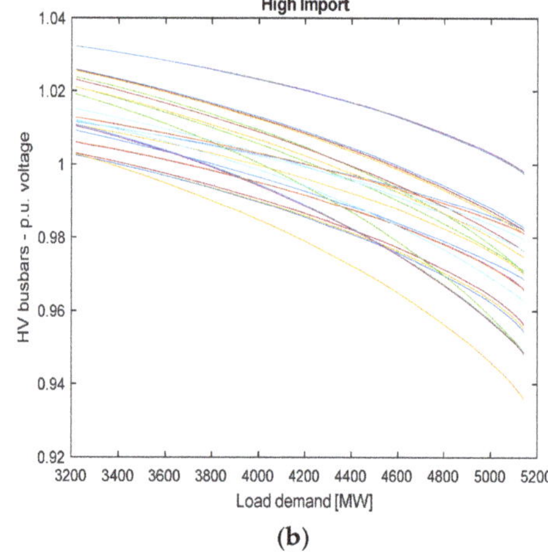

(a)   (b)

Figure 18. HV voltage profile of the *Island* DP (**a**), HV voltage profile of the *High Import* DP (**b**).

The results are summarized in the following table (Table 22, where the improvements are highlighted): for all the other DPs, the Priolo compensators can be kept off, as they already present reasonable loadability margins.

Table 22. Loadability margins for the selected DPs for 2050 with a full RES reactive contribution at Priolo as a synchronous compensator.

Dispatching Profile	Final Load (MW)	Loadability (MW)	Loadability (%)	Loadability without Priolo (%)
High Import	5144	1924	60.0	24.0
Island	6007	1718	40.0	35.0

A second analysis was carried out to better investigate the problem, where the power plants in Priolo were completely dismantled and the connection busses employed for the connection of wind turbine parks, hence taking advantage of the already existing infrastructures. The new wind turbines were connected to the grid through PWM converters, able to provide reactive power. The capability has again been extended up to the maximum available to maximize the reactive support (previously defined as "full inverters capability"). The loadability margins were recalculated: the results are shown in Table 23.

Table 23. Loadability margins for the selected DPs for 2050 with a full RES reactive contribution of Priolo as wind park.

Dispatching Profile	Final Load (MW)	Loadability (MW)	Loadability (%)	Loadability without Priolo (%)
High Import	5625	2405	74.6	24.0
Island	6036	1747	41.0	35.0

Thanks to the larger reactive capability provided by the converters, the *High Import* DP can reach a satisfactory loadability level equal to 75%, while in the *Island* one is increased by few MW up to a final 41%. Anyway, this latest value can be considered suitable for grid stability and the analysis can now be considered settled.

It has been noticed that when Sicily is importing power from continental Italy, it is facing a high-power demand or it is even not connected, its loadability margin, and hence its voltage stability is reduced. The capability of the static converters has been extended up to the maximum available, supporting the voltage stability to a satisfactory level. For the *Island* and *High Import* DPs the local reactive compensation at the Priolo substation allows increasing the loadability margin. Sicily does not present a good loadability margin in the base case, with both the combination of synchronous generators and RES plants equipped with the current reactive settings [53,54]. However, the presence of RES plant with "full inverters capability", along with specific compensation in local substations, improves the voltage stability and guarantees the voltage security of the power system.

6. Conclusions

This paper reports the results of a dynamic analysis carried out by EnSiEL for WP1 of the OSMOSE project (Task 1.4.3) concerning the feasibility of some of the scenarios and DPs identified for 2050 in Task 1.1 and Task 1.2, in the particular case of the power system of Sicily, one of the two main islands of Italy.

In particular, EnSiEL has assessed some typical perturbations of power systems, e.g., loss of a large generator, slow increase in loads, or contingencies of branches. These simulations were performed on a dedicated large grid model of the electrical network of Sicily, provided by Terna, the Italian TSO. This grid model was updated according to both the capacities defined by the scenarios provided by Task 1.1 and the system-wide balancing of energy supply and demand identified area by area by Task 1.2. Therefore, Task 1.4.3 has evaluated if the outputs of the above-mentioned Tasks are feasible from a dynamic point of view, identified possible lack of stability, and, in such a case, suggested possible countermeasures, picking up any flexibility resource that can be useful.

The analysis has regarded angle stability under large and small perturbations and voltage stability. Table 24 reports the conclusions and indications drawn for each DP with respect to the three types of stability.

Table 24. Conclusions summary.

Type of Stability	Is Stability Guaranteed?	Does the Operational Condition Need Specific Additional Flexibilitiy Tools?	Solutions Identified/Notes
High Export DP			
Large perturbation angle stability	YES	YES	DSR [1] and BESS FCR [2] are required to guarantee the normal operating conditions
Small perturbation angle stability	YES	NO	-
Voltage stability	YES	YES	Extend the reactive capability of RES units
High Import DP			
Large perturbation angle stability	NO	YES	DSR and BESS FCR are required
Small perturbation angle stability	YES	YES	SI [3] by FCWTs [4] and PSS tuning are required
Voltage stability	NO	YES	Extend the reactive capability of RES units + Local compensation in Priolo substation
High Load DP			
Large perturbation angle stability	YES	NO	-
Small perturbation angle stability	YES	NO	-
Voltage stability	NO	YES	Extend the reactive capability of RES units
Island DP			
Large perturbation angle stability	NO	YES	DSR and BESS FCR are required
Small perturbation angle stability	YES	NO	-
Voltage stability	NO	YES	Extend the reactive capability of RES units + Local compensation in Priolo substation
Low Load DP			
Large perturbation angle stability	NO	YES	DSR and BESS FCR are required
Small perturbation angle stability	NO	YES	SI by FCWTs and PSS tuning are required
Voltage stability	YES	YES	Extend the reactive capability of RES units
Lines out of service DP			
Large perturbation angle stability	NO	YES	DSR and BESS FCR are required
Small perturbation angle stability	NO	YES	SI by FCWTs and PSS tuning are required
Voltage stability	YES	YES	Extend the reactive capability of RES units

[1] Demand-Side Response; [2] Frequency Containment Reserve; [3] Synthetic Inertia; [4] Full-Converter Wind Turbines.

Results show that large-perturbation angle stability is guaranteed in all the DPs and for all events, except for a very unlikely operating condition that results in a high import power loss. However, it is possible to achieve stability in such a critical operating condition, taking into account the contribution of BESSs and/or DSR, which results to be mandatory. As for the post fault steady state, the contribution to frequency regulation of BESSs and/or DSR is generally sufficient to keep frequency in the range 49.9–50.1 Hz. In only two cases, this has resulted not been possible; however, this constraint can be managed by subsequent control actions.

Regarding small-perturbation angle stability, for the DPs in which RES support frequency control is not active, in some cases, low damped electromechanical modes are present. In particular, the interarea mode can be stabilized thanks to the contribution of SI provided by wind farms and by suitably tuning the PSSs of the relevant generators. Therefore, thanks to the availability of SI, the Sicilian grid shows strong small perturbation stability features.

Finally, simulations assessing the voltage stability were first performed considering only the synchronous generator's reactive contribution and then the RES support as well. Because of the high penetration of RESs in 2050, high voltage instability is observed unless flexibility provided by RES is considered. Indeed, exploiting the reactive contribution of RESs voltage stability is guaranteed in all the studied operating conditions.

Despite that the analysis provided in this paper covers many stability issues, considering a complete set of flexibility options with a detailed model of a real power system, further analyses should be conducted to confirm the conclusions we have obtained. First of all, grid portions different from the one of Sicily should be considered. Moreover, further stability aspects should be analyzed, focusing on time scales lower than the one of electromechanical phenomena considered in this work. Indeed, the high penetration of converters may introduce stability issues not taken into account till today by conventional power system analysis and this will be related to high-frequency dynamics, such as the ones associated with the switching of converters. Future works will be thus dedicated to extend our stability analysis by considering power systems different from the one of Sicily and using models able to capture the dynamics at time scales lower than the electrochemical phenomena.

Author Contributions: Conceptualization: A.B.; methodology: A.B., F.C., F.D., V.I., F.N. and A.V.; data curation: A.V.; software: J.A.A., F.C., F.D., V.I., T.P. and A.V.; validation: J.A.A., F.C., T.P. and A.V.; writing—original draft preparation: J.A.A., F.C., T.P. and A.V.; writing—review and editing: A.B., F.D., V.I. and F.N.; supervision: A.B. and F.N. All authors have read and agreed to the published version of the manuscript.

Funding: This work is part of the OSMOSE project. It has received funding from the European Union's Horizon 2020 research and innovation program under grant agreement n. 773406.

Institutional Review Board Statement: Not applicable.

Informed Consent Statement: Not applicable.

Data Availability Statement: Not applicable.

Conflicts of Interest: The authors declare no conflict of interest. The funders had no role in the design of the study; in the collection, analyses, or interpretation of data; in the writing of the manuscript, or in the decision to publish the results.

References

1. OSMOSE—Optimal System-Mix of Flexibility Solutions for European Flexibility. Available online: https://www.osmose-h2020.eu (accessed on 26 April 2022).
2. Liu, H.; Yang, S.; Yuan, X. Inertia Control Strategy of DFIG-Based Wind Turbines Considering Low-Frequency Oscillation Suppression. *Energies* **2022**, *15*, 29. [CrossRef]
3. Wang, J.; Xu, Y.; Wu, X.; Huang, J.; Zhang, X.; Yuan, H. Enhanced Inertial Response Capability of a Variable Wind Energy Conversion System. *Energies* **2021**, *14*, 8132. [CrossRef]

4. Li, Y.; Xu, Z.; Wong, K.P. Advanced Control Strategies of PMSG-Based Wind Turbines for System Inertia Support. *IEEE Trans. Power Syst.* **2017**, *32*, 3027–3037. [CrossRef]

5. Xu, Y.; Chen, P.; Zhang, X.; Yang, D. An Improved Droop Control Scheme of a Doubly-Fed Induction Generator for Various Disturbances. *Energies* **2021**, *14*, 7980. [CrossRef]

6. Bolzoni, A.; Terlizzi, C.; Perini, R. Analytical Design and Modelling of Power Converters Equipped with Synthetic Inertia Control. In Proceedings of the 2018 20th European Conference on Power Electronics and Applications (EPE'18 ECCE Europe), Riga, Latvia, 17–21 September 2018; pp. 1–10.

7. Fang, J.; Zhang, R.; Li, H.; Tang, Y. Frequency Derivative-based Inertia Enhancement by Grid-Connected Power Converters with a Frequency-Locked-Loop. *IEEE Trans. Smart Grid* **2018**, *10*, 4918–4927. [CrossRef]

8. Conte, F.; Massucco, S.; Silvestro, F. Frequency control services by a building cooling system aggregate. *Electr. Power Syst. Res.* **2016**, *141*, 137–146. [CrossRef]

9. Tindemans, S.H.; Strbac, G. Low-Complexity Decentralized Algorithm for Aggregate Load Control of Thermostatic Loads. *IEEE Trans. Ind. Appl.* **2021**, *57*, 987–998. [CrossRef]

10. Trovato, V.; Sanz, I.M.; Chaudhuri, B.; Strbac, G. Advanced Control of Thermostatic Loads for Rapid Frequency Response in Great Britain. *IEEE Trans. Power Syst.* **2017**, *32*, 2106–2117. [CrossRef]

11. Lin, Y.; Barooah, P.; Meyn, S.; Middelkoop, T. Experimental evaluation of frequency regulation from commercial building HVAC systems. *IEEE Trans. Smart Grid* **2015**, *6*, 776–783. [CrossRef]

12. Cheng, B.; Powell, W.B. Co-Optimizing Battery Storage for the Frequency Regulation and Energy Arbitrage Using Multi-Scale Dynamic Programming. *IEEE Trans. Smart Grid* **2018**, *9*, 1997–2005. [CrossRef]

13. Zhao, H.; Wu, Q.; Hu, S.; Xu, H.; Rasmussen, C.N. Review of energy storage system for wind power integration support. *Appl. Energy* **2015**, *137*, 545–553. [CrossRef]

14. Conte, F.; Massucco, S.; Schiapparelli, G.P.; Silvestro, F. Day-ahead and intra-day planning of integrated BESS-PV systems providing frequency regulation. *IEEE Trans. Sustain. Energy* **2020**, *11*, 1797–1806. [CrossRef]

15. Berizzi, A.; Bovo, C.; Ilea, V.; Merlo, M.; Miotti, A.; Zanellini, F. Decentralized reactive power control of wind power plants. In Proceedings of the 2012 IEEE International Energy Conference and Exhibition, ENERGYCON 2012, Florence, Italy, 9–12 September 2012.

16. Christakou, K.; Tomozei, D.C.; Bahramipanah, M.; Le Boudec, J.Y.; Paolone, M. Primary voltage control in active distribution networks via broadcast signals: The case of distributed storage. *IEEE Trans. Smart Grid* **2014**, *5*, 2314–2325. [CrossRef]

17. Ilea, V.; Bovo, C.; Falabretti, D.; Merlo, M.; Arrigoni, C.; Bonera, R.; Rodolfi, M. Voltage control methodologies in active distribution networks. *Energies* **2020**, *13*, 3293. [CrossRef]

18. Ma, K.; Yao, T.; Yang, J.; Guan, X. Residential power scheduling for demand response in smart grid. *Int. J. Electr. Power Energy Syst.* **2016**, *78*, 320–325. [CrossRef]

19. Bañales, S.; Dormido, R.; Duro, N. Smart Meters Time Series Clustering for Demand Response Applications in the Context of High Penetration of Renewable Energy Resources. *Energies* **2021**, *14*, 3458. [CrossRef]

20. Wang, J.; Padullaparti, H.; Ding, F.; Baggu, M.; Symko-Davies, M. Voltage Regulation Performance Evaluation of Distributed Energy Resource Management via Advanced Hardware-in-the-Loop Simulation. *Energies* **2021**, *14*, 6734. [CrossRef]

21. Stephen, A.A.; Musasa, K.; Davidson, I.E. Voltage Rise Regulation with a Grid Connected Solar Photovoltaic System. *Energies* **2021**, *14*, 7510. [CrossRef]

22. Szultka, A.; Szultka, S.; Czapp, S.; Karolak, R.; Andrzejewski, M.; Kapitaniak, J.; Kulling, M.; Bonk, J. Voltage Profiles Improvement in a Power Network with PV Energy Sources—Results of a Voltage Regulator Implementation. *Energies* **2022**, *15*, 723. [CrossRef]

23. Ding, Z.; Huang, X.; Liu, Z. Active Exploration by Chance-Constrained Optimization for Voltage Regulation with Reinforcement Learning. *Energies* **2022**, *15*, 614. [CrossRef]

24. Li, W.; Tang, M.; Zhang, X.; Gao, D.; Wang, J. Operation of Distributed Battery Considering Demand Response Using Deep Reinforcement Learning in Grid Edge Control. *Energies* **2021**, *14*, 7749. [CrossRef]

25. Weibezahn, J.; Göke, L.; Orrù, L. OSMOSE D1.1: European Long-Term Scenarios Description. 2019. Available online: https://www.osmose-h2020.eu/download/d1-1-european-long-term-scenarios-description (accessed on 29 March 2022).

26. Weibezahn, J.; Göke, L.; Poli, D.; Protard, V.; Bourien, Y.M.; Ažman, G. OSMOSE D1.2: Flexibility Cost and Operational Data Outlook. 2020. Available online: https://www.osmose-h2020.eu/download/d1-2-flexibility-cost-and-operational-data-outlook (accessed on 29 March 2022).

27. Bourmaud, J.-Y.; Lhuillier, N.; Orlic, D.; Kostic, M.; Heggarty, T.; Grisey, N. OSMOSE D1.3: Optimal Mix of Flexibility. 2022. Available online: https://www.osmose-h2020.eu/wp-content/uploads/2022/04/D1.3-Optimal-Mix-of-Flexibility-1.pdf (accessed on 29 March 2022).

28. ENSIEL. Available online: https://www.consorzioensiel.it (accessed on 28 February 2022).

29. Heggarty, T.; Bourmaud, J.Y.; Girard, R.; Kariniotakis, G. Quantifying power system flexibility provision. *Appl. Energy* **2020**, *279*, 115852. [CrossRef]

30. Berizzi, A.; Ilea, V.; Vicario, A.; Conte, F.; Massucco, S.; Adu, J.A.; Nucci, C.A.; Pontecorvo, T. Stability analysis of the OSMOSE scenarios: Main findings, problems, and solutions adopted. In Proceedings of the 2021 AEIT International Annual Conference (AEIT), Milan, Italy, 4–8 October 2021; pp. 1–6.

31. DIgSILENT GmbH. *DIgSILENT PowerFactory, User Manual*; Version 20; DIgSILENT GmbH: Gomaringen, Germany, 2019.

32. Kundur, P. *Power System Stability And Control*; McGraw-Hill: New York, NY, USA, 1994.
33. Adrees, A.; Milanović, J.V.; Mancarella, P. Effect of inertia heterogeneity on frequency dynamics of low-inertia power systems. *IET Gener. Transm. Distrib.* **2019**, *13*, 2951–2958. [CrossRef]
34. IEEE/NERC. Task Force on Short-Circuit and System Performance Impact of Inverter Based Generation. In *Impact of Inverter Based Generation on Bulk Power System Dynamics and Short-Circuit Performance*; IEEE: Piscataway, NJ, USA, 2018.
35. Eftekharnejad, S.; Vittal, V.; Heydt, G.T.; Keel, B.; Loehr, J. Small signal stability assessment of power systems with increased penetration of photovoltaic generation: A case study. *IEEE Trans. Sustain. Energy* **2013**, *4*, 960–967. [CrossRef]
36. Map of the Sicilian 380–220 kV System. Available online: http://www.regione.sicilia.it (accessed on 28 February 2022).
37. EC Directive (EU). 2018/2001 of the European Parliament and of the Council of 11 December 2018 on the promotion of the use of energy from renewable sources. *Off. J. Eur. Union* **2018**, *2001*, 1–128.
38. TERNA S.p.A. Grid Code Annexes 11, "Criteri Generali per la Taratura delle Protezione delle Reti a Tensione Uguale o Superiore a 110 kV". 2018. Available online: https://download.terna.it/terna/0000/0105/28.pdf (accessed on 29 March 2022). (In Italian)
39. TERNA S.p.A. GAUDI' Portal. Available online: https://www.terna.it/it/sistema-elettrico/gaudi (accessed on 28 February 2022).
40. Western Electricity Coordinating Council Modeling and Validation Work Group. *WECC Renewable Energy Modeling Task Force WECC Solar Plant Dynamic Modeling Guidelines*; World Environment Center: Washington, DC, USA, 2014.
41. TERNA S.p.A. Grid Code Annexes 12, "Criteri di Taratura dei Relè di Frequenza del Sistema Elettrico e Piano di Alleggerimento". 2018. Available online: https://download.terna.it/terna/0000/1127/88.pdf (accessed on 29 March 2022). (In Italian).
42. Berizzi, A.; Bosisio, A.; Ilea, V.; Marchesini, D.; Perini, R.; Vicario, A. Analysis of Synthetic Inertia Strategies from Wind Turbines for Large System Stability. In *IEEE Transactions on Industry Applications*; IEEE: New York, NY, USA, 2022. [CrossRef]
43. Wang, B.; Ni, J.; Geng, J.; Lu, Y.; Dong, X. Arc flash fault detection in wind farm collection feeders based on current waveform analysis. *J. Mod. Power Syst. Clean Energy* **2017**, *5*, 211–219. [CrossRef]
44. Bolzoni, A.; Perini, R. Experimental validation of a novel angular estimator for synthetic inertia support under disturbed network conditions. In Proceedings of the 2019 21st European Conference on Power Electronics and Applications (EPE'19 ECCE Europe), Genova, Italy, 3–5 September 2019; pp. 1–10. [CrossRef]
45. Hu, J.; Huang, Y.; Wang, D.; Yuan, H.; Yuan, X. Modeling of grid-connected DFIG-based wind turbines for dc-link voltage stability analysis. *IEEE Trans. Sustain. Energy* **2015**, *6*, 1325–1336. [CrossRef]
46. Adu, J.A.; Napolitano, F.; Nucci, C.A.; Diego Rios Penaloza, J.; Tossani, F. A DC-Link Voltage Control Strategy for Fast Frequency Response Support. In Proceedings of the 2020 IEEE 20th Mediterranean Electrotechnical Conference (MELECON), Palermo, Italy, 16–18 June 2020; pp. 470–475. [CrossRef]
47. Adu, J.A.; Rios Penaloza, J.D.; Napolitano, F.; Tossani, F. Virtual Inertia in a Microgrid with Renewable Generation and a Battery Energy Storage System in Islanding Transition. In Proceedings of the SyNERGY MED 2019—1st International Conference on Energy Transition in the Mediterranean Area, Cagliari, Italy, 28–30 May 2019.
48. ENTSO-E. Demand Connection Code. Available online: https://www.entsoe.eu/major-projects/network-code-development/demand-connection (accessed on 20 June 2020).
49. TERNA S.p.A. Codice di trasmissione, Dispacciamento, Sviluppo e Sicurezza Della Rete. 2018. Available online: https://download.terna.it/terna/0000/0107/31.pdf (accessed on 29 March 2022). (In Italian)
50. ENTSO-E. *Frequency Stability Evaluation Criteria for the Synchronous Zone of Continental Europe: Requirements and Impacting Factors*; ENTSO-E: Brussels, Belgium, 2016.
51. Machowski, J.; Bialek, J.W.; Bumby, J.R. *Power System Dynamics: Stability and Control*; John Wiley & Sons: Hoboken, NJ, USA, 2008; ISBN 9780470725580.
52. Gonzalez-Longatt, F.; Rueda, J. *PowerFactory Applications for Power System Analysis*; Springer: Berlin/Heidelberg, Germany, 2014; ISBN 978-3-319-12957-0.
53. TERNA S.p.A. Grid Code Annexes 17, "Centrali Eoliche—Condizioni Generali di Connessione Alle Reti AT. Sistemi di Protezione, Regolazione e Controllo". 2019. Available online: https://download.terna.it/terna/0000/0105/34.pdf (accessed on 29 March 2022). (In Italian)
54. TERNA S.p.A. Grid Code Annexes 68, "Centrali Fotovoltaiche—Condizioni Generali di Connessione Alle Reti AT. Sistemi di Protezione, Regolazione e Controllo". 2019. Available online: https://download.terna.it/terna/0000/0105/84.pdf (accessed on 29 March 2022). (In Italian)
55. CEI 0–16. *Regola Tecnica di Riferimento per la Connessione di Utenti Attivi e Passivi Alle Reti AT ed MT Delle Imprese Distributrici di Energia Elettrica*; Comitato Elettrico Italiano: Milano, Italy, 2019. (In Italian)

Article

Forced Oscillation Grid Vulnerability Analysis and Mitigation Using Inverter-Based Resources: Texas Grid Case Study

Khaled Alshuaibi [1], Yi Zhao [1], Lin Zhu [2], Evangelos Farantatos [2,*], Deepak Ramasubramanian [2], Wenpeng Yu [1] and Yilu Liu [1,3]

[1] Department of Electrical Engineering and Computer Science, The University of Tennessee, Knoxville, TN 37996, USA; kalshuai@vols.utk.edu (K.A.); yzhao77@utk.edu (Y.Z.); wyu10@utk.edu (W.Y.); liu@utk.edu (Y.L.)

[2] Electric Power Research Institute, Palo Alto, CA 94304, USA; lzhu@epri.com (L.Z.); dramasubramanian@epri.com (D.R.)

[3] Oak Ridge National Laboratory, Oak Ridge, TN 37830, USA

[*] Correspondence: efarantatos@epri.com

Abstract: Forced oscillation events have become a challenging problem with the increasing penetration of renewable and other inverter-based resources (IBRs), especially when the forced oscillation frequency coincides with the dominant natural oscillation frequency. A severe forced oscillation event can deteriorate power system dynamic stability, damage equipment, and limit power transfer capability. This paper proposes a two-dimension scanning forced oscillation grid vulnerability analysis method to identify areas/zones in the system that are critical to forced oscillation. These critical areas/zones can be further considered as effective actuator locations for the deployment of forced oscillation damping controllers. Additionally, active power modulation control through IBRs is also proposed to reduce the forced oscillation impact on the entire grid. The proposed methods are demonstrated through a case study on a synthetic Texas power system model. The simulation results demonstrate that the critical areas/zones of forced oscillation are related to the areas that highly participate in the natural oscillations and the proposed oscillation damping controller through IBRs can effectively reduce the forced oscillation impact in the entire system.

Keywords: forced oscillation; inverter-based resources (IBRs); grid vulnerability analysis; active power modulation

Citation: Alshuaibi, K.; Zhao, Y.; Zhu, L.; Farantatos, E.; Ramasubramanian, D.; Yu, W.; Liu, Y. Forced Oscillation Grid Vulnerability Analysis and Mitigation Using Inverter-Based Resources: Texas Grid Case Study. *Energies* **2022**, *15*, 2819. https://doi.org/10.3390/en15082819

Academic Editor: Abu-Siada Ahmed

Received: 24 February 2022
Accepted: 11 April 2022
Published: 12 April 2022

Publisher's Note: MDPI stays neutral with regard to jurisdictional claims in published maps and institutional affiliations.

1. Introduction

A forced oscillation in power systems is usually excited by an external periodic disturbance from a cyclic load, equipment failure, poor control design, or the mechanical oscillation of a generator during abnormal operation conditions [1–4]. With the increasing load variability and high penetration of renewable energy resources with intermittent nature, such as wind generation [5–7], forced oscillation events are becoming more frequent and more severe. In addition, in today's deregulated and competitive market, more buyers and wind producers are joining the electricity market, and as a result, the strategic behavior of electricity consumers and renewable power producers might increase forced oscillation events [8,9]. Multiple forced oscillation events which were sustained for minutes to hours have been reported in the U.S., China, Canada etc. Forced oscillation can deteriorate power system stability, damage equipment, limit power transfer capability, and reduce power quality [10,11]. A recent example of a forced oscillation event occurred on 11 January 2019, in the U.S. Eastern Interconnection due to a faulty input to the steam turbine controller of a generator. The forced oscillation frequency was close to the natural oscillation mode of 0.25 Hz, and the forced oscillation lasted for 18 min before the source was removed [12].

Unlike forced oscillation, natural oscillation is the inherent oscillation due to the physical properties of the power grid. The damping of natural oscillation is becoming lower because of the high penetration of IBRs and the retirement of conventional synchronous generators with power system stabilizers (PSSs) [13–15]. Resonance can occur when the forced oscillation frequency is close to the frequency of a poorly damped natural oscillation mode. As a result, the oscillation amplitude can be significantly amplified and can propagate across the entire power system [16].

The most effective strategy to suppress the forced oscillation is to locate and remove the source from the power system [17,18]. However, accurate source identification may be a time-consuming process after the event is detected. Power oscillation damping controllers through IBRs and other devices have been proven to enhance the damping of a natural oscillation quickly and effectively [19–21]. Meanwhile, it is a feasible measure to suppress the forced oscillation energy to a safe level using a damping controller before the source is removed. Few papers have investigated the mitigation of a forced oscillation via a damping controller. In [22–24], power static synchronous compensator with energy storage (E-STATCOM), voltage source converter-based high voltage direct current (VSC-HVDC), and battery energy storage systems were investigated to suppress the forced oscillation by modulating their active or reactive power. Only [24] reports the testing of a large system among all the aforementioned strategies to mitigate forced oscillation. In [25], a forced oscillation damping controller with an event-triggered control strategy was proposed and tested on the IEEE 14-machine South-East Australian model.

Different from the previous work, this paper proposes a general method for forced oscillation grid vulnerability analysis and mitigation through IBRs in large-scale power grids. To reduce the forced oscillation effect in the entire system, a forced oscillation damping controller at an effective IBR actuator can be activated before the exact forced source is identified and removed. First, critical areas/zones under different specified forced oscillation frequencies that can excite the most severe forced oscillation across a large-scale power system can be identified by using a detailed location and frequency scanning method. These identified critical areas/zones under the specified forced oscillation frequencies are also effective locations as the actuators of forced oscillation damping controllers when the forced oscillation sources are narrowed down inside or close to the critical areas/zones. Secondly, a forced oscillation damping controller through active power modulation control via IBRs was designed to reduce the impact of the forced oscillation event before the forced source is removed. Finally, the 2000-bus synthetic Texas grid model in [26] is used as the study system to verify the proposed method.

The major contributions of this paper can be summarized as follows: (1) a two-dimension scanning forced oscillation grid vulnerability analysis method is proposed to identify the forced oscillation frequency and areas/zones that are critical to forced oscillation. (2) An effective IBR actuator location to forced oscillation damping controller is identified. (3) A forced oscillation damping controller is designed to reduce the impact of a forced oscillation event to a safe level and provide the system operator sufficient time to locate and remove the forced source.

The rest of this paper is outlined as follows: the forced oscillation grid vulnerability analysis, actuator selection of the forced oscillation controller, and forced oscillation mitigation methods are described in Section 2; the simulation results of the Texas grid case study are discussed in Section 3; and Section 4 gives the conclusion.

2. Forced Oscillation Vulnerability Analysis and Mitigation

2.1. Framework of the Algorithm

Figure 1 shows the framework of the proposed method to identify the critical areas/zones under different forced oscillation frequency events and the damping control of the forced oscillation.

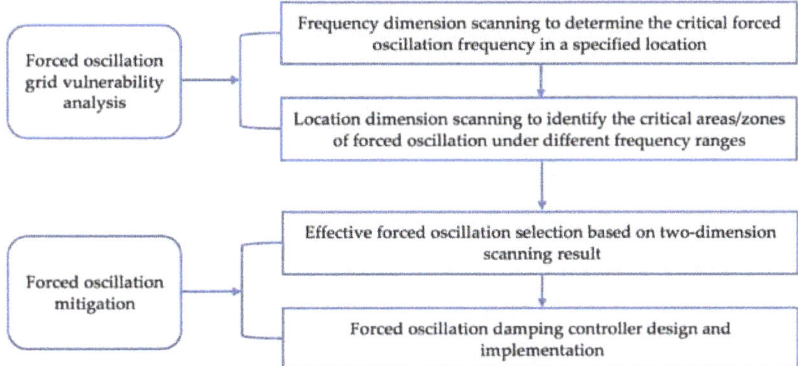

Figure 1. Framework of the forced oscillation identification and mitigation procedure.

To mimic the forced oscillation events in the simulation model, a sinusoidal signal with a specified frequency was added to the active power set point of the governor model at a synchronous generator. A two-dimensional (frequency and location) scanning method was utilized for the forced oscillation vulnerability analysis to identify the critical areas/zones and the oscillation frequencies that could excite the most severe forced oscillation. The frequency dimension scanning was performed by changing the forced oscillation frequency in a specific range (e.g., from 0.1 Hz to 1.5 Hz) at a fixed source location. A high-voltage bus frequency deviation in each area/zone was used as the indicator of the forced oscillation energy in each area/zone. The location dimension scanning was performed by changing the forced oscillation source location across the system under different specified forced oscillation frequencies. The critical frequency and critical locations that can excite the maximum bus frequency deviation were determined through frequency dimension scanning and location dimension scanning.

Once the critical areas/zones under different forced oscillation frequencies were identified, a local forced oscillation damping controller can be designed and implemented at IBRs in these areas as candidates to suppress the forced oscillation event. The ideal location of the damping controller/actuator would be the same as the external source to reduce the forced oscillation energy. However, when the forced source location can be quickly narrowed down in the relative area that is inside or close to the critical areas/zones with the forced damping controller, the damping controller can be activated to reduce the forced oscillation impact before the accurate forced source is identified and removed. In other words, the identified critical areas/zones are the effective locations of the actuator with a forced oscillation damping controller.

2.2. Forced Oscillation Vulnerability Analysis Method

Forced oscillation can be mimicked in simulations by changing the active power set point of a generator governor model or changing the voltage set point of the exciter model [24]. In this case study, a sinusoidal signal was added to the governor active power set point, as shown in Figure 2. The steam turbine-governor TGOV1 is used as an example, and the governor model parameter is shown in Table 1. The sinusoidal waveform is defined as:

$$\Delta P(t) = A \sin(2\pi f t) \tag{1}$$

where f is the forced oscillation frequency, A is the amplitude of the sinusoidal wave, and $\Delta P(t)$ is the active power that is added to the governor active power set point.

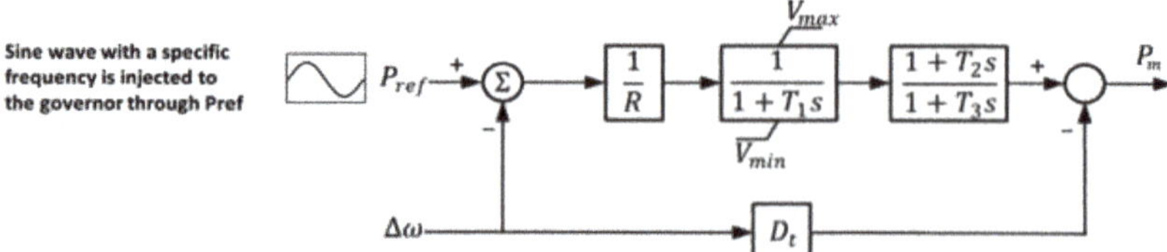

Figure 2. TGOV1 governor turbine model with an external sinusoidal signal.

Table 1. TGOV1 governor parameter.

R	T1	Vmax	Vmin	T2	T3	Dt
20	0.1	99	−99	0.1	0.1	0

In the frequency dimension and location dimension scanning, the sinusoidal wave amplitude was adjusted to generate the same peak–peak forced oscillation energy for all sources. In this way, the bus frequency deviation can be compared fairly under all the disturbances with the same perturbation energy. The peak–peak forced oscillation energy was set to be such that it could excite a severe forced oscillation in the system. In this paper, the maximum frequency deviation over the ±36 mHz governor deadband was considered a severe forced oscillation. Figure 3 shows an example of the generator's active power output when the forced oscillation frequency is 0.4 Hz. The peak–peak forced oscillation energy here is 100 MW.

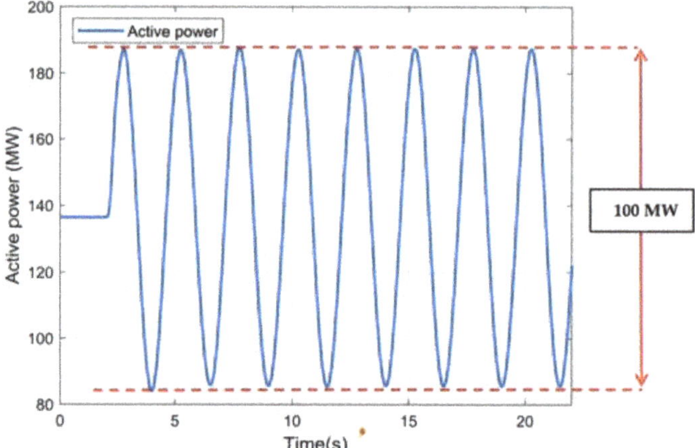

Figure 3. Active power output in MW of a generator with a 0.4 Hz sinusoidal wave added to the governor reference power set point.

The frequency dimension scanning was performed by changing the forced oscillation frequency with a small, fixed step size in a specified range of frequencies at a fixed source location. For each specified oscillation frequency, the bus frequency signals at the high voltage substations in each zone/area were recorded to calculate their peak–peak frequency deviation. In this paper, the maximum peak–peak bus frequency deviation in each zone was used as an indicator of the severity of a forced event impact in each zone. By comparing the peak–peak bus frequency deviation in all these zones/areas, the maximum value will be chosen to represent the peak–peak frequency deviation under this specified oscillation frequency. The peak–peak frequency deviations for the different frequencies are sorted in ascending order, and the frequency with the largest frequency deviation is determined as the most critical forced oscillation frequency at this fixed source location.

Location dimension scanning was conducted by changing the source location through the entire grid with the same forced energy. The peak–peak frequency deviations under different forced sources were then compared to identify the critical location to the forced oscillation event. The generator with the largest capacity in each area/zone was selected to add the sinusoidal signal at its governor active power set point. The above-mentioned frequency dimension scanning procedure was repeated at each selected generator. For each selected forced oscillation source, the maximum peak–peak bus frequency deviation under each specified forced oscillation frequency can be obtained. After the location dimension scanning, the peak–peak frequency deviations under each specified forced oscillation frequency at different forced oscillation sources were sorted in ascending order, and the source location with the largest frequency deviation was identified as the most critical location for the forced oscillation event with the specified forced oscillation frequency in the power system.

Based on the above procedure, the identified critical locations under different frequencies are the ones that can excite the largest frequency deviation in the system with the forced oscillation source.

2.3. Actuator Selection of Forced Oscillation Controller via IBRs

Based on the above two-dimension scanning results, the critical locations to the forced oscillation event under different specified oscillation frequencies can be further considered as the candidate effective actuator location of the oscillation damping controller to reduce the critical forced oscillation impact in the whole system.

Since the oscillation frequency and the relatively large area with the forced source can be quickly estimated, the effective candidate actuators (IBRs) that are close to or in the forced source areas can be activated to reduce the forced source impact before any accurate forced source is located and removed.

2.4. Forced Oscillation Mitigation Method

The proposed forced oscillation controller in [24], is implemented through a utility-scale wind/solar plant. The IBRs were modeled using the generic models that were developed by the Western Electricity Coordinating Council (WECC) Renewable Energy Modeling Task Force [27]. The proposed controller is a droop controller with the frequency deviation of its local high voltage bus as input. The controller's output is added at Paux to modulate the active power output of the IBRs.

3. Texas Power System Case Study

3.1. Synthetic Texas Power System Model Description

The 2000-bus synthetic Texas power grid model that was developed by Texas A&M was used in this paper for the forced oscillation study [26]. The model consists of 432 in-service machines with a maximum generation capacity of 84,996 MW that partially supply a total load of 67,109 MW. In these 432 machines, there are 81 renewable resources with a maximum capacity of 11,833 MW and an actual generation of 8875 MW. In addition, there are 2345 transmission lines and four high voltage levels: 500, 230, 161, and 115 kV. As shown in Figure 4, the synthetic Texas power grid consists of eight areas, and each area includes one or more zones with a total of 28 zones. Area 5 (North Central) is the load center which receives large power that is transmitted from Area 7 (Coast) and Area 8 (East). The renewables are located in Areas 1 to 5, with the renewable penetration level based on the maximum MW capacity of the whole system being 13.92%.

Figure 4. 2000-bus synthetic Texas power grid diagram [28].

PSSs, governors, and exciters are implemented at each synchronous generator. For the dynamic models of the renewables, the generic models that were developed by the WECC were used, and in particular, the generator/converter model (REGCAU2) and the electrical control model (REECCU1). Small-signal analysis using the DSAtools/SSAT tool, showed that two natural oscillation modes exist in the system. One is a 0.67 Hz oscillation mode between Area 4 and Area 7, with a damping ratio of 5.1%. The other is a 0.60 Hz oscillation mode between Areas 1, 2, 3, 4, 5, and parts of Area 8 and Area 7, with a damping ratio of 6.31%. Table 2 shows the generators with a high participation factor of the two natural oscillation modes in different zones and it was calculated using the small-signal analysis tool in the DSAtools. It is noted that Area 4 has the highest participation factor for the natural oscillation mode of 0.67 Hz.

Table 2. Participation factor for natural 0.67 Hz and 0.60 Hz oscillation.

Generator Location	Generator Bus NO.	Participation Factor (0.67 Hz)	Participation Factor (0.60 Hz)
Area 4 Zone 19	4192	1.00	0.14
Area 4 Zone 21	4058	0.72	0.01
Area 4 Zone 20	4115	0.40	0.22
Area 8 Zone 8	8080	0.19	0.77
Area 1 Zone 9	1051	0.18	1.00
Area 5 Zone 12	5063	0.18	0.01
Area 2 Zone 11	2023	0.18	0.76
Area 5 Zone 18	5319	0.17	0.73

3.2. Grid Vulnerability Analysis

This section presents the grid vulnerability analysis in the Texas system using the proposed two-dimensional scanning method. The purpose of the scanning is to compare the forced oscillation energy at different frequencies and forced source locations in order to identify the areas that can excite the largest frequency deviation. In this study, the forced oscillation starts at t = 2 s and is simulated for 20 s. The conventional generator (coal, oil, nuclear, and hydro) with the maximum capacity at each zone in the Texas system was selected as the source of the forced oscillation. To excite sufficient frequency deviations in the system, the oscillation energy at each source location was set to 100 MW peak–peak for all the events.

3.2.1. Frequency Dimension Scanning

The frequency dimension scanning procedure involves changing the forced oscillation frequency at a fixed source location. In this study, the oscillation frequency ranges from 0.1 Hz to 1.5 Hz with a step size of 0.05 Hz. A high voltage bus frequency response at each zone was selected to calculate its peak–peak bus frequency deviation by using its last 1.5 oscillation cycles. By comparing the peak–peak bus frequency deviation among the buses in different zones, the maximum peak–peak bus frequency deviation under each specified forced oscillation frequency was identified.

Figure 5 shows the maximum peak–peak bus frequency deviation under different frequencies when the forced oscillation source was located at Area 4 Zone 19. Figure 6 illustrates the frequency response of the selected high voltage buses at each zone in the Texas system when the 0.67 Hz oscillation source was located at Area 4 Zone 19. As shown in Figure 5, when the oscillation frequency goes to around 0.67 Hz (the natural oscillation frequency), the most severe oscillation deviation can be observed in the system than other forced oscillation frequencies.

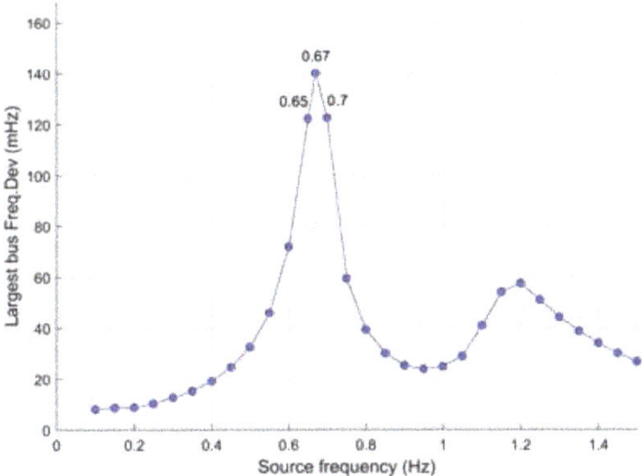

Figure 5. Frequency dimension scanning results when the forced source at Area 4 Zone 19.

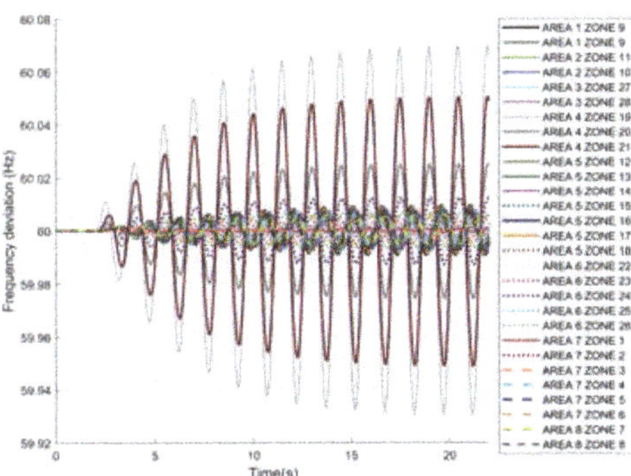

Figure 6. Frequency response at different zones when the 0.67 Hz forced oscillation source is located at Area 4 Zone 19.

3.2.2. Location Dimension Scanning

The location dimension scanning is to vary the location of the forced oscillation source across the system with different forced oscillation frequencies. Figure 7 shows the location scanning results of the forced oscillation source with the top five frequency deviations. Regardless of the forced source location, compared to other forced oscillation frequency ranges, the maximum frequency deviation can always be observed in the Texas synthetic system when the forced oscillation frequency is equal or close to the natural oscillation at 0.67 Hz. This proves that the forced oscillation impact on the power system stability is magnified when the forced oscillation frequency coincides with the natural oscillation frequency. What's more, when the forced oscillation source was located at zones that have high participation of the natural 0.67 Hz oscillation (see Table 2), a larger frequency deviation can be observed in the system than other forced oscillation sources.

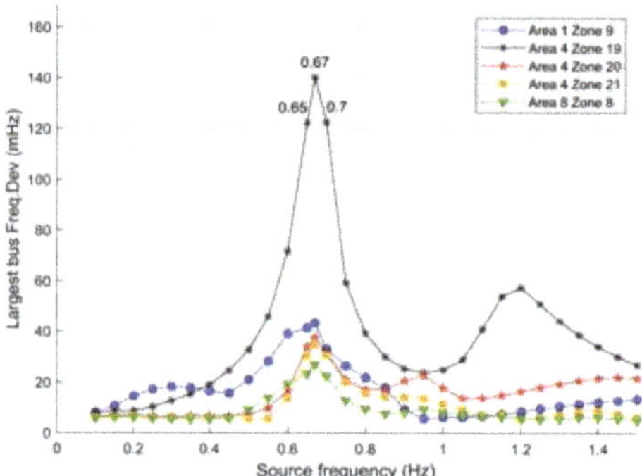

Figure 7. Location dimension scanning results of the forced oscillation source with the top five frequency deviations.

Table 3 shows an example of the location dimension scanning results under the critical 0.67 Hz forced oscillation. It can be observed that Area 4 Zone 19 is the critical source location that can excite the maximum peak–peak bus frequency deviation in the system. This is consistent with the participation factor sorting of 0.67 Hz natural oscillation in Table 2.

Table 3. Location dimension scanning results when the critical forced oscillation frequency is 0.67 Hz.

Source Location	Area—Zone with Largest Bus Frequency Deviation	Peak–Peak Bus Frequency Deviation (mHz)
Area 4 Zone 19	Area 4 Zone 19	140.090
Area 1 Zone 9	Area 4 Zone 19	43.622
Area 4 Zone 20	Area 4 Zone 19	37.798
Area 4 Zone 21	Area 4 Zone 19	34.996
Area 8 Zone 8	Area 4 Zone 19	30.780
Area 5 Zone 12	Area 4 Zone 19	27.221
Area 7 Zone 4	Area 4 Zone 19	27.125
Area 2 Zone 11	Area 4 Zone 19	26.910
Area 7 Zone 3	Area 4 Zone 19	25.536
Area 5 Zone 18	Area 4 Zone 19	24.451
Area 7 Zone 2	Area 4 Zone 19	23.359
Area 3 Zone 28	Area 3 Zone 28	23.172
Area 8 Zone 7	Area 4 Zone 19	19.345
Area 7 Zone 5	Area 4 Zone 19	18.532
Area 5 Zone 14	Area 4 Zone 19	18.099
Area 5 Zone 16	Area 4 Zone 19	17.938
Area 6 Zone 22	Area 4 Zone 19	17.764
Area 5 Zone 13	Area 4 Zone 19	17.663
Area 5 Zone 17	Area 4 Zone 19	15.185
Area 6 Zone 26	Area 4 Zone 19	13.886
Area 7 Zone 6	Area 4 Zone 19	12.948
Area 6 Zone 25	Area 4 Zone 19	11.598
Area 6 Zone 23	Area 4 Zone 19	9.652
Area 7 Zone 1	Area 7 Zone 4	9.191

3.3. Actuator Selection of Forced Oscillation Controller via IBRs

Based on the two-dimension scanning, the critical areas/zones under different frequency range are summarized in Table 4. This table is useful for selecting the effective actuators to install the oscillation damping controller and reduce the forced oscillation in the power system. Based on Table 4, IBRs in Area 1, Zone 9 were effective in suppressing oscillations from 0.10 Hz to 0.35 Hz; IBRs in Area 4, Zone 19 were effective in suppressing oscillations around 0.4 Hz and from 0.5 Hz to 0.85 Hz and 1.10 Hz to 1.3 Hz; and IBRs in Area 3, Zone 28 were effective in suppressing oscillations around 0.45 Hz and from 0.90 Hz to 1.05 Hz and 1.35 Hz to 1.50 Hz. The IBRs with a large capacity in these areas are recommended to implement forced oscillation damping controllers.

Table 4. The critical areas/zones under different frequency range based on two-dimensional scanning.

Source Location	Oscillation Frequency (Hz)	Area—Zone with Largest Bus Frequency Deviation	Peak–Peak Bus Frequency Deviation (mHz)
1-4 Area 1 Zone 9	0.1	Area 7 Zone 4	8.00
Area 1 Zone 9	0.15	Area 7 Zone 4	10.53
Area 1 Zone 9	0.2	Area 7 Zone 4	14.67
Area 1 Zone 9	0.25	Area 7 Zone 4	17.33
Area 1 Zone 9	0.3	Area 7 Zone 4	18.35
Area 1 Zone 9	0.35	Area 7 Zone 4	18.00
Area 4 Zone 19	0.4	Area 4 Zone 19	19.19
Area 3 Zone 28	0.45	Area 3 Zone 28	25.22
Area 4 Zone 19	0.5	Area 4 Zone 19	32.63
Area 4 Zone 19	0.55	Area 4 Zone 19	45.95
Area 4 Zone 19	0.6	Area 4 Zone 19	71.88
Area 4 Zone 19	0.65	Area 4 Zone 19	122.30
Area 4 Zone 19	0.67	Area 4 Zone 19	140.09
Area 4 Zone 19	0.7	Area 4 Zone 19	122.58
Area 4 Zone 19	0.75	Area 4 Zone 19	59.45
Area 4 Zone 19	0.8	Area 4 Zone 21	39.37
Area 4 Zone 19	0.85	Area 4 Zone 21	30.16
Area 3 Zone 28	0.9	Area 3 Zone 28	30.35
Area 3 Zone 28	0.95	Area 3 Zone 28	31.79
Area 3 Zone 28	1	Area 3 Zone 28	31.78
Area 3 Zone 28	1.05	Area 3 Zone 28	33.20
Area 4 Zone 19	1.1	Area 4 Zone 19	40.93
Area 4 Zone 19	1.15	Area 4 Zone 19	54.10
Area 4 Zone 19	1.2	Area 4 Zone 19	57.49
Area 4 Zone 19	1.25	Area 4 Zone 19	51.09
Area 4 Zone 19	1.3	Area 4 Zone 19	44.23
Area 3 Zone 28	1.35	Area 3 Zone 28	46.49
Area 3 Zone 28	1.4	Area 3 Zone 28	48.79
Area 3 Zone 28	1.45	Area 3 Zone 28	51.21
Area 3 Zone 28	1.5	Area 3 Zone 28	53.53

When the forced oscillation frequency is monitored, the actuators that are effective to damp the oscillations can be identified from Table 4. Once the forced source is narrowed down within a relatively large area, the effective actuators (IBRs) that are close to or in the forced source areas can be activated to reduce the impact of the forced source energies before any accurate forced source is located.

3.4. Forced Oscillation Control via IBRs

This section demonstrates the performance of the forced oscillation damping controller. The aim of the controller is to reduce the impact of the forced oscillation until the forced oscillation source is located and removed. The controller performance is validated under the most serious resonance case, i.e., when the forced source is at Area 4 Zone 19 with the forced oscillation frequency coinciding with the natural oscillation mode 1 (0.67 Hz). Based on the results in Tables 3 and 4, the effective location for a forced oscillation damping controller is at Area 4 Zone 19. The 243.6 MVA IBR at bus 4183 in Area 4 Zone 19 is then selected as the actuator and the feedback signal is the 230 kV bus 4070 that is connected with the IBR in Area 4 Zone 19.

In this paper, the droop controller gain was set to be −190. Figure 8 depicts the Area 4 Zone 19 frequency deviation improvement and the active power output of the actuator when the forced source at Area 4 Zone 19 before and after implementing the control. Table 5 illustrates the peak–peak frequency deviation improvement of the selected high voltage buses at each zone in the Texas system after implementing the controller. The largest deviation of the entire system can be reduced to less than ±36 mHz (66.6 mHz) with −190 control gain and 17% limiter. The peak–peak actuator output is 50.05 MW which is around ±16.44% of the actual active power of the actuator of 152.25 MW. Table 5 shows when the actuator is at the same zones of the forced source, the peak –peak bus frequency deviation is reduced at all the zones in the system.

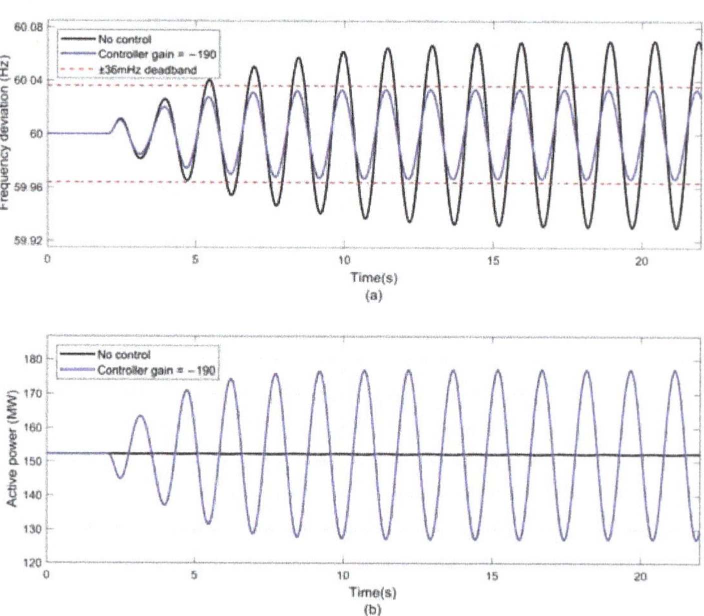

Figure 8. (**a**) Area 4 Zone 19 frequency deviation improvement. (**b**) Active power output of the actuator at Area 4 Zone 19 when the forced source is at Area 4 Zone 19.

Considering that the governors in the system have a 36 m Hz deadband before being activated, the forced oscillation damping controller is helpful to ensure the system response is limited in the deadband of governors, and reduces the forced oscillation impact on the entire system.

Figure 9 illustrates Area 4 Zone 19 frequency deviation improvement and the active power output of the actuator when the forced source at Area 4 Zone 20 before and after implementing the control. In this case, the peak–peak forced oscillation energy is increased to 160 MW in order for the peak–peak bus frequency deviation to exceed the deadband of the governor (±36 mHz). Area 4 Zone 19 had the largest peak–peak frequency deviation, and it can be reduced to 40.51 mHz. The peak–peak actuator output was 30.73 MW which is around ±10.1% of the actual active power of the actuator of 152.25 MW.

Table 5. The peak–peak bus frequency deviation improvement before and after implementing the forced oscillation damping controller.

Area-Zone	Peak–Peak Bus Frequency Deviation (mHz)		Bus Frequency Deviation Reduction (mHz)
	No Control	With Control	
Area 1 Zone 9	13.00	5.99	7.01
Area 2 Zone 11	18.41	8.53	9.88
Area 2 Zone 10	14.74	6.76	7.98
Area 3 Zone 27	13.44	6.19	7.25
Area 3 Zone 28	7.83	3.62	4.21
Area 4 Zone 19	140.09	66.60	73.49
Area 4 Zone 20	51.11	23.64	27.47
Area 4 Zone 21	101.95	47.10	54.85
Area 5 Zone 12	16.07	7.43	8.65
Area 5 Zone 13	12.95	5.99	6.95
Area 5 Zone 14	15.29	7.08	8.21
Area 5 Zone 15	12.01	5.55	6.46
Area 5 Zone 16	18.39	8.52	9.87
Area 5 Zone 17	17.52	8.11	9.41
Area 5 Zone 18	16.24	7.51	8.72
Area 6 Zone 22	13.46	6.24	7.22
Area 6 Zone 23	8.19	3.78	4.40
Area 6 Zone 24	25.01	11.57	13.44
Area 6 Zone 25	7.85	3.63	4.23
Area 6 Zone 26	13.53	6.27	7.26
Area 7 Zone 1	1.27	0.54	0.73
Area 7 Zone 2	14.98	6.94	8.03
Area 7 Zone 3	13.10	6.07	7.03
Area 7 Zone 4	15.47	7.17	8.30
Area 7 Zone 5	13.27	6.15	7.12
Area 7 Zone 6	11.20	5.19	6.01
Area 8 Zone 7	11.97	5.54	6.43
Area 8 Zone 8	19.28	8.93	10.35

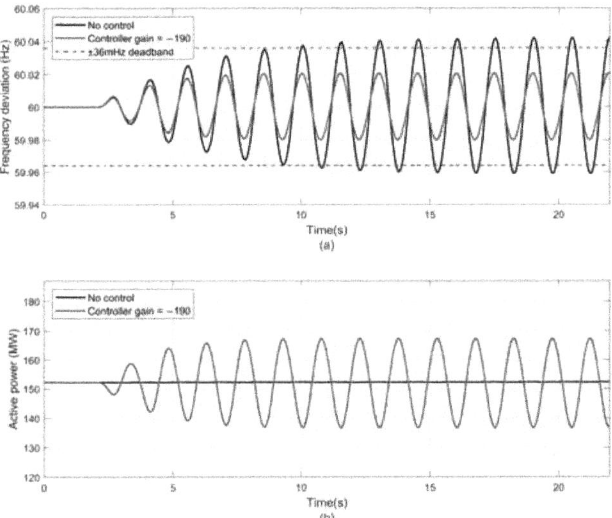

Figure 9. (a)Area 4 Zone 19 frequency deviation improvement. (**b**) Active power output of the actuator at Area4 Zone 19 when the forced source at Area 4 Zone 20.

4. Conclusions

A two-dimension scanning forced oscillation grid vulnerability analysis method in large-scale power grids was proposed in this paper. Based on the grid vulnerability analysis results, the effective location for forced oscillation damping controller can be selected. Active power modulation control through IBRs at the effective location with local measurement was also designed to mitigate forced oscillation under serious forced oscillation cases.

The key findings of this paper are summarized below:

- Based on frequency dimension scanning, forced oscillation at or close to the natural oscillation excites the largest peak–peak frequency deviation in the entire system.
- When the forced oscillation source is located at the areas that have high participation of the natural oscillations, the forced oscillation will be further magnified throughout the system and significantly impact the system stability.
- Active power modulation control through IBRs with local measurement was effective in reducing the frequency deviation that was caused by the forced oscillation. This will allow the operator to have sufficient time to locate and disconnect the forced source.
- The proposed forced oscillation controller can reduce all the zones' frequency deviation to a safety level in the synthetic Texas power grid when the actuator is close to the forced source. The largest peak–peak bus frequency deviation can be reduced by more than 50%, to 66.6 mHz (in the range of governor deadband ± 36 mH) with appropriate control gain.

Future work may include the investigation of the impact of forced oscillation through reactive power perturbation using the proposed two-dimension scanning method. The forced oscillation damping controller through reactive power modulation of IBRs will be designed to mitigate the forced oscillation.

Author Contributions: Conceptualization, L.Z., E.F., D.R. and Y.L.; methodology, K.A., Y.Z. and L.Z.; software, K.A., Y.Z. and W.Y.; validation, K.A. and Y.Z.; formal analysis, K.A. and Y.Z.; investigation, K.A. and Y.Z.; resources, K.A., Y.Z. and W.Y.; data curation, K.A., Y.Z. and W.Y.; writing—original draft preparation, K.A. and Y.Z.; writing—review and editing, L.Z., E.F., D.R. and Y.L.; visualization, K.A. and W.Y.; supervision, L.Z., E.F., D.R. and Y.L.; project administration, L.Z., E.F., D.R. and Y.L.; funding acquisition, E.F. All authors have read and agreed to the published version of the manuscript.

Funding: This work was primarily supported by Electric Power Research Institute (EPRI) and partly supported by National Science Foundation under the Award Number 1839684 and 1941101. This work also made use of Engineering Research Center Shared Facilities supported by the Engineering Research Center Program of the National Science Foundation and DOE under NSF Award Number EEC-1041877 and the CURENT Industry Partnership Program.

Data Availability Statement: Not applicable.

Conflicts of Interest: The authors declare no conflict of interest.

References

1. *Reliability Guideline: Forced Oscillation Monitoring and Mitigation*; The North American Electric Reliability Corporation (NERC): Atlanta, GA, USA, 2017. Available online: https://www.nerc.com/comm/PC_Reliability_Guidelines_DL/Reliability_Guideline_-_Forced_Oscillations_-_2017-07-31_-_FINAL.pdf (accessed on 27 December 2021).
2. Ghorbaniparvar, M. Survey on forced oscillations in power system. *J. Mod. Power Syst. Clean Energy* **2017**, *5*, 671–682. [CrossRef]
3. Power System Oscillatory Behaviors: Sources, Characteristics, and Analyses; North American SynchroPhasor Initiative (NASPI), USA, Technical Report, May 2017. Available online: https://www.pnnl.gov/main/publications/external/technical_reports/PNNL-26375.pdf (accessed on 27 December 2021).
4. Khan, M.A.; Pierre, J.W. Detection of Periodic Forced Oscillations in Power Systems Using Multitaper Approach. *IEEE Trans. Power Syst.* **2019**, *34*, 1086–1094. [CrossRef]
5. Zuhaib, M.; Rihan, M.; Saeed, M.T. A novel method for locating the source of sustained oscillation in power system using synchrophasors data. *Prot. Control Mod. Power Syst.* **2020**, *5*, 30. [CrossRef]
6. Ye, H.; Liu, Y.; Zhang, P.; Du, Z. Analysis and Detection of Forced Oscillation in Power System. *IEEE Trans. Power Syst.* **2017**, *32*, 1149–1160. [CrossRef]
7. Su, C.; Hu, W.; Chen, Z.; Hu, Y. Mitigation of power system oscillation caused by wind power fluctuation. *IET Renew. Power Gener.* **2013**, *7*, 639–651. [CrossRef]
8. Srinivasan, D.; Trung, L.T.; Singh, C. Bidding and cooperation strategies for electricity buyers in power markets. *IEEE Syst. J.* **2014**, *10*, 422–433. [CrossRef]
9. Xiao, D.; AlAshery, M.K.; Qiao, W. Optimal price-maker trading strategy of wind power producer using virtual bidding. *J. Mod. Power Syst. Clean Energy Early Access.* **2021**, 1–3. [CrossRef]

10. Surinkaew, T.; Emami, K.; Shah, R.; Islam, S.; Mithulananthan, N. Forced Oscillation in Power Systems with Converter Controlled-Based Resources–A Survey with Case Studies. *IEEE Access* **2021**, *9*, 150911–150924. [CrossRef]

11. Estevez, P.G.; Marchi, P.; Galarza, C.; Elizondo, M. Non-Stationary Power System Forced Oscillation Analysis Using Synchrosqueezing Transform. *IEEE Trans. Power Syst.* **2021**, *36*, 1583–1593. [CrossRef]

12. *Eastern Interconnection Oscillation Disturbance January 11, 2019 Forced Oscillation Event*; The North American Electric Reliability Corporation (NERC): Atlanta, GA, USA, 2019. Available online: https://www.nerc.com/pa/rrm/ea/Documents/January_11_Oscillation_Event_Report.pdf (accessed on 27 December 2021).

13. Rogers, G. *Power System Oscillations*; Kluwer Academic Publishers: New York, NY, USA, 2000.

14. Venkatraman, A.; Markovic, U.; Shchetinin, D.; Vrettos, E.; Aristidou, P.; Hug, G. Improving Dynamic Performance of Low-Inertia Systems Through Eigensensitivity Optimization. *IEEE Trans. Power Syst.* **2021**, *36*, 4075–4088. [CrossRef]

15. Tan, A.; Lin, X.; Sun, J.; Lyu, R.; Li, Z.; Peng, L.; Khalid, M.S. A Novel DFIG Damping Control for Power System with High Wind Power Penetration. *Energies* **2016**, *9*, 521. [CrossRef]

16. Sarmadi, S.A.N.; Venkatasubramanian, V. Inter-Area Resonance in Power Systems from Forced Oscillations. *IEEE Trans. Power Syst.* **2016**, *31*, 378–386. [CrossRef]

17. Chevalier, S.; Vorobev, P.; Turitsyn, K. A Bayesian Approach to Forced Oscillation Source Location Given Uncertain Generator Parameters. *IEEE Trans. Power Syst.* **2019**, *34*, 1641–1649. [CrossRef]

18. Feng, S.; Zheng, B.; Jiang, P.; Lei, J. A Two-Level Forced Oscillations Source Location Method Based on Phasor and Energy Analysis. *IEEE Access* **2018**, *6*, 44318–44327. [CrossRef]

19. Zhao, Y.; Zhu, L.; Xiao, H.; Liu, Y.; Farantatos, E.; Patel, M.; Darvishi, A.; Fardanesh, B. An Adaptive Wide-Area Damping Controller via FACTS for the New York State Grid Using a Measurement-Driven Model. In Proceedings of the 2019 IEEE Power & Energy Society General Meeting (PESGM), Atlanta, GA, USA, 8 August 2019; pp. 1–5. [CrossRef]

20. Zhu, Y.; Liu, C.; Wang, B.; Sun, K. Damping control for a target oscillation mode using battery energy storage. *J. Mod. Power Syst. Clean Energy* **2018**, *6*, 833–845. [CrossRef]

21. Saadatmand, M.; Gharehpetian, C.B.; Moghassemi, A.; Guerrero, J.M.; Siano, P.; Alhelou, H.H. Damping of Low-Frequency Oscillations in Power Systems by Large-Scale PV Farms: A Comprehensive Review of Control Methods. *IEEE Access* **2021**, *9*, 72183–72206. [CrossRef]

22. Feng, S.; Wu, X.; Jiang, P.; Xie, L.; Lei, J. Mitigation of Power System Forced Oscillations: An E-STATCOM Approach. *IEEE Access* **2018**, *6*, 31599–31608. [CrossRef]

23. Xu, Y.; Bai, W.; Zhao, S.; Zhang, J.; Zhao, Y. Mitigation of forced oscillations using VSC-HVDC supplementary damping control. *Electr. Power Syst. Res.* **2020**, *184*, 106333. [CrossRef]

24. Lin, Z.; Yu, W.; Jiang, Z.; Zhang, C.; Zhao, Y.; Dong, J.; Wang, W.; Liu, Y.; Farantatos, E.; Ramasubramanian, D.; et al. A Comprehensive Method to Mitigate Forced Oscillations in Large Interconnected Power Grids. *IEEE Access* **2021**, *9*, 22503–22515.

25. Surinkaew, T.; Shah, R.; Nadarajah, M.; Muyeen, S.M. Forced Oscillation Damping Controller for an Interconnected Power System. *IET Gener. Transm. Distrib.* **2020**, *14*, 339–347. [CrossRef]

26. ACTIVSg2000: 2000-Bus Synthetic Grid on Footprint of TEXAS. Available online: https://electricgrids.engr.tamu.edu/electric-grid-test-cases/activsg2000/ (accessed on 20 December 2021).

27. *Generic Solar Photovoltaic System Dynamic Model Specification*; Prepared by Western Electricity Coordinating Council (WECC) Renewable Energy Modeling Task Force: Salt Lake City, UT, USA, 2012. Available online: https://www.wecc.org/_layouts/15/WopiFrame.aspx?sourcedoc=/Reliability/WECC%20Solar%20PV%20Dynamic%20Model%20Specification%20-%20September%202012.pdf&action=default&DefaultItemOpen=1 (accessed on 21 February 2022).

28. ERCOT Load Zones. Available online: https://www.lonestartransmission.com/what-we-do/ercot-load-zones.html (accessed on 23 February 2022).

Article

Frequency Dynamics in Fully Non-Synchronous Electrical Grids: A Case Study of an Existing Island

Mariano G. Ippolito, Rossano Musca *, Eleonora Riva Sanseverino and Gaetano Zizzo *

Engineering Department, University of Palermo, 90128 Palermo, Italy;
marianogiuseppe.ippolito@unipa.it (M.G.I.); eleonora.rivasanseverino@unipa.it (E.R.S.)
* Correspondence: rossano.musca@unipa.it (R.M.); gaetano.zizzo@unipa.it (G.Z.)

Abstract: The operation of a power system with 100% converter-interfaced generation poses several questions and challenges regarding various aspects of the design and the control of the system. Existing literature on the integration of renewable energy sources in isolated systems mainly focuses on energy aspects or steady-state issues, and only a few studies examine the dynamic issues of autonomous networks operated with fully non-synchronous generation. A lack of research can be found in particular in the determination of the required amount of grid-forming power, the selection of the number and rated power of the units which should implement the grid-forming controls, and the relative locations of the grid-forming converters. The paper aims to address those research gaps starting from a theoretical point of view and then by examining the actual electrical network of an existing island as a case study. The results obtained from the investigations indicate specific observations and design opportunities, which are essential for securing the synchronization and the stability of the grid. Possible solutions for a fully non-synchronous operation of autonomous systems, in terms of dynamic characteristics and frequency stability, are presented and discussed.

Keywords: microgrids; frequency control; grid-forming; 100% converter-interfaced generation; virtual synchronous machine

Citation: Ippolito, M.G.; Musca, R.; Riva Sanseverino, E.; Zizzo, G. Frequency Dynamics in Fully Non-Synchronous Electrical Grids: A Case Study of an Existing Island. *Energies* **2022**, *15*, 2220. https://doi.org/10.3390/en15062220

Academic Editor: Ferdinanda Ponci

Received: 7 February 2022
Accepted: 17 March 2022
Published: 18 March 2022

Publisher's Note: MDPI stays neutral with regard to jurisdictional claims in published maps and institutional affiliations.

1. Introduction

The continuous integration of generated energy from renewable energy sources is significantly changing the operation paradigm of electrical power systems. This transition process raises questions and challenges regarding the dynamic characteristics of the system which expand beyond the classical definitions of stability, introducing the perspective of power systems fully operated by 100% converter-interfaced energy generation [1–8]. In [3], the viability of an all converter-interfaced generation system operating at constant frequency is investigated. The study considers grid-forming converters to enable the grid operation at constant frequency, also taking into account power sharing between the different sources in the system. The work in [4] presents a short-term voltage stability assessment of the Great Britain synchronous area, demonstrating how the application of the grid-forming control can improve the voltage stability of the system under 100% power electronics interfaced generation. The work in [5] investigates the dynamic behavior of the all-island Irish transmission system in the assumed scenario of 100% non-synchronous generation. The study considers a virtual synchronous generator control scheme for the grid-forming converters. The work in [6] formulates a small-signal model of isolated electrical networks with 100% converter-interfaced generation, identifying the critical factors for the determination of the converter-interfaced generation sources to be designated as grid-forming. In [7], the influence of the load dynamics on converter-dominated isolated power systems is examined, proposing control strategies to achieve operational scenarios with 100% penetration of converter-based generation. The work in [8] presents a dynamic analysis of an isolated power system where a single grid-forming unit is sized with a minimum required power capacity to ensure the stability of the system. The perspective of fully non-synchronous generation can be particularly relevant for autonomous power systems, such as microgrids and electrical networks of geographical islands. In this case, the availability of renewable energy sources and the integration of energy storage systems can accelerate the change, moving from a traditional synchronous nature of the system toward a complete non-synchronous generation scenario [9–16]. The operation of an electrical system involving exclusively generation sources interfaced through power electronics is expected to be characterized by specific constraints and requirements, pushing for a re-thinking of the overall characteristics of the system and new ways of designing them. In this context, the specific capabilities which can be offered by the power converters are recognized as key enablers for a successful transition to systems with 100% converter-interfaced generation. Most of power converters are integrated in the systems with conventional grid-following controls. In a system with only non-synchronous generation sources, the converters are requested to provide all those services and supportive actions which have been traditionally provided by synchronous machines. Grid-following controls can provide voltage and frequency support to the system, with the implementation of specific additional controls. However, for a system with 100% interfaced generation, the grid-following control concept cannot sustain a stable and self-sufficient operation of the system. The power converters are then required to have the specific capability of determining independently voltage and frequency, and thus providing an inherent synchronization mechanism to the system. Power converters with this particular capability belong to the category of grid-forming: the basic characteristics of these controls is in fact to form the voltage with controlled magnitude and angle, without the need for an already "formed" grid. Grid-forming controls offer a wide range of fundamental features and applications, and for that reason, this emerging control concept has attracted significant attention in recent years in both academia and industry.

In general, it is possible to notice that there are some works in the literature dealing with the topic of 100% converter-dominated power systems. These works usually refer to standard benchmarking systems, and only very few refer to existing electrical networks. There is therefore a need for more studies and analyses performed on actual power systems. Another research gap is the dimensioning of the proportion between grid-following and grid-forming in a fully non-synchronous electrical grid. In some works, only the interactions between grid-following and grid-forming are studied, but the determination of the required grid-forming power and the selection of the units which should implement the grid-forming controls are not addressed. In addition, the location of the grid-forming converters are typically fixed in the system, and the aspects related to the positions of the units with grid-forming capabilities are not fully investigated. In this context, the oscillatory stability and the damping characteristics of autonomous systems operated exclusively with converter-interfaced generation are also other aspects which require further and more specific investigations.

This work presents a study of the power–frequency dynamics for autonomous power systems with 100% interfaced generation. To ensure a stable and reliable operation of these systems, it is fundamental to assess which generation sources should be designated to be grid-forming and which specific characteristics must be realized by the converters controls. The design of a power system completely operated by interfaced generation has not been yet fully explored. The main contribution of the paper is in proposing and describing a possible approach for the design of autonomous power networks, with the definition of a few basic principles specifically related to the power–frequency dynamics. According to the given design principles, it is possible to determine how much grid-forming power is required in an autonomous system, and the total amount of inertial effect which should be synthetically provided by the interfaced generation sources. The next step of the design would be then to determine how many converters should implement grid-forming capabilities and where they should be located in the system. A possible approach to address these points is presented in the paper, and it is based on the identification of the most relevant factors which can have a critical impact on the power–frequency dynamics of the system. These aspects are identified by the formulation of the small-signal model of an autonomous system, including multiple grid-forming units interconnected with each other. The main aspects identified in the analysis are number and rated powers of the grid-forming converters, location and mutual electrical distances between them, and certainly relevant control parameters, such as the inertia time constant and the virtual impedance. The assessment of the additional design requirements is realized through the examination of the identified critical aspects for a specific case study: concepts and investigation factors derived in the preliminary analysis are applied to the existing power network of a Mediterranean island. The system of the island is considered in different configurations, assuming a future scenario where the load demand is completely supplied by converter-interfaced generation. The system is analyzed with a positive sequence RMS simulation model, focusing on the interactions between the power–frequency controls of the grid-forming converters. The capability of realizing a quick and successful synchronization between the oscillating grid-forming converters is recognized in the analysis as the essential aspect for the power–frequency dynamics of the system, suggesting the potential risk of instability and indicating at the same time the opportunity for specific system design and configurations.

2. Grid-Forming Models for Phasor RMS Simulations

The control schemes which belong to the category of grid-forming are various. Different control schemes and variations have been in fact studied and proposed in the literature [17–24]. Two of the most common grid-forming controls are the virtual synchronous machine (VSM) and the droop-based control. The VSM can be found in a variety of control schemes and variations [25–29], and it is based on the emulation of the swing dynamics of the synchronous machines. The droop-based control was originally designed for converter-dominated microgrids and it provides grid-forming functionalities through droop regulators for voltage and frequency [30–32]. The grid-forming control considered in this work is represented in Figure 1. The angle control can implement either a virtual synchronous machine scheme (Figure 2a) or a power-synchronization control (Figure 2b). In the diagram, u_c and δ_c are, respectively, magnitude and angle of the converter voltage; u_{ref} and p_{ref} are the references for terminal voltage and active power; u and p are the voltage and the power measured at the converter output; ω_n is the rated angular frequency; H is the time constant of the integral action responsible for the realization of the inertial effect, while R is the gain applied on the frequency deviation realizing the droop control for primary frequency reserve; K_u and T_u are, respectively, gain and time constant of the given voltage control; T_{pf} and T_{uf} are the time constants of the low-pass filters applied on the measurements.

The considered grid-forming control consists of two main parts, the angle control loop and the AC voltage controller. The angle control provides the reference angle δ_c for the internal frame transformations of the converter, and it realizes the fundamental capabilities of the grid-forming control. For the VSM, the power-angle control implements the emulation of the swing dynamics of synchronous machines, and it includes a droop control on the frequency error. The integrator time constant realizes synthetically the inertial effect physically provided by synchronous machines. The coefficient of the droop control is instead equivalent to the frequency droop R of the turbine/governor connected to synchronous machines. In the droop-based control, the power-angle control implements a simple droop control on the deviation of the measured active power from the reference value, realizing the intrinsic synchronization mechanism without the provision of any inertial effect. It can be then easily demonstrated that the two considered formulations of virtual synchronous machine and power-synchronization control become equivalent when considering $H = 0$ and $K_i = \omega_n R$. The AC voltage controller provides the reference magnitude u_c of the voltage at the converter output for the internal generation of the modulation index, and it implements a simple droop control on the deviation of the voltage measured at the converter terminal from the reference value. Since the simulation model considered in this work is a dynamic model formulated in the phasor RMS domain, the output variables u_c and δ_c respectively determined by the two main control loops directly provide the real and imaginary parts of the controlled voltage expressed in the common *dq0* rotating frame of the system. The converter control also contains an intrinsic dependence on the local measurement of electrical quantities such as voltage, current, and power. For that, low-pass filters (LPF) are often added to filter out the noise, to avoid attaching to harmonics, and to limit signal jumps. The measurements of voltage and active power are included in the mathematical model passing the measured electrical quantity through a first-order LPF with a given cut-off frequency. From the point of view of model implementation, the equations describing the mathematical model of the overall power system are typically referred to as a common per unit system with base power $S_b = 100$ MVA. The equations describing the model of the interfaced generation source are instead usually expressed per unit of the rated power S_r of the converter. It is, therefore, necessary to have a base change as conversion between the two per unit systems.

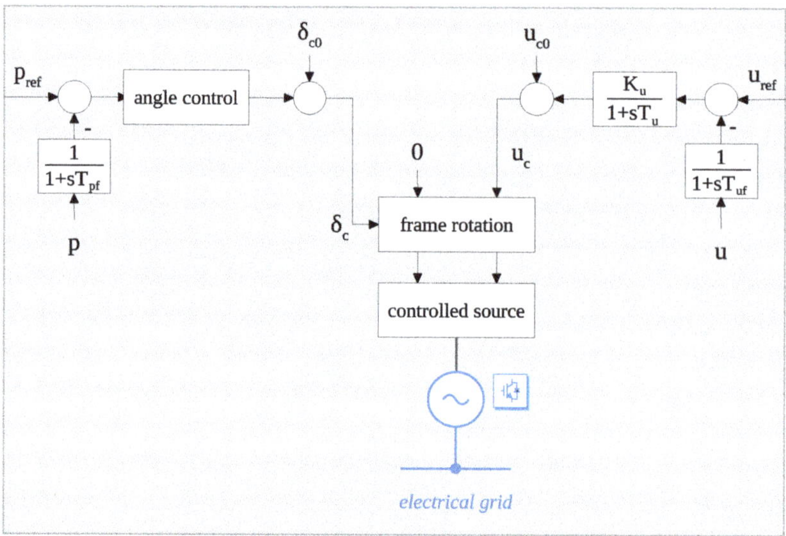

Figure 1. Block diagram of the grid-forming control.

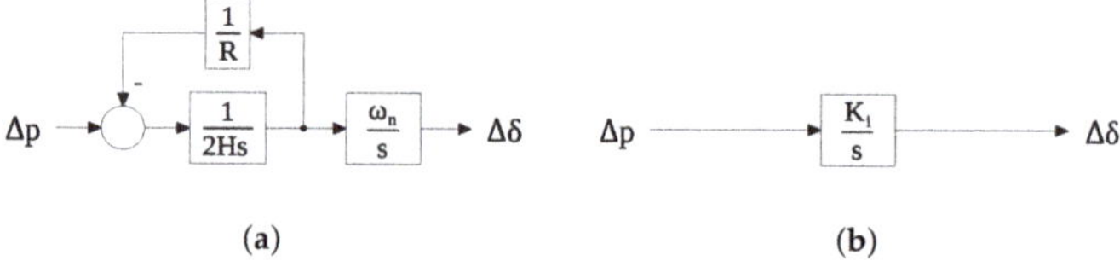

(a)   (b)

Figure 2. Detail of the angle control block: (**a**) Virtual synchronous machine; (**b**) Power-synchronization control.

The considered grid-forming model does not represent the DC side of the converter and it does not include fast inner control loops for current and voltage regulation. These assumptions were considered in the work since they are not expected to have a significant impact on the frequency dynamics and generally on the slow transients of the system, which are the main focus of the presented investigations.

The phasor RMS model used in this work is developed in the power systems analysis tool NEPLAN [33]. The validation of the model to the specific purposes of the study is made comparing the phasor RMS implementation with a detailed EMT dynamic model developed in the specialized toolbox Simscape of MATLAB/Simulink [34]. The results are reported for comparison in Figure 3. It can be seen that a difference exists between the RMS and the EMT implementations. However, the results of the simulation in the two time domains indicate a substantial match for the considered events and time scales. For investigations of frequency dynamics and slow transients, the phasor RMS models of grid-forming converters appear therefore to be appropriate.

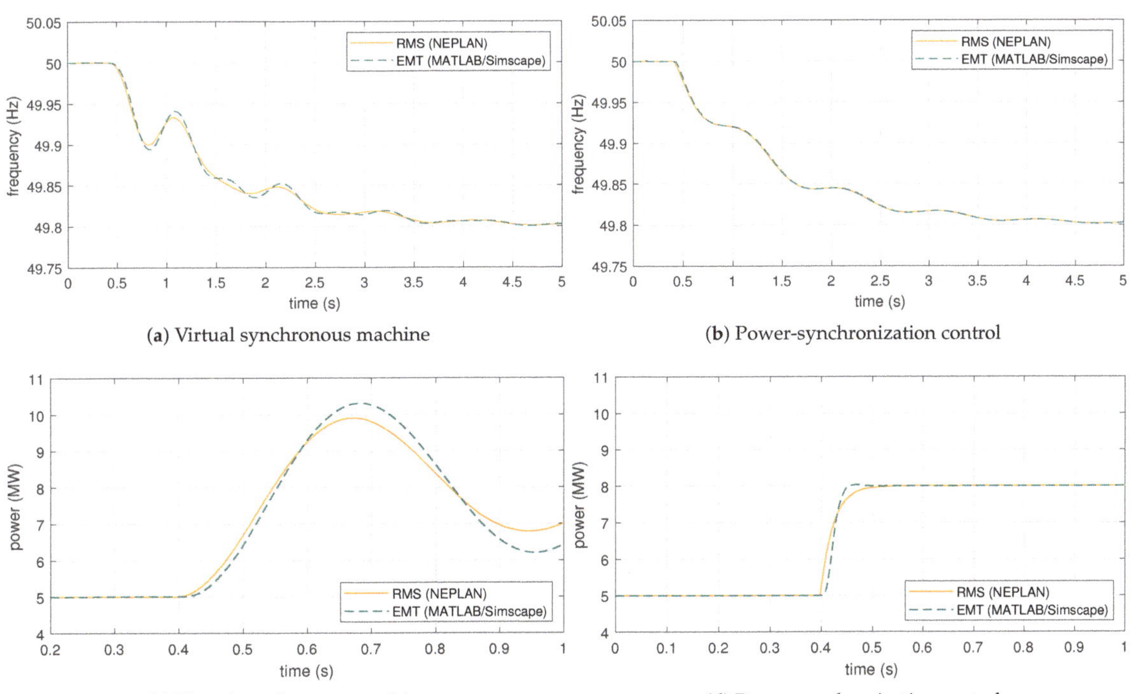

(**a**) Virtual synchronous machine

(**b**) Power-synchronization control

(**c**) Virtual synchronous machine

(**d**) Power-synchronization control

Figure 3. Comparison and validation of grid-forming simulation models: (**a**,**b**) frequency transients caused by the application of a power imbalance; (**c**,**d**) step in the active power reference of the grid-forming controls.

3. Frequency Dynamics of a Fully Non-Synchronous Autonomous System

The main purpose of this section is the identification of the factors which can have a critical impact on the frequency dynamics of an autonomous power system with 100% converter-interfaced generation. These factors can be identified referring to a small-signal model of the system and considering the representative case of coupled oscillators. The general case of multiple grid-forming converters interconnected between each other is reduced to a simple equivalent representation, focusing the analytical approach for the derivation of specific considerations about the frequency dynamics of a fully non-synchronous autonomous system operated with grid-forming technologies.

3.1. Identification and Study of Critical Factors

The general case of an autonomous system with multiple oscillating grid-forming units can be studied as a system composed by pairs of coupled oscillators (Figure 4a), and then focusing the attention on a generic pair *i-j* of oscillating sources (Figure 4b).

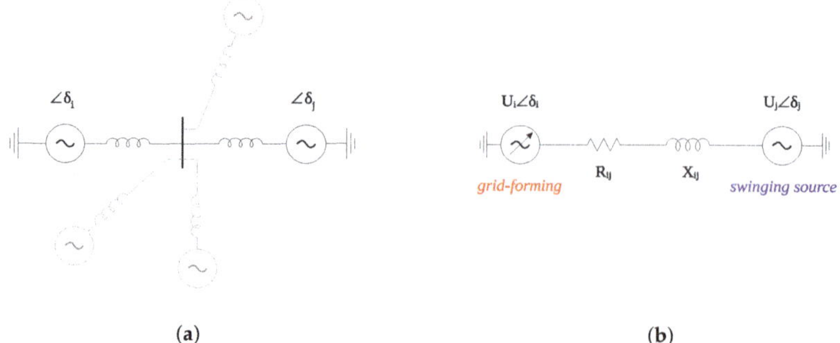

(**a**)

(**b**)

Figure 4. Autonomous systems with multiple grid-forming units: (**a**) Outline of a generic multiple oscillators electric system; (**b**) Pair of coupled oscillators belonging to the multiple oscillators system.

The active power exchanged between i and j can be expressed as:

$$p_{ij} = U_i U_j Y_{ij} \cos(\delta - \theta_{ij}) \tag{1}$$

where the difference between the two phase angles δ_i and δ_j is indicated as δ for compact notation. In (1), θ_{ij} is the angle of the complex impedance $\dot{Z}_{ij}$ representing the network interconnection between the two oscillators:

$$\dot{Z}_{ij} = R_{ij} + jX_{ij} = Z_{ij}e^{\theta_{ij}} \tag{2}$$

This representation is general and it takes into account the resistive-inductive nature of the grid interconnections which typically characterize small autonomous electric systems. It is also important to note that the total impedance $\dot{Z}_{ij}$ between the coupled oscillators i and j can include the impedances of the physical components interconnecting the two elements and the impedances which can be virtually realized by the corresponding control systems. For the study of small-signal deviations, the expression of the active power in (1) can be linearized around an initial steady-state operating point characterized by δ_0 as follows:

$$\Delta p_{ij} = \left. \frac{\partial p_{ij}}{\partial \delta} \right|_{\delta_0} \Delta\delta = U_i U_j Y_{ij} \left(-\sin\delta_0 \cos\theta_{ij} + \cos\delta_0 \sin\theta_{ij} \right) \Delta\delta \tag{3}$$

and then:

$$\Delta p_{ij} = U_i U_j Y_{ij} \sin(\theta_{ij} - \delta_0)\Delta\delta = K_s\Delta\delta \tag{4}$$

The factor K_s can be referred to as the synchronizing coefficient between the oscillators i and j, and it can be regarded as the expression of the elastic restoring torques between pairs of coupled oscillating elements which enforce the synchronization [35].

The mathematical representation of the two oscillators in Figure 4b can be formulated with different degrees of complexity and detail. Since the purpose of the work is the investigation of the frequency dynamics of the system, an analytical approach can be focused on the equations governing the power-angle control of the considered sources. Proceeding in this way, the system in Figure 4b can be described by the following mathematical system of differential-algebraic equations:

$$\frac{d\Delta\omega_j}{dt} = \frac{1}{2H_j}\left(\Delta p_{\text{ref}j} - \Delta p_{ji}\frac{S_b}{S_{rj}} - \frac{1}{R_j}\Delta\omega_j\right) \tag{5}$$

$$\frac{d\Delta\delta_j}{dt} = \omega_n\Delta\omega_j \tag{6}$$

$$\frac{d\Delta\omega_i}{dt} = \frac{1}{2H_i}\left(\Delta p_{\text{ref}i} - \Delta p_{ij}^m\frac{S_b}{S_{ri}} - \frac{1}{R_i}\Delta\omega_i\right) \tag{7}$$

$$\frac{d\Delta\delta_i}{dt} = \omega_n\Delta\omega_i \tag{8}$$

$$\frac{d\Delta p_{ij}^m}{dt} = \frac{1}{T_{\text{pf}}}\left(\Delta p_{ij} - \Delta p_{ij}^m\right) \tag{9}$$

$$\Delta p_{ij} = K_s\left(\Delta\delta_i - \Delta\delta_j\right) \tag{10}$$

$$\Delta p_{ji} = K_s\left(\Delta\delta_j - \Delta\delta_i\right) \tag{11}$$

where the active power exchanges Δp_{ij} and the factor K_s are given by (4). In the mathematical representation of the oscillator i, Equation (9) has been additionally considered. This equation represents the low-pass filter of the active power measurement included in the grid-forming control (Figure 1). The mathematical model also considers the required conversion between the common per unit system with base power S_b = 100 MVA and the local per unit system of the element model with rated power S_r as base power. For the study of the system given in Figure 4b, it is possible in fact to note that the oscillator i represents the specific grid-forming converter, while the oscillator j describes a generic swinging element characterized by a first-order dynamic. For the sake of a more straightforward notation, the quantities related to the generic swinging element will be denoted with the subscript g, while for the grid-forming converter the subscript c will be used. It is then possible to express the small-signal model of the system (5)–(11) in a state-space formulation where the matrix A and the state vector Δx are given by:

$$A = \begin{bmatrix} -\frac{1}{2H_g R_g} & -\frac{K_s}{2H_g}\frac{S_b}{S_g} & 0 & \frac{K_s}{2H_g}\frac{S_b}{S_g} & 0 \\ \omega_n & 0 & 0 & 0 & 0 \\ 0 & 0 & -\frac{1}{2H_c R_c} & 0 & -\frac{1}{2H_c} \\ 0 & 0 & \omega_n & 0 & 0 \\ 0 & -\frac{K_s}{T_{\text{pf}}}\frac{S_b}{S_c} & 0 & \frac{K_s}{T_{\text{pf}}}\frac{S_b}{S_c} & -\frac{1}{T_{\text{pf}}} \end{bmatrix} \qquad \Delta x = \begin{bmatrix} \Delta\omega_g \\ \Delta\delta_g \\ \Delta\omega_c \\ \Delta\delta_c \\ \Delta p^m \end{bmatrix} \tag{12}$$

The first two rows of the matrix A correspond to the dynamic representation of the generic swinging element, while the remaining rows correspond to the small-signal model of the angle control of the grid-forming element (Figure 4b). From the coefficients of the matrix, it can be immediately observed that they depend on:

- rated power of the grid-forming converter (S_c);
- inertial effect of the grid-forming angle control (H_c);
- frequency droop of the grid-forming angle control (R_c);
- time constant of the low-pass filter on the active power measurement (T_{pf});
- the characteristics of the interconnections (K_s);
- strength of the network (S_g).

The computation of the eigenvalues of the matrix A can be performed to provide a deeper insight on the possible impact of the factors listed before. In particular, the calculation of the eigenvalues can be iteratively repeated for a parametric analysis with a sweep of the identified relevant parameters. For the analysis, the parameter S_g characterizing the strength of the network is fixed to three values: infinite bus, 1000 MVA, and 10 MVA. The last two values can cover two possible conditions of autonomous electrical systems, high strength (relatively large network, several oscillating sources) and low strength (small network, limited number of oscillating sources). The infinite bus case is considered only for the sake of comparison. All the other parameters are related to the grid-forming converter: for them, a sweep in a typical range of values is considered. The results of the computations are summarized in Figure 5.

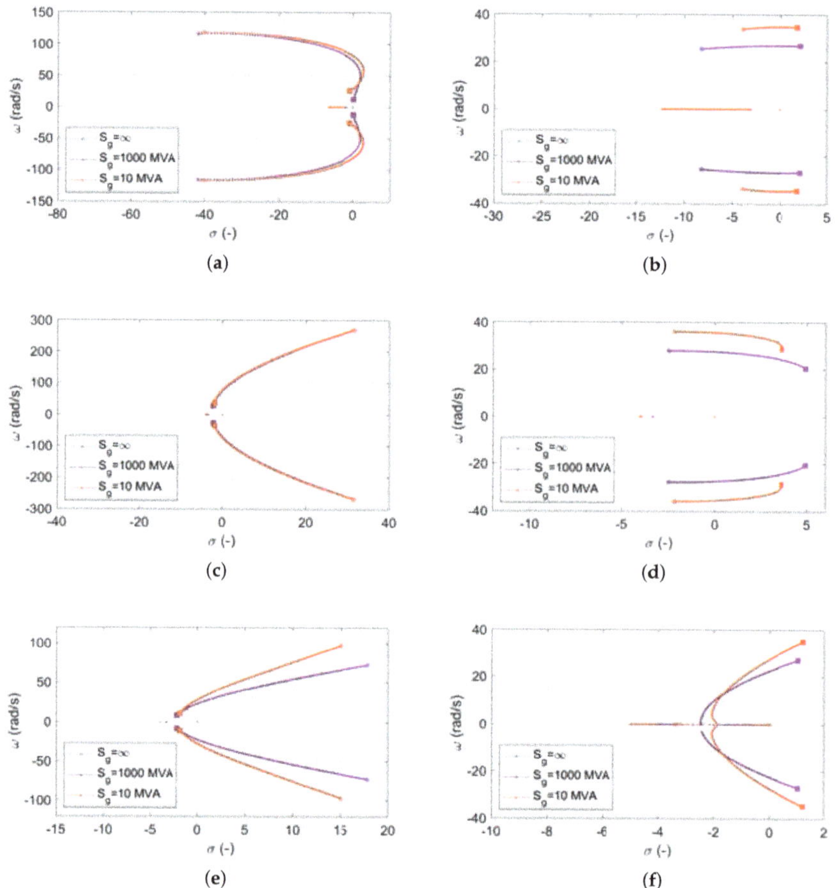

Figure 5. Plot of system eigenvalues for different parametric sweeps: starting from stars, ending to squares. (**a**) Inertia H_c (0.01–10 s); (**b**) Droop R_c (0.01–0.1 pu); (**c**) Converter rated power S_c (0.1–10 MVA); (**d**) Measurement time constant T_{pf} (0–0.05 s); (**e**) Impedance magnitude (0.1–10 pu); (**f**) Impedance angle (0–90 degrees).

From the results, a first general observation is that the strength of the grid S_g has the expected impact on the impedance magnitude and angle, and also on the sensitivities of the grid-forming droop and of the measurement LPF time constant. The sensitivities of inertial effect and converter rated power are instead marginally affected by the value of S_g. From the plot of Figure 5a, it is possible to notice that an increase of the inertial effect has a negative impact on the stability of the system. It can be however observed that a further increase of the inertia H_c of the grid-forming eliminates the instability phenomenon, as is indicated by the eigenvalues of the system coming back in the left-hand side of the plane.

From the plots of Figure 5b,d, it can be seen that increasing the grid-forming droop R_c and the time constant of the active power measurement filter T_{pf} can definitely determine the instability of the system. From the plot of Figure 5c, it can be immediately observed that small grid-forming converters can make the system unstable: since the grid-forming must act as a reference for the system, a converter with a small rated power might not be strong enough to contribute positively to the frequency dynamics. From the plot of Figure 5e, it is possible to notice that an increase of the impedance modulus has a positive impact on the stability of the considered oscillators system. This result suggests that a grid-forming converter close to another oscillating source might experience difficulties for successfully operating in stable conditions. The plot of Figure 5f examines the sensitivity of the resistive-inductive nature of the network interconnections, sweeping from a purely resistive grid (impedance angle 0°) to a purely inductive grid (impedance angle 90°). It can be observed that the presence of a resistive component has a positive impact on the stability of the system, contributing to a relevant increase of the damping.

In summary, the small-signal model of the controls governing the frequency dynamics of a multiple oscillators system and the parametric analysis of the selected parameters indicate a specific impact on the dynamics of fully non-synchronous autonomous networks, suggesting a potential instability of the system under particular conditions. Certainly, combinations and variations of the identified critical factors can determine specific results and further considerations. This aspect will be addressed in the analysis of the case study discussed in Section 5.

The root cause of the observed instability can be ultimately identified in the time constant T_{pf} of the LPF applied on the active power measurement of the converter. This is a known issue with LPF and active power control in grid-forming converters [36–39]. For a better illustration of this issue, the simple case of a single oscillator connected to an infinite bus can be used as illustrative example (Figure 6).

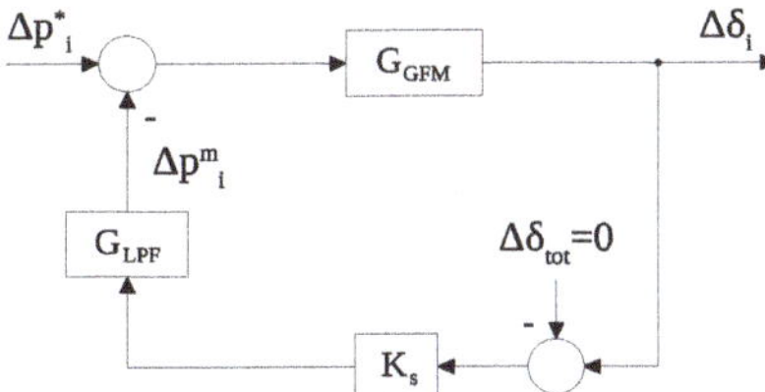

Figure 6. Small-signal block diagram for the study of a single oscillator dynamics.

In the diagram, the feedback pathway includes the representation of the network interconnection with the coefficient K_s and the effects of the LPF on the active power measurement. The components of synchronizing and damping coefficients due to network interconnection and low-pass filter can be isolated employing a similar approach to the one followed in the case of synchronous machines for the change in the electrical torque due to field flux variations caused by angle changes [40]. The open-loop transfer function between Δp^m and $\Delta\delta$ in the feedback pathway of the model can be expressed by:

$$\frac{\Delta p^m}{\Delta\delta} = K_s\, G_{LPF} = K_s \frac{\omega_f}{\omega_f + s} \tag{13}$$

where $\omega_f = 1/T_{pf}$.

The transfer function in (13) can be regarded as the variation of Δp^m caused by a change in the angle $\Delta\delta$. The expression in (13) can be rearranged multiplying numerator and denominator by $\omega_f - s$, resulting in:

$$\Delta p^m = K_s \frac{\omega_f^2}{\omega_f^2 - s^2}\Delta\delta - K_s \frac{\omega_f}{\omega_f^2 - s^2}s\Delta\delta \tag{14}$$

Recalling that the derivative of the angle is the frequency through $s\Delta\delta = \omega_{\mathrm{n}}\Delta\omega$ and substituting $s = j\omega$ in (14), the variation of Δp^{m} caused by a change in the frequency $\Delta\omega$ can be expressed by:

$$\frac{\Delta p^{\mathrm{m}}}{\Delta\omega} = -\omega_{\mathrm{n}}K_{\mathrm{s}}\frac{\omega_{\mathrm{f}}}{\omega_{\mathrm{f}}^2 + \omega^2} \qquad (15)$$

which represents the component of the damping coefficient due to network interconnection and low-pass filter. The term given in (15) immediately indicates that the contribution provided by the feedback component is negative, causing therefore a reduction of the damping provided by the whole control. The reduction is clearly due to the presence of the LPF on the feedback pathway of the system, and it is affected by the cut-off frequency ω_{f} of the filter. The magnitude of the reduction is also determined by the given oscillation frequency ω and by the coefficient K_{s}. Recalling the dependence of the synchronizing coefficients, it can be observed that small rated powers and close electrical distances would result in severe damping reduction, possibly threatening the stability of the system. These two factors have, therefore, an additional specific impact on the power–frequency dynamics, referring in particular to the damping characteristics of the system.

3.2. Discussion about Stability and Damping

The factors identified in the previous sections are summarized in Table 1. The number of oscillating grid-forming units connected to the system, their mutual distances and their rated powers are main aspects which can affect the dynamic interactions during a frequency transient. The characteristics of the grid-forming controls governed by specific parameters such as the inertia time constant and the virtual impedances are also factors to consider, since they certainly affect the power–frequency dynamics of the system. As observed in the previous analysis, all these factors can have a critical impact on the system stability. The aspects related to damping and oscillatory stability of power systems dominated by power converters are addressed in several works [36–39,41,42]. In [41], it is remarked that grid-forming controls should be properly implemented, as they could result in a poorly damped closed-loop system. In [36], the effects related to LPF within the active power loop of the converter control are analytically investigated and recognized as a potential source of a critical lack of damping, leading to instability and loss of synchronization. Other works and research propose specific solutions to this problem [37–39]. More generally, the issue of damping provision by power converters is widely discussed in several papers and technical reports [42–51]. While this article focuses on the frequency stability of an autonomous system operated with multiple grid-forming converters, other various aspects related to the dynamic interactions between grid-forming converters are addressed from different points of view and hierarchical control levels [52–62].

Table 1. Investigation factors with critical impact.

Key Factor	Description
Number and size	Share of grid-forming converters and rated powers
Location	Electrical distances and impedances
Control parameters	Inertial effect and virtual impedance

4. Design Principles for Fully Non-Synchronous Autonomous Systems

4.1. Reference Hypothesis

The design of an autonomous system completely operated with interfaced generation is an aspect which has been explored only in very few cases, often in the context of pilot projects and experimental test cases. In general, the design objectives for an autonomous system with 100% converter-interfaced generation can be various and are dependent on the given characteristics of the system. In this work, a general design approach is proposed. The method is a three-step procedure and it is focused on fundamental aspects of the frequency dynamics, targeting both steady-state and transient frequency performance as design principles. As an initial step, it is fundamental to define a specific design hypothesis for the system under planning. The purpose of this design objective is the determination of a reference power imbalance Δp^*, which will be used in the next steps of the design process. The definition of a reference incident might be made according to several methods and criteria. For instance, a possible design hypothesis for Δp^* could be the loss of the biggest generation unit in the system, adopting an N-1 criterion. Another possibility could be to assign Δp^* in the range of 3–5% of the maximum system load: a power imbalance in that range is in fact already considered as a large disturbance for the system [63]. More advanced techniques might be based on probabilistic methods, including considerations about the availability of the generation sources interfaced by the converters. In this work, the approach based on the designation of the reference power imbalance as percentage of the maximum load of the system will be considered, fixing Δp^* to 5% of the maximum load.

4.2. Total Required Grid-Forming

One first point which should be surely addressed is how much grid-forming power is required in an autonomous system. This point can be subjected to several different constraints and addressed from several points of view. For instance, some works indicate that there is a minimum amount of non-synchronous generation sources with grid-forming capabilities required to guarantee the stability of the system [5,8,64]. To keep the design process simple, the determination of the total required grid-forming is exclusively derived for the realization of the given steady-state conditions on frequency dynamics, leaving the consideration related to transient performance to the other design step. The design principle will be then focused, in this case, on the realization of proper active power sharing between the generation sources. This will also give an idea of the possible proportion between grid-forming and grid-following units in the system. For the determination of the total rated power, which must summed up by all the converters with grid-forming capabilities, it is necessary to fix two design targets: the frequency deviation at steady-state Δf_s^* which should correspond to the reference power imbalance Δp^*, and the frequency droop gain R which will govern the power sharing of the grid-forming sources. The target value of the frequency deviation at steady-state Δf_s^* can be fixed according to the grid code specifications existing for the autonomous system under design. The value of the frequency droop R is instead typically assigned in the range from 1 to 10% of the rated power of the element: if the per unit frequency droop will be assumed equal for all the generation sources participating in the primary reserve with the grid-forming control, a given power imbalance in the system will be thus shared between them proportionally. For the design, it is assumed that the other interfaced generation sources are operated as grid-following without implementing any frequency droop control. This is a conservative assumption, since in many cases grid-following units can participate in the primary reserve while operating in frequency sensitive mode. For an autonomous system where the generation sources are limited, it is, however, reasonable to think that some units will not be always available for participation in the primary frequency control. For fixed values of the reference power imbalance Δp^* and of the target frequency deviation at steady-state Δf_s^*, the total required power S_{tot}^* can be expressed as:

$$S_{\text{tot}}^* = \frac{f_n R \Delta p^*}{\Delta f_s^*} \tag{16}$$

Using (16), it is possible to determine the minimum total required amount of grid-forming converters to match the steady-state performance design. The number of grid-forming converters and the subdivision of the total required power between them are aspects which can be determined according to additional constraints and further considerations. For that, some insights will be provided by the investigations of the critical factors for frequency dynamics described in Section 5.

4.3. Total Required Inertia

Another point which should also be considered in the design concerns the transient performance of the power–frequency dynamics of the system. The main aspects are the total required amount of inertial effect, the maximum/minimum allowed instantaneous frequency deviation and the maximum allowed absolute frequency rate. These aspects are interrelated, since the inertial effect taking place in the very first instants of the transient affect both the frequency rate and the instantaneous deviation of the frequency. The design principle can be then focused on the realization of a predefined inertial effect, synthetically provided by the interfaced generation sources which will be operated as grid-forming. For the design, it is assumed that the other interfaced generation sources operated as grid-following will not implement any synthetic inertia control. Like the case of primary reserve, this assumption is conservative because grid-following units can generally provide synthetic inertia through specific additional controls. For the determination of the total amount of inertial effect which must be provided by the converters with grid-forming capabilities, the approach can be based on the concepts of the center of inertia and aggregated swing dynamics. Contrarily to large interconnected power systems, in the case of autonomous networks, it is reasonable to assume that the dynamics of all the oscillating elements can be concentrated at the center of inertia, with a unique coherent frequency for the whole system. Two alternative design targets can be then fixed: the total required inertia can be in fact determined either considering the maximum allowed frequency rate or the maximum/minimum allowed instantaneous frequency deviation. While for large interconnected systems both targets would likely require a complex and elaborated design procedure, for small autonomous power systems the design can be easily kept simple in both cases: if the target is either the limitation of the frequency rate or the containment of the instantaneous frequency deviation, the possibility of assuming the dynamics of the system concentrated at the center of inertia makes both design targets viable. In this work, the total required amount of inertial effect will be determined assuming the maximum frequency rate as design target. The alternative of considering the instantaneous frequency deviation is discarded: since the frequency dynamics of a multiple grid-forming autonomous system can be essentially characterized by a first-order dynamics, no significant differences between instantaneous and steady-state frequency deviations are expected, and therefore the inertia would only have a limited effect on this design target. For fixed values of the reference power imbalance Δp^* and of the target frequency rate $\rho^*_{\max}$, the total required inertial effect can be expressed in terms of total inertia constant H^*_{tot} as:

$$H^*_{tot} = \frac{f_n \Delta p^*}{2\rho^*_{\max} S^*_{tot}} \tag{17}$$

Using (17), it is possible to determine the minimum required amount of inertial effect which should be provided by grid-forming converters to meet the specific transient performance design. It is worth noting that the inertial effect synthetically provided by grid-forming converters has also a particular impact on the synchronization mechanism between the oscillating grid-forming generation sources, and consequently on the oscillatory characteristics of the system. For that, some additional insights will be provided by the investigations of the critical factors for frequency dynamics described in Section 5.

5. Case Study: Application to the Existing Power System of a Mediterranean Island

The system considered as case study is the medium voltage network of Pantelleria, an island in the Mediterranean sea. The total demand of the island strongly depends on the period of the year, and it varies from a minimum of around 4 MW to a maximum of around 8 MW. The load demand is currently supplied by a thermal generation power plant, composed by diesel generators and located close to the urban center. The island is therefore dependent on external sources of energy. The power is delivered to the loads through four main feeders, all departing from the thermal power plant. The total lengths of the medium voltage feeders span from 14 km for the longest lines to 4 km for the shortest. The lines are mainly cables, with resistance in the range 0.33–0.47 Ω/km, reactance 0.23–0.31 Ω/km and capacitance 0.1–0.24 μF/km. There are only a few photovoltaic plants installed in the system, mainly in the urban center: the total active power is around 300 kW, so the photovoltaic plants represent just a minor generation share over the total load of the island. A schematic outline of the system is shown in Figure 7.

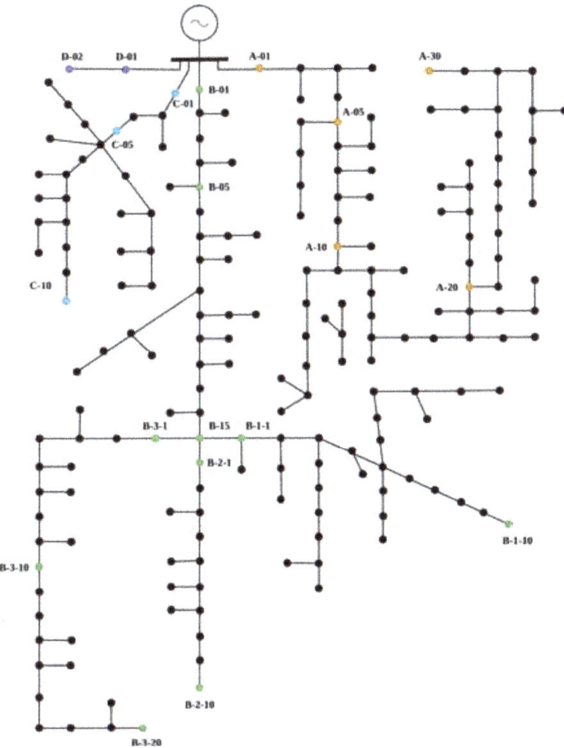

Figure 7. Current network scheme of Pantelleria island, with selected nodes in the 4 feeders.

The autonomous system of the island is then imagined in future operating conditions, characterized by the complete phase out of all synchronous machines of the diesel power plant and the corresponding integration of the necessary amount of renewable power generation to guarantee the load supply. In the considered scenario, the electric system of the island would be therefore powered by 100% converter-interfaced generation. This scenario is then formulated in several different potential combinations and configurations. The interfaced generation sources integrated in the system of the island will vary in number and rated power, they will be assumed to be installed at different locations across the system, and they will either implement a grid-following or a grid-forming control, considering also different tuning of the control parameters. All the particular combinations and configurations will be adapted depending on the aspects to be investigated. For an efficient management of all the various system configurations and simulation cases, the handling of the calculations is automated with a stand-alone application written in C#. The utility accesses externally the functionalities of the powers systems analysis software NEPLAN [33] by calling the available APIs of the software. The application makes then possible to manage all the elements and the models of the system in a custom-defined and automatic way, run consecutive simulations programmatically and post-process the results for statistics. For the simulations, the power system of the island is represented with a positive sequence RMS dynamic model. The models of the converter-interfaced generation sources are implemented according to the control structure presented in Section 2, and they are developed as user-written dynamic models with SYMDEF (*SYMbolic DEFinition*), proprietary modeling language of NEPLAN. In all the considered configurations, the power system is simulated in perturbed conditions, characterized by an uniform disturbance identically applied to all the loads of the network. For that, a step change of $\Delta p_d = 1\%$ in the initial active power of all the loads is applied in the time-domain simulations. The value of the load step is selected as a relatively small perturbation for the system, considering that a load change in the range 3–5% is already regarded as a large disturbance [63]. Moreover, the approach of considering a uniform distributed disturbance for all the loads allows to investigate the effects of different percentages and locations of non-synchronous generation sources with inertial and damping capabilities on the system dynamics [65]. For the simulation model, the parameters have the following default values: $H = 2$ s, $R = 0.05$ pu, $K_u = 1$ pu, $T_u = 0.001$ s, $X_v = 0.1$ pu, $\omega_f = 100$ rad/s.

5.1. System Design for Frequency Dynamics

The design of the power system of Pantelleria island for a future scenario with 100% converter-interfaced generation can be made according to the methods introduced in Section 4. The maximum load of the system is $P_{max} = 8.1$ MW, observed in the summer season due to the influx of tourists. The reference power imbalance Δp^* can be then calculated as 5% of the maximum load, resulting in $\Delta p^* = 0.05 \cdot P_{max} = 400$ kW. For the frequency control, it is decided to fix the frequency droop to 5% for all the generation sources participating in the primary reserve. The aggregated frequency droop R is thus equal to $R = 5/100 = 0.05$ pu, in per unit of the given grid-forming converter rated power. The design targets are the steady-state frequency deviation, fixed to $\Delta f_s^* = 50 - 49 = 1$ Hz, and the maximum absolute value of the frequency rate, fixed to $\rho_{max} = 5$ Hz/s. Applying (16) with all the previous values for the design targets, it is possible to calculate the total amount of rated power S_{tot}^* which should implement grid-forming capabilities as:

$$S_{tot}^* = \frac{f_n R \Delta p^*}{\Delta f_s^*} = \frac{50 \times 0.05 \times 0.4}{1} = 1\,\text{MVA} \tag{18}$$

Applying (17), it is possible to calculate the total amount of inertial effects in terms of inertia constant H_{tot}^* which should be guaranteed in the system approximately as:

$$H_{tot}^* = \frac{f_n \Delta p^*}{2 \rho_{max}^* S_{tot}^*} = \frac{50 \times 0.4}{2 \times 5 \times 1} = 2\,\text{s} \tag{19}$$

The calculated value of S_{tot}^* represents the minimum total amount of grid-forming converters required to guarantee the steady-state frequency performance, while the calculated value represents the total inertial effect which should be provided by the grid-forming converters to meet the specific transient performance assumed as design target.

5.2. Impact of Size

The impact of the size of grid-forming converters is investigated with the following system configuration: only two interfaced generation sources are operated as grid-forming units, while all the remaining interfaced generation sources are controlled as grid-following. The two grid-forming units are connected at the nodes A-01 and A-05, at locations close to the same feeder (Figure 7). Different combinations of the rated powers of the two grid-forming converters are examined. The first combination considers two big grid-forming units, both with rated power $S_r = 4$ MVA. The second combination considers instead two relatively small grid-forming units, with rated power $S_r = 1$ MVA each. Finally, the third combination considers a big and a small grid-forming units, with $S_r = 4$ MVA and $S_r = 1$ MVA, respectively. In this analysis case, the size of $S_r = 4$ MVA is considered to be a big generation unit in comparison with the total load of the system ($P_{max} = 8.1$ MW), while the size of $S_r = 1$ MVA is assumed to be instead a small generation unit as conventionally equal to or smaller than about 10% of the total load. The simulation of the described system configurations give the results shown in Figure 8. The results indicate that small grid-forming units in an autonomous system might lead to a detrimental oscillatory instability. In fact, it can be noticed that two small units are not capable of realizing a successful synchronization in the system. In other words, the small sized grid-forming units are not strong enough to impose the frequency of the voltage at their terminals, and having comparable sizes, they will "compete" with each other for synchronism without success. It can also be observed that it is not only the rated power of the single oscillating grid-forming converter which matters, but it is rather the ratio between the rated powers of the grid-forming units involved in the oscillations which can have a significant impact. In the case of one big unit and one small unit, the system response is in fact unstable. The phenomenon of the oscillatory instability is mitigated and eventually compensated for by bigger sizes of the grid-forming converters: for the given system configurations with two big sized grid-forming units, the system response is stable. In that case, the oscillating units are strong enough to sustain a reference frequency for the terminal voltage, thus contributing to a successful synchronization process.

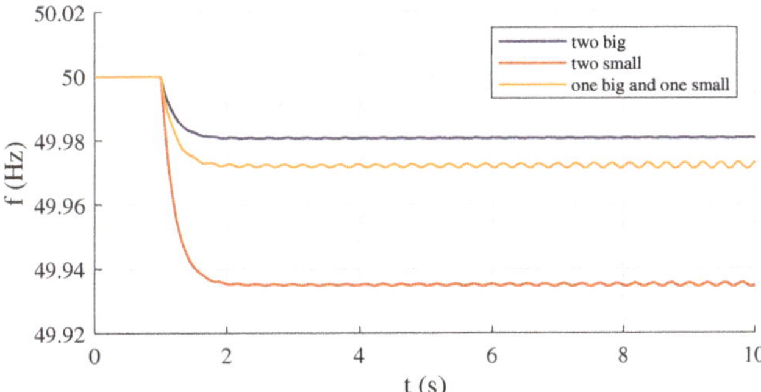

Figure 8. Impact of size of grid-forming units.

5.3. Impact of Number

The impact of the number of grid-forming converters is investigated with the following two configurations. In the first configuration, only four relatively big converter-interfaced generation sources are connected to the network, representing the whole generation of the system. All four converter-interfaced generation sources implement grid-forming capabilities, and they have a rated power of $S_r = 2$ MVA each. The four grid-forming units are assumed to be connected in the middle of the four feeders of the network, located not too close to each other. In the second configuration, 80 small converter-interfaced generation units are distributed all across the autonomous system, with a rated power of $S_r = 100$ kVA each. All the generation units are controlled with the grid-forming scheme: almost every medium voltage substation of the network has, therefore, a generation source with grid-forming control. In both configurations, the grid-forming units cover the total load demand of the island, providing also the same amount of inertial effect. The two configurations are then equivalent from the point of view of generated power, total rated power and total inertial effect. The results of the time-domain simulations are shown in Figure 9. It can be immediately observed that an high number of small grid-forming converters puts the system at the edge of the stability, generating sustained oscillations in the transient response after a perturbation. The system does not experience the oscillatory instability in this case because the distribution of oscillating grid-forming units across the grid is so dense that they are very close with each other, and therefore able to reach the synchronism. If instead the grid-forming capabilities are provided by a few units concentrated in the network, the dynamic behavior of the system is stable and exhibits better transient performance. The results confirm then the observations made about the impact of the rated power of grid-forming units, indicating the opportunity of having only a few big generation units operated as grid-forming, instead of many small oscillating grid-forming distributed in the system.

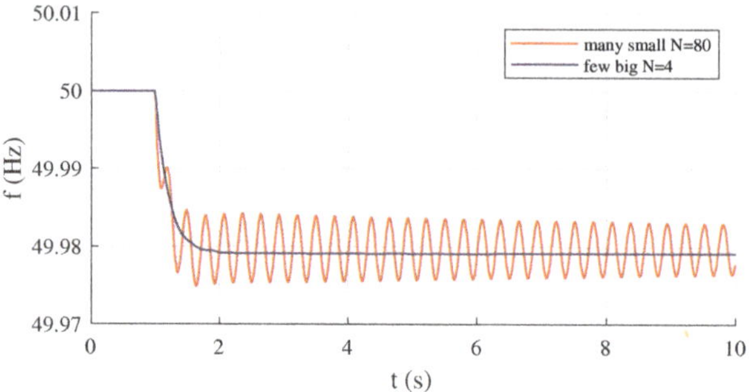

Figure 9. Impact of number of grid-forming units.

5.4. Impact of Location

The impact of electrical distances and mutual locations between grid-forming converters is investigated with the following system configurations. The first configuration considers three grid-forming units installed at the most close locations in the grid: the selected nodes with the grid-forming controls are then A-01, B-01 and C-01 (Figure 7). The second configuration considers the same three grid-forming units installed at the most distant locations between each other: the selected nodes are in this case A-30, B-3-20 and C-10. The third configuration considers two of the three grid-forming units installed close to each other, while the third grid-forming unit is connected far from the other two. The selected nodes are in this case A-01, B-01 and C-10. The fourth and last configuration considers two of the three grid-forming units installed at very distant locations from each other, while the third grid-forming unit is connected in an intermediate position between the other two. The selected nodes are in this case A-30, B-01 and C-10. In all the configurations, the grid-forming units have a rated power of S_r = 2 MVA each. The simulation of the four system configurations give the results shown in Figure 10. The results indicate that the location of the grid-forming units can have a critical impact on the oscillatory stability of the system. It can be in fact observed that when two or more oscillating grid-forming converters are electrically close in the network, they might not be capable to realize the synchronization, "competing" with each other through sustained oscillations and eventually leading to the instability of the system. In the case of grid-forming units electrically far from each other, they can successfully reach the synchronism and the response of the system is stable. It can be additionally observed that the case of a grid-forming converter in an intermediate position between the other two grid-forming distant from each other shows some initial oscillations at the beginning of the frequency transient (yellow line in Figure 10), indicating already an alteration of the power–frequency dynamics of the system. It is clear that the location of generation sources is subject to several constraints, and it is not a factor of free decision in the system operation. However, the understanding and the considerations about the impact of electrical distances between grid-forming converters might be useful in supporting a more careful selection and design of the converters implementing a grid-forming control in autonomous power systems.

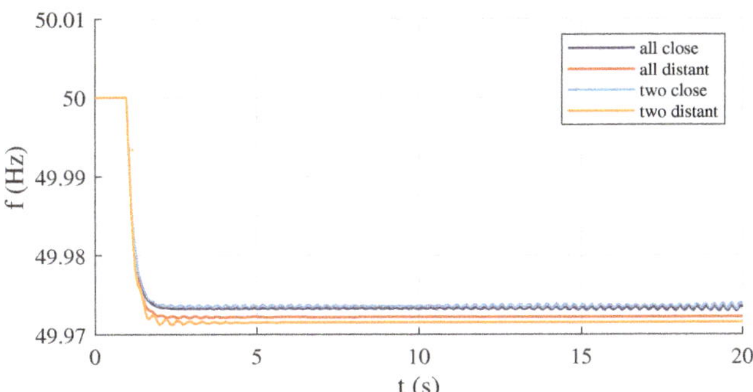

Figure 10. Impact of location of grid-forming units.

5.5. Impact of Inertial Effect

The most relevant parameter of the considered grid-forming control is the time constant of the first integrator. This parameter is in fact responsible for the realization of the inertial effect provided by the control, and it also significantly affects the inherent synchronization mechanism between grid-forming units. The impact of the inertial effect provided by the control is investigated considering again the case of three grid-forming units connected at the most close locations in the system, and performing a parametric analysis by simulating the system with different values of the inertia time constant H. The considered values are the inertia constant of the base case ($H = 2$ s), and then four times smaller ($H = 0.5$ s) and four times bigger ($H = 8$ s) inertia constants. The simulation results are shown in Figure 11. The base case with $H = 2$ s corresponds to the instability observed in the previous section, when the grid-forming units are electrically near and they swing against each other without reaching a successful synchronization. For the same system configuration, if the inertia time constant of all the grid-forming converters is reduced by reducing the constant H, the control actions become faster and more damped, realizing a swift synchronization between the oscillating units. With the reduction of the inertial effect, the implemented control approaches the zero-inertia grid-forming model proposed in [20,28], and it would also correspond to other inertia-less grid-forming schemes, such as the power-synchronization control or droop-based controls without low-pass filters [21,66]. In the simulated case, a reduced inertia time constant of $H = 0.5$ s can already ensure a stable transient operation of the autonomous system. If instead the inertial effect is increased by increasing the time constant H of the integrator, the phenomenon of the oscillatory instability becomes mitigated for as well. For the simulations, a value of $H = 8$ s has been considered. In this case, the increased inertial effect makes the control actions slower, introducing a sort of temporal margin which gives flexibility in the synchronization process between the grid-forming units. The response of the system in this case would be still oscillatory, with the tendency to stabilize. From the results, it appears the possible existence of a range of inertia time constant, where the oscillating grid-forming units with given rated powers engage themselves increasing oscillations, eventually leading to the instability of the system.

Given the recognized relevance of the inertial effect on the frequency dynamics of the system, a more detailed analysis is dedicated to the impact of this factor, considering four additional simulation cases. The configuration used to investigate the impact of different rated powers of the grid-forming units is here analyzed again, assuming different values of the inertia time constant. Two grid-forming units are then connected at relatively close locations in the grid (nodes A-01 and A-05 in Figure 7), considering different values and combinations of the rated powers. The results of the time-domain simulations are summarized in Table 2. To complete the previous considerations, it can be further observed that the reduction of the inertial effect is more effective when applied to grid-forming unit with relatively big rated power, or, in a specular way, that the increase of the inertial effect producing a temporal lag as margin for synchronization is more effective when applied to grid-forming units with relatively small rated power.

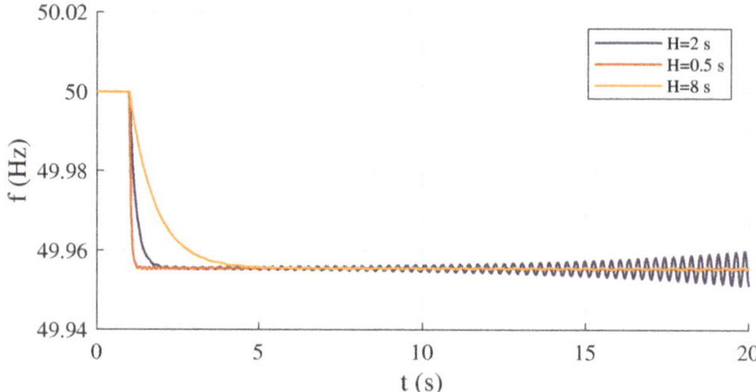

Figure 11. Impact of inertial effect in grid-forming converters.

Table 2. Impact of rated powers and inertial effect.

S_r (MVA)		System Response		
GFM_1	GFM_2	$H_1 = H_2$	$H_1 \ll H_2$	$H_1 \gg H_2$
1	1	unstable	marginally stable	marginally stable
4	4	stable	stable	stable
1	4	unstable	marginally stable	stable
4	1	unstable	stable	marginally stable

It is worth observing that the proposed methodology for the determination of the total amount of inertial effect can be effectively used for the realization of the desired design target, with the containment of the frequency rate within a given maximum value, but the oscillatory characteristics and the power–frequency control of the system can be also significantly affected. For that, more elaborated design techniques of the inertial effect could be applied.

5.6. Impact of Virtual Impedance

The virtual impedance of a grid-forming scheme is another control parameter such as the inertia time constant which can have a significant impact on the dynamic response of the system [67–69]. As with what was achieved for the inertia time constant, the impact of the virtual impedance is investigated considering the case of three grid-forming units connected at the most close locations in the system, performing a parametric analysis with different values of the virtual impedance X_v implemented in the control. The simulation results are reported in Figure 12. The base case with $X_v = 0.1$ pu corresponds to the phenomenon observed studying the impact of location, when the grid-forming units are electrically near and they swing against each other eventually leading to the instability of the system. If the virtual impedance X_v is increased, the grid-forming converters become virtually more distant, the positive effect of long electrical distances is emulated, and the power–frequency dynamics of the system becomes stable. In this case, the increased virtual impedance makes the couplings between the oscillating grid-forming units looser, enabling a more flexible synchronization process. According to the previous considerations, the increase of X_v determines in fact a reduction of the synchronizing coefficients K_s. This reduction however is not critical for the frequency dynamics of the system, but it rather introduces a sort of elasticity which allows the grid-forming units to successfully reach the synchronism after a perturbation. The difference observed in the steady-state frequency between the two cases with different virtual impedances is related to the self-regulating effect of the loads: the implementation of a virtual impedance modifies in fact the voltage control realized at the terminal of the converters, and consequently the loads produce different changes in the power according to the different values of the voltage.

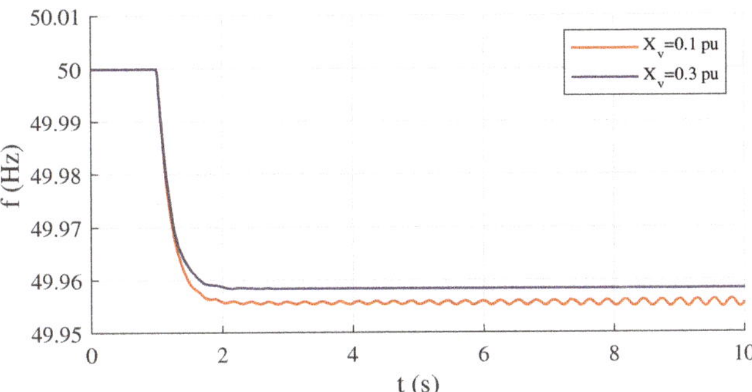

Figure 12. Impact of virtual impedance in grid-forming converters.

5.7. Impact of Control Strategy

The possible contribution of converter-interfaced generation sources implementing grid-supporting controls is investigated considering two critical cases identified in the previous analyses. As observed, the realization of the inertial effect in grid-forming controls through an internal synthetic oscillator can significantly affect the power–frequency dynamics of an autonomous system. If the inertial effect is reduced, the control action of the grid-forming converters becomes faster and all the oscillating units can rapidly reach a successful synchronization between each other. Implementing inertia-less grid-forming schemes such as droop-based controls, the frequency dynamics of the system would be clearly stabilized. However, a given amount of inertia might still be required. A possible alternative for the realization of the inertial effect would be to call the other grid-following units to participate in the power–frequency dynamics of the system. This category of converters control can in fact realize the synthetic inertia in different ways, i.e., emulating the principle of synchronous machines through derivative-based methods on the frequency acquired by the phase-locked loop, but in any case without the need for including a grid-independent internal oscillator in the control. This possibility is here investigated for two different system configurations.

In the first configuration, the system includes only two grid-forming units, with relatively small rated powers and connected at close locations in the network. The two units have a rated power of $S_r = 2$ MVA each, and they are connected at close locations to the same feeder (A-01 and A-05 in Figure 7). As seen, small rated powers and short electrical distances can have a critical impact on the frequency dynamics of the system. In contrast to the previous simulation cases, the grid-following units are assumed this time to be capable of supporting the grid with specific frequency services. For that, they implement a derivative-based control to synthetically provide an inertial effect in the dynamic response. The analysis is performed for different values of the inertia provided by the grid-supporting units. The simulation results are shown in Figure 13. It can be immediately observed that the exclusion of the inertial action from the grid-forming controls has a positive effect on the power–frequency dynamics of the system, allowing a successful synchronization between the oscillating grid-forming units. The provision of inertial effect by the grid-supporting units does not undermine the oscillatory stability of the system, but it rather contributes positively to obtain the required performance in the frequency dynamics.

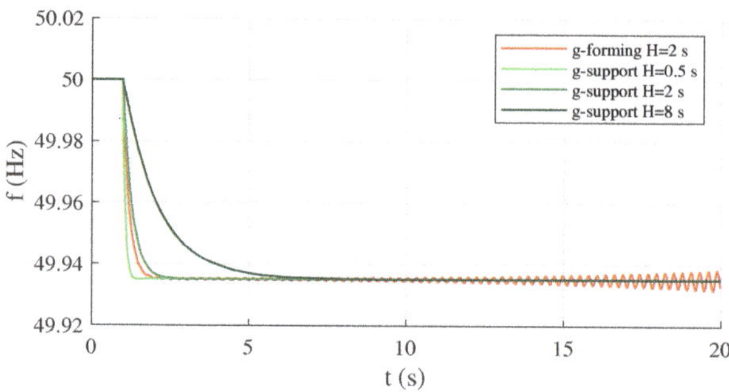

Figure 13. Contribution of grid-supporting controls with two grid-forming converters.

In the second configuration, many small grid-forming units are distributed across the system, amounting to the 30% of the total required generation. As seen, a high number of grid-forming with small rated powers is a critical condition for the frequency dynamics of the system. The system includes only one grid-forming unit relatively bigger than the other, connected to guarantee an appropriate primary reserve. The small units have a rated power of $S_r = 100$ kVA each. Similarly to the previous case, the interfaced generation sources implement a grid-supporting control scheme, with the capability of providing synthetic inertia to the system. The analysis is performed for different values of the inertia provided by the grid-supporting units. Even if this configuration is less likely than the first one, it is reported in the analysis for the sake of completeness. The simulation results are shown in Figure 14. The results basically confirm the opportunity of including inertia-less grid-forming units in the system. In this case, the grid-forming units are assigned the task of realizing an effective synchronization process through fast control actions. The task of providing the required inertial effect is left instead to the grid-supporting units, ensuring a stable frequency dynamics of the system.

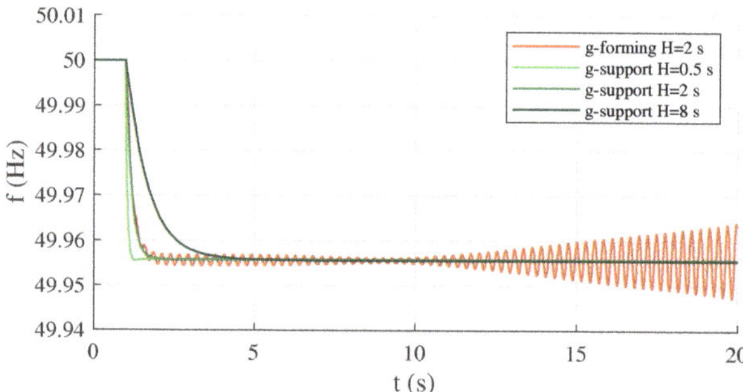

Figure 14. Contribution of grid-supporting controls with many grid-forming converters.

6. Conclusions

The power–frequency dynamics of autonomous systems with 100% converter-interfaced generation has been analyzed, starting from a theoretical point of view and then examining the existing electrical network of Pantelleria island as a case study. For a fully non-synchronous operation of the system, basic design principles for the determination of the required amount of grid-forming power are proposed and applied to the case study. The considerations obtained from the investigations and the possible solutions for a fully non-synchronous operation of autonomous systems can be summarized as follows. The selection of a few big units with grid-forming capabilities appears to be generally preferable. The few selected grid-forming must be ideally reliable sources with relatively high rated powers, such as storage systems or large wind power plants. Small grid-forming converters might not be able to effectively synchronize with each other, since they would not be strong enough to impose the frequency at their terminals. Many small grid-forming units distributed across the system might introduce sustained oscillations and experience critical issues in the synchronization process, competing for synchronism with each other and in the worst cases leading to the instability of the system. For the units implementing the grid-forming control, electrical distances and relative locations in the grid, fast control actions and the reduction of the inertial effect are all recognized as essential aspects for securing the synchronization and the stability of the system. For a good system design, grid-forming units should implement fast control actions with small inertial time constants, while the required inertial effect could be provided by grid-following units participating in the frequency control.

It is finally worth underlying the importance of this kind of study from the perspective of a full and effective integration of renewable energy sources on small islands. Indeed, studies in the literature are mainly focused on energy aspects or steady-state issues (power losses, voltage regulation, etc.) related to the integration of renewable energy sources in isolated systems, and only a few authors discuss the dynamic issues of fully non-synchronous autonomous networks. The proposed study aims to address some of the research gaps discussed in the introduction, and it demonstrates that much effort must still be made in the definition of an effective way to size and design autonomous systems dominated by converter-interfaced generation, while preserving their stability and security of supply.

Author Contributions: Conceptualization, methodology, collection and processing of data, R.M. and G.Z.; Model development and simulations, R.M.; Writing—original draft, R.M.; Writing—review and editing, G.Z.; Supervision, M.G.I. and E.R.S. All authors have read and agreed to the published version of the manuscript.

Funding: This research received no external funding.

Conflicts of Interest: The authors declare no conflict of interest.

References

1. Ackermann, T.; Prevost, T.; Vittal, V.; Roscoe, A.J.; Matevosyan, J.; Miller, N. Paving the Way: A Future without Inertia Is Closer Than You Think. *IEEE Power Energy Mag.* **2017**, *15*, 61–69. [CrossRef]
2. Matevosyan, J.; Badrzadeh, B.; Prevost, T.; Quitmann, E.; Ramasubramanian, D.; Urdal, H.; Huang, S.; Vital, V.; O'Sullivan, J.; Quint, R. Grid-Forming Inverters: Are They the Key for High Renewable Penetration? *IEEE Power Energy Mag.* **2019**, *17*, 89–98. [CrossRef]
3. Ramasubramanian, D.; Farantatos, E.; Ziaeinejad, S.; Mehrizi-Sani, A. Operation paradigm of an all converter interfaced generation bulk power system. *IET Gener. Transm. Distrib.* **2018**, *12*, 4240–4248. [CrossRef]
4. Ndreko, M.; Rüberg, S.; Winter, W. Grid forming control scheme for power systems with up to 100% power electronic interfaced generation: A case study on Great Britain test system. *IET Renew. Power Gener.* **2020**, *14*, 1268–1281. [CrossRef]
5. Chen, J.; Liu, M.; Milano, F.; O'Donnell, T. 100% Converter-Interfaced generation using virtual synchronous generator control: A case study based on the irish system. *Electr. Power Syst. Res.* **2020**, *187*, 106475. [CrossRef]
6. Kouzelis, K.; Musca, R. A Study of Power-Frequency Dynamics in Isolated Power Networks with 100% Converter-Interfaced Generation. In Proceedings of the 19th Wind Integration Workshop, Virtual Conference, 11–12 November 2020.
7. Gouveia, J.; Moreira, C.L.; Peças Lopes, J.A. Influence of Load Dynamics on Converter-Dominated Isolated Power Systems. *Appl. Sci.* **2021**, *11*, 2341. [CrossRef]
8. Gouveia, J.; Moreira, C.L.; Peças Lopes, J.A. Grid-Forming Inverters Sizing in Islanded Power Systems—A stability perspective. In Proceedings of the 2019 International Conference on Smart Energy Systems and Technologies (SEST), Porto, Portugal, 9–11 September 2019.
9. Bongiorno, M.; Favuzza, S.; Ippolito, M.G.; Musca, R.; Navia, M.S.N.; Sanseverino, E.R.; Zizzo, G. An Analysis of the Intertial Response of Small Isolated Power Systems in Presence of Generation from Renewable Energy Sources. In Proceedings of the IEEE 4th International Forum on Research and Technologies for Society and Industry, Palermo, Italy, 10–13 September 2018.
10. Delille, G.; François, B.; Malarange, G. Dynamic Frequency Control Support by Energy Storage to Reduce the Impact of Wind and Solar Generation on Isolated Power System's Inertia. *IEEE Trans. Sustain. Energy* **2012**, *3*, 931–939. [CrossRef]
11. Horne, J.; Flynn, D.; Littler, T. Frequency Stability Issues for Islanded Power Systems. In Proceedings of the IEEE PES Power Systems Conference and Exposition, New York, NY, USA, 10–13 October 2004.
12. Yuan, C.; Xie, P.; Yang, D.; Xiao, X. Transient Stability Analysis of Islanded AC Microgrids with a Significant Share of Virtual Synchronous Generators. *Energies* **2018**, *11*, 44. [CrossRef]

13. Bongiorno, M.; Favuzza, S.; Ippolito, M.G.; Musca, R.; Navia, M.S.N.; Sanseverino, E.R.; Zizzo, G. System Stability of a Small Island's Network with Different Levels of Wind Power Penetration. In Proceedings of the IEEE 4th International Forum on Research and Technologies for Society and Industry, Palermo, Italy, 10–13 September 2018.

14. Eskandari, M.; Li, L.; Moradi, M.; Siano, P. A nodal approach based state-space model of droop-based autonomous networked microgrids. *Sustain. Energy Grids Netw.* **2019**, *18*, 100216. [CrossRef]

15. Krismanto, A.; Mithulananthan, N.; Krause, O. Stability of Renewable Energy based Microgrid in Autonomous Operation. *Sustain. Energy Grids Netw.* **2018**, *13*, 134–147. [CrossRef]

16. Oureilidis, K.O.; Demoulias, C. A decentralized impedance-based adaptive droop method for power loss reduction in a converter-dominated islanded microgrid. *Sustain. Energy Grids Netw.* **2016**, *5*, 39–49. [CrossRef]

17. Unruh, P.; Nuschke, M.; Strauß, P.; Welck, F. Overview on Grid-Forming Inverter Control Methods. *Energies* **2020**, *13*, 2589. [CrossRef]

18. Weise, B.; Korai, A.; Constantin, A. Comparison of Selected Grid-Forming Converter Control Strategies for Use in Power Electronic Dominated Power Systems. In Proceedings of the 18th Wind Integration Workshop, Dublin, Ireland, 16–18 October 2019.

19. Zhong, Q.C.; Weiss, G. Synchronverters: Inverters that mimic synchronous generators. *IEEE Trans. Ind. Electron.* **2011**, *58*, 1259–1267. [CrossRef]

20. Roscoe, A.J.; Yu, M.; Dyśko, A.; Booth, C.; Ierna, R.; Zhu, J.; Urdal, H. A VSM (Virtual Synchronous Machine) Convertor Control Model Suitable for RMS Studies for Resolving System Operator Owner Challenges. In Proceedings of the 15th Wind Integration Workshop, Vienna, Austria, 15–17 November 2016.

21. Zhang, L.; Harnefors, L.; Nee, H.P. Power-Synchronization Control of Grid-Connected Voltage-Source Converters. *IEEE Trans. Power Syst.* **2010**, *25*, 809–820. [CrossRef]

22. Rocabert, J.; Luna, A.; Blaabjerg, F.; Rodriguez, P. Control of Power Converters in AC Microgrids. *IEEE Trans. Power Electron.* **2012**, *27*, 4734–4749. [CrossRef]

23. Tayyebi, A.; Groß, D.; Anta, A.; Kupzog, F.; Dörfler, F. Interactions of Grid-Forming Power Converters and Synchronous Machines—A Comparative Study. *arXiv* **2019**, arXiv:1902.10750.

24. Zuo, Y.; Yuan, Z.; Sossan, F.; Zecchino, A.; Cherkaoui, R.; Paolone, M. Performance assessment of grid-forming and grid-following converter-interfaced battery energy storage systems on frequency regulation in low-inertia power grids. *Sustain. Energy Grids Netw.* **2021**, *27*, 100496. [CrossRef]

25. Bevrani, H.; Ise, T.; Miura, Y. Virtual synchronous generators: A survey and new perspectives. *Electr. Power Syst. Res.* **2014**, *54*, 244–254. [CrossRef]

26. Mohammed, O.O.; Otuoze, A.O.; Salisu, S.; Ibrahim, O.; Rufa'i, N.A. Virtual Synchronous Generator: An Overview. *Niger. J. Technol.* **2019**, *38*, 153–164. [CrossRef]

27. D'Arco, S.; Suul, J.A.; Fosso, O.B. A Virtual Synchronous Machine implementation for distributed control of power converters in SmartGrids. *Electr. Power Syst. Res.* **2015**, *122*, 180–197. [CrossRef]

28. Yu, M.; Roscoe, A.J.; Booth, C.D.; Dysko, A.; Ierna, R.; Zhu, J.; Urdal, H. Use of an Inertia-less Virtual Synchronous Machine within Future Power Networks with High Penetrations of Converters. In Proceedings of the 2016 Power Systems Computation Conference (PSCC), Genoa, Italy, 20–24 June 2016.

29. Hesse, R.; Turschner, D.; Beck, H.P. Micro grid stabilization using the Virtual Synchronous Machine (VISMA). In Proceedings of the International Conference on Renewable Energies and Power, Valencia, Spain, 15–17 April 2009.

30. Pogaku, N.; Prodanovic, M.; Green, T.C. Modeling, Analysis and Testing of Autonomous Operation of an Inverter-Based Microgrid. *IEEE Trans. Power Electron.* **2007**, *22*, 613–625. [CrossRef]

31. Brabandere, K.D.; Bolsens, B.; den Keybus, J.V.; Woyte, A.; Driesen, J.; Balmans, R. A voltage and frequency droop control method for parallel inverters. *IEEE Trans. Power Electron.* **2007**, *22*, 1107–1115. [CrossRef]

32. Khaledia, A.; Golkar, M.A. Analysis of droop control method in an autonomous microgrid. *J. Appl. Res. Technol.* **2017**, *15*, 371–377. [CrossRef]

33. NEPLAN Power Systems Analysis Software. Available online: https://www.neplan.ch/en-products/ (accessed on 2 February 2022).

34. MATLAB Scientific Computation Software. Available online: https://mathworks.com/products/matlab.html (accessed on 2 February 2022).

35. Dörfler, F.; Bullo, F. Synchronization in complex networks of phase oscillators: A survey. *Automatica* **2014**, *50*, 1539–1564. [CrossRef]

36. Pan, D.; Wang, X.; Liu, F.; Shi, R. Transient Stability of Voltage-Source Converters With Grid-Forming Control: A Design-Oriented Study. *IEEE J. Emerg. Sel. Top. Power Electron.* **2020**, *8*, 1019–1033. [CrossRef]

37. Liu, J.; Miura, Y.; Ise, T. Comparison of Dynamic Characteristics between Virtual Synchronous Generator and Droop Control in Inverter-Based Distributed Generators. *IEEE Trans. Power Electron.* **2016**, *31*, 3600–3611. [CrossRef]

38. Pan, R.; Sun, P. Extra transient block for virtual synchronous machine with better performance. *IET Gener. Transm. Distrib.* **2020**, *14*, 1186–1196. [CrossRef]

39. Qoria, T.; Gruson, F.; Colas, F.; Denis, G.; Prevost, T.; Guillaud, X. Inertia Effect and Load Sharing Capability of Grid Forming Converters Connected to a Transmission Grid. In Proceedings of the 15th IET International Conference on AC and DC Power Transmission, Coventry, UK, 5–7 February 2019.

40. Kundur, P.; Balu, N.J.; Lauby, M.G. *Power System Stability and Control*; McGraw-Hill: New York, NY, USA, 1994.
41. Harnefors, L.; Hinkkanen, M.; Riaz, U.; Rahman, F.M.M.; Zhang, L. Robust Analytic Design of Power-Synchronization Control. *IEEE Trans. Ind. Electron.* **2019**, *66*, 5810–5819. [CrossRef]
42. *High Penetration of Power Electronic Interfaced Power Sources and the Potential Contribution of Grid Forming Converters*; ENTSO-E Technical Report; ENTSO-E: Brussels, Belgium, January 2020.
43. Liu, J.; Miura, Y.; Ise, T. A Comparative Study on Damping Methods of Virtual Synchronous Generator Control. In Proceedings of the 21st European Conference on Power Electronics and Applications (EPE19), Genova, Italy, 2–6 September 2019.
44. Gao, B.; Xia, C.; Chen, N.; Cheema, K.M.; Yang, L.; Li, C. Virtual Synchronous Generator Based Auxiliary Damping Control Design for the Power System with Renewable Generation. *Energies* **2017**, *10*, 1146. [CrossRef]
45. Pipelzadeh, Y.; Chaudhuri, B.; Green, T.C. Control Coordination Within a VSC HVDC Link for Power Oscillation Damping: A Robust Decentralized Approach Using Homotopy. *IEEE Trans. Control Syst. Technol.* **2013**, *21*, 1270–1279. [CrossRef]
46. Ndreko, M.; van der Meer, A.A.; Rawn, B.G.; Gibescu, M.; van der Meijden, M.A.M.M. Damping Power System Oscillations by VSC-Based HVDC Networks: A North Sea Grid Case Study. In Proceedings of the International Workshop on Large-Scale Integration of Wind Power into Power Systems as well as on Transmission Networks for Offshore Wind Power Plants, London, UK, 22–24 October 2013.
47. Dong, S.; Chen, Y.C. Adjusting Synchronverter Dynamic Response Speed via Damping Correction Loop. *IEEE Trans. Energy Convers.* **2017**, *32*, 608–619. [CrossRef]
48. Ebrahimi, M.; Khajehoddin, S.A.; Karimi-Ghartemani, M. An Improved Damping Method for Virtual Synchronous Machines. *IEEE Trans. Sustain. Energy* **2019**, *10*, 1491–1500. [CrossRef]
49. Bongiorno, M.; Ippolito, M.G.; Musca, R.; Zizzo, G. Damping Provision by Different Virtual Synchronous Machine Schemes. In Proceedings of the 20th International Conference on Environment and Electrical Engineering and 4th Industrial and Commercial Power Systems Europe (EEEIC/I&CPS Europe), Madrid, Spain, 9–12 June 2020.
50. Bongiorno, M.; Musca, R.; Zizzo, G. Grid Forming Converters in Weak Grids: The Case of a Mediterranean Island. In Proceedings of the 18th Wind Integration Workshop, Dublin, Ireland, 16–18 October 2019.
51. Huang, L.; Xin, H.; Wang, Z. Damping Low-Frequency Oscillations Through VSC-HVDC Stations Operated as Virtual Synchronous Machines. *IEEE Trans. Power Electron.* **2019**, *34*, 5803–5818. [CrossRef]
52. Bao, F.; Guo, J.; Wang, W.; Wang, B. Cooperative Control Strategy of Multiple VSGs in Microgrid Based on Adjacent Information. *IEEE Access* **2021**, *9*, 125603–125615. [CrossRef]
53. Zhang, L.; Zheng, H.; Wan, T.; Shi, D.; Lyu, L.; Cai, G. An integrated control algorithm of power distribution for islanded microgrid based on improved virtual synchronous generator. *IET Renew. Power Gener.* **2021**, *15*, 2674–2685. [CrossRef]
54. Liu, X.; Gong, R. A control strategy of microgrid-connected system based on VSG. In Proceedings of the 2020 IEEE International Conference on Power, Intelligent Computing and Systems (ICPICS), Shenyang, China, 28–30 July 2020. [CrossRef]
55. Kim, J.Y.; Jeon, J.H.; Kim, S.K.; Cho, C.; Park, J.H.; Kim, H.M.; Nam, K.Y. Cooperative control strategy of energy storage system and microsources for stabilizing the microgrid during islanded operation. *IEEE Trans. Power Electron.* **2010**, *25*, 3037–3048. [CrossRef]
56. Choopani, M.; Hosseinian, S.H.; Vahidi, B. New transient stability and LVRT improvement of multi-VSG grids using the frequency of the center of inertia. *IEEE Trans. Power Syst.* **2020**, *35*, 527–538. [CrossRef]
57. Choopani, M.; Hosseinain, S.H.; Vahidi, B. A novel comprehensive method to enhance stability of multi-VSG grids. *Int. J. Electr. Power Energy Syst.* **2019**, *104*, 502–514. [CrossRef]
58. Zhenao, S.; Fanglin, Z.; Xingchen, C. Study on a Frequency Fluctuation Attenuation Method for the Parallel Multi-VSG System. *Front. Energy Res.* **2021**, *9*, 310. [CrossRef]
59. Du, W.; Chen, Z.; Schneider, K.P.; Lasseter, R.H.; Nandanoori, S.P.; Tuffner, F.K.; Kundu, S. A Comparative Study of Two Widely Used Grid-Forming Droop Controls on Microgrid Small-Signal Stability. *IEEE J. Emerg. Sel. Top. Power Electron.* **2020**, *8*, 963–975. [CrossRef]
60. Chen, M.; Zhou, D.; Blaabjerg, F. Active Power Oscillation Damping Based on Acceleration Control in Paralleled Virtual Synchronous Generators System. *IEEE Trans. Power Electron.* **2021**, *36*, 9501–9510. [CrossRef]
61. Guo, J.; Chen, Y.; Liao, S.; Wu, W.; Zhou, L.; Xie, Z.; Wang, X. Analysis and Mitigation of Low-Frequency Interactions Between the Source and Load Virtual Synchronous Machine in an Islanded Microgrid. *IEEE Trans. Ind. Electron.* **2021**, *69*, 3732–3742. [CrossRef]
62. Beires, P.P.; Moreira, C.L.; Peças Lopes, J.A. Grid-forming inverters replacing Diesel generators in small-scale islanded power systems. In Proceedings of the 2019 IEEE Milan PowerTech, Milan, Italy, 23–27 June 2019.
63. Fang, J.; Li, H.; Tang, Y.; Blaabjerg, F. Distributed Power System Virtual Inertia Implemented by Grid-Connected Power Converters. *IEEE Trans. Power Electron.* **2018**, *33*, 8488–8499. [CrossRef]
64. MIGRATE Deliverable D1.1. Report on Systemic Issues. December 2016. Available online: https://ec.europa.eu/research/participants/documents/downloadPublic?documentIds=080166e5af08ecd7&appId=PPGMS (accessed on 2 February 2022).
65. Poolla, B.K.; Groß, D.; Dörfler, F. Placement and Implementation of Grid-Forming and Grid-Following Virtual Inertia and Fast Frequency Response. *IEEE Trans. Power Syst.* **2019**, *34*, 3035–3046. [CrossRef]
66. D'Arco, S.; Suul, J. Equivalence of Virtual Synchronous Machines and Frequency-Droops for Converter-Based MicroGrids. *IEEE Trans. Smart Grid* **2014**, *5*, 394–395. [CrossRef]

67. Rodríguez-Cabero, A.; Roldán-Pérez, J.; Prodanovic, M. Virtual Impedance Design Considerations for Virtual Synchronous Machines in Weak Grids. *IEEE J. Emerg. Sel. Top. Power Electron.* **2020**, *8*, 1477–1489. [CrossRef]
68. Suul, J.A.; D'Arco, S.; Rodriguez, P.; Molinas, M. Extended stability range of weak grids with Voltage Source Converters through impedance-conditioned grid synchronization. In Proceedings of the 11th IET International Conference on AC and DC Power Transmission, Birmingham, UK, 10–12 February 2015.
69. Wang, X.; Taul, M.G.; Wu, H.; Liao, Y.; Blaabjerg, F.; Harnefors, L. Grid-Synchronization Stability of Converter-Based Resources—An Overview. *IEEE Open J. Ind. Appl.* **2020**, *1*, 115–134. [CrossRef]

Article

A Perturbation-Based Methodology to Estimate the Equivalent Inertia of an Area Monitored by PMUs

Guido Rossetto Moraes [1,2,*], **Valentin Ilea** [3], **Alberto Berizzi** [3], **Cosimo Pisani** [4], **Giorgio Giannuzzi** [4] and **Roberto Zaottini** [4]

[1] Department of Electrical Engineering, Federal University of Santa Catarina, Florianópolis 88040-900, Brazil
[2] INESC P&D Brasil, Santos 11055-300, Brazil
[3] Department of Energy, Politecnico di Milano, 20156 Milan, Italy; valentin.ilea@polimi.it (V.I.); alberto.berizzi@polimi.it (A.B.)
[4] Terna Rete Italia SpA, 00138 Rome, Italy; cosimo.pisani@terna.it (C.P.); giorgio.giannuzzi@terna.it (G.G.); roberto.zaottini@terna.it (R.Z.)
* Correspondence: guido.moraes@posgrad.ufsc.br

Abstract: This paper proposes a novel methodology to estimate equivalent inertia of an area, observed from its boundary buses where Phasor Measurement Units (PMUs) are assumed to be installed. The areas are divided according to the measurement points, and the methodology proposed can obtain the equivalent dynamic response of the area dependent of or independent of coherency of the generators inside, which is the first contribution of this paper. The methodology is divided in three parts: estimating the frequency response, estimating the power imbalance and estimating inertia through the solution of the swing equation by Least-Squares Method (LSM). The estimation of the power imbalance is the second contribution of this paper, enabling the study of areas that contain perturbations and attending the limitation of methods of the literature that rely on assumptions of slow mechanical power. It can be further divided in three steps: accounting the total power injected, estimating an equivalent load behavior and estimating an equivalent mechanical power. The quality of results is proved with test systems of different sizes, simulating different types of perturbations.

Keywords: power system stability; inertia estimation; PMU

Citation: Rossetto Moraes, G.; Ilea, V.; Berizzi, A.; Pisani, C.; Giannuzzi, G.; Zaottini, R. A Perturbation-Based Methodology to Estimate the Equivalent Inertia of an Area Monitored by PMUs. *Energies* **2021**, *14*, 8477. https://doi.org/10.3390/en14248477

Academic Editor: Teuvo Suntio

Received: 18 October 2021
Accepted: 9 December 2021
Published: 15 December 2021

Publisher's Note: MDPI stays neutral with regard to jurisdictional claims in published maps and institutional affiliations.

1. Introduction

The increasing penetration of Renewable Energy Sources (RES) in power systems is bringing new challenges to Transmission System Operators (TSOs) worldwide. RES-based generators are connected to the grid by means of converters that electrically decouple their inertial response, if any, from the system. Hence, equivalent inertia is decreasing, causing higher and faster frequency excursions following power imbalances, which may result in frequency instability. To mitigate this impact, a requirement of synthetic compensation of inertia is being adopted by many TSOs [1,2]. However, RES are intermittent, such that assessing equivalent inertia is a rising need [3,4].

A possibility to estimate equivalent inertia is provided by the recently deployed and disseminated PMUs, devices capable of providing measurements of frequency, current and voltage (magnitude and phase angles) in real time in a precise and synchronized way, with a resolution of 10–60 samples per second [5]. Due to the easiness of working with phasors, PMUs have been used in many different applications, such as analysis of perturbations and oscillations [6–8], topology monitoring and parameter estimation [9]. Regarding inertia estimation, literature normally divides the topic in small perturbation studies [8,10,11], large perturbation studies [12–14] and studies under ambient conditions [15,16].

In the field of large perturbation studies, a particular effort has been devoted in recent years to characterize areas including multiple loads and generators. However, most papers assume either monitoring generating units individually or monitoring coherent groups of generators. In [17], a robust Kalman Filter is proposed to estimate the mean frequency, but full dynamical observability is required. In [18], an approach that accommodates frequency and voltage variations through curve-fitting is proposed, but accuracy highly depend on the percentage of generating units monitored. In [19], the inertia of an area as perceived by a particular bus of the system is evaluated through an autoregressive moving average exogenous input model, but in all testes, generators are assumed as monitored.

However, monitoring generating units individually may be hard to ensure in real systems. Despite recent efforts on developing low-cost solutions [20–22], most commercial PMUs are still expensive [23,24], such that most countries still count with a limited number of PMUs installed. Hence, difficulties for TSOs are many: not all generators individually monitored, coherent groups always changing due to the impact of intermittent RES-based generation, lack of observability of medium-voltage and low-voltage grids, and limited or slow communication.

Taking these practical aspects into consideration, this paper proposes a novel way of looking at the problem: instead of defining an area through the traditional criteria of coherency adopted in the literature, an area may be defined according to the location of available PMUs in the network. Whether considering units already installed or planning the placement of new PMUs, the proposed requirements to define an area based on phasor measurements are just two: to monitor the total power injected by the area in the system, and to monitor the dynamics of the Center of Inertia (COI) of the area. Whenever these requirements are met, an area can be defined, and a dynamic equivalent can be obtained.

The methodology proposed can build a dynamic equivalent of an area in three major steps: estimating the frequency response, estimating the power imbalance and estimating inertia. To estimate the frequency response, the methodology reduces the system around the measurement points available and adapts part of the method proposed in [25] to obtain equivalents at the retained points. The many equivalents are then combined in one, which represents the dynamics of the COI of the area. To estimate power imbalance of the area, the methodology proposes estimating the behavior of the total load, total losses and total mechanical power of the area, seen from its boundary buses. Finally, inertia is estimated solving the swing equation related to the dynamic equivalent of the area through LSM.

Results are obtained with data provided by simulations with the well-known 11-bus Kundur's test system [26] and with the benchmark test system proposed in [27]. The first part of the results section presents a study validating the methodology, with details about its steps and insights on the behavior of the dynamic equivalents. A switch of a synchronous generator by a Wind Power (WP) generator is simulated, and the reduction of inertia following the event is estimated with the methodology proposed. The second part of the results section presents different study cases evaluating the performance of each step and its impact on the final inertia estimated.

Thus, the main contribution of this paper is the innovative manner of defining areas based on practical assumptions. Second, the methodology derived is capable of estimating inertia regardless the coherency of generators inside the area of study. Finally, the methodology can deal with perturbations inside or outside the area considered.

The remaining of this paper is organized as follows: Section 2 presents a brief introduction on the dynamic behavior of an area, Section 3 presents the methodology, Section 4 presents the results and Section 5 presents the conclusions.

2. The Dynamic Behavior of an Area

The dynamics of a synchronous machine *i* following a perturbation is ruled by the Swing Equation,

$$\frac{2H_i}{2\pi f_0} \frac{d^2\delta_i(t)}{dt^2} = \Delta P_i(t) = P_{m_i}(t) - P_{g_i}(t), \qquad [pu] \qquad (1)$$

where H_i is the constant of inertia, δ_i is the rotor angle, $\Delta P_i(t)$ is the total power imbalance, P_{m_i} is the mechanical power and P_{g_i} is the electrical power produced by machine i. Additionally, the term $\frac{d^2\delta_i(t)}{dt^2}$ may be changed to $\frac{df_i(t)}{dt}$ since

$$f_i = \frac{1}{2\pi}\frac{d\delta_i(t)}{dt} + f_0, \qquad [Hz] \tag{2}$$

where f_0 is the nominal frequency of the system.

To evaluate the dynamic behavior of a multi-machine system, it is possible to write Equation (1) for every machine that is connected to the grid. Alternatively, it is also possible to study the behavior of the system with an equivalent, based on the concept of the COI, a rotational analogy of the center of mass of an object. The dynamic behavior of every synchronous generator tends to follow the behavior of the COI of the system [26]. This concept may be particularized to any group of generators, restricted to a regional area of interest, for example. The restriction of the area of study is a matter of deciding the extension of the data set [10].

The frequency associated with the COI (also called mean frequency), is a weighted average of the frequencies of every machine of the area, determined by

$$f_{COI_a}(t) = \frac{\sum_{i=1}^{n} H_i S_{b_{Gi}} f_{G_i}(t)}{\sum_{i=1}^{n} H_i S_{b_{Gi}}} \tag{3}$$

where H_i is the inertia, $S_{b_{Gi}}$ is the nominal power and f_{G_i} is the electrical frequency of each generator G_i in Area a.

The constant of inertia at the COI of Area a is

$$H_{COI_a} = \sum_{i=1}^{n} H_i S_{b_{Gi}} \tag{4}$$

Furthermore, the dynamic behavior of an area may be studied from the point of view of its boundary bus (or boundary buses). Consider the left part of Figure 1 as a generic representation of an area of a power system. An equivalent of Area a seen from its boundary bus (BB) is represented in the right part of the figure, where M_a represents the moving masses of the area, P_{mov_a} is the moving power, G_{eq_a} is an equivalent generator and P_{e_a} is the power exiting the area. To represent G_{eq_a}, the second-order model is assumed in Figure 1, where x'_a is the transient reactance and $\mathbf{E}_a$ is the internal voltage of the machine.

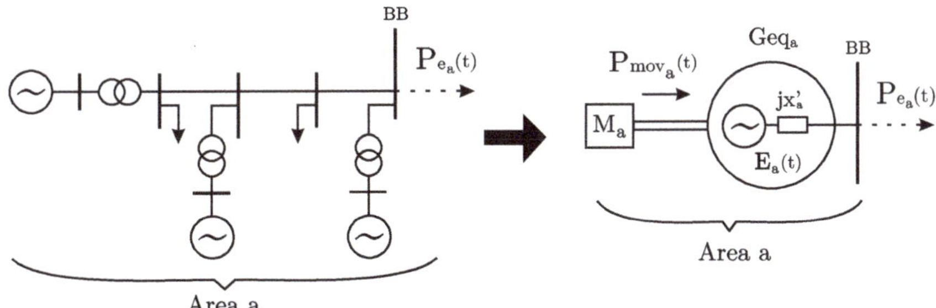

Figure 1. Equivalent system of an area seen from its boundary bus.

The moving power of Area a is defined here as:

$$P_{mov_a}(t) \triangleq P_{m_a}(t) - P_{l_a}(t) - P_{losses_a}(t) \triangleq \sum_{i=1}^{n} P_{m_i}(t) - \sum_{l=1}^{n_l} P_{l_l} - \sum_{m=1}^{n_b} P_{losses_m} \tag{5}$$

where P_{m_a} denotes the total mechanical power, P_{l_a} denotes the total load power and P_{losses_a} denotes the total losses of area a; P_{m_i} denotes the mechanical power of each of the

$i = 1, \dots, n$ generators, P_{l_l} denotes the active power consumed at each of the $l = 1, \dots, n_l$ load buses and P_{losses_m} are the losses in each of the $m = 1, \dots, n_b$ branches inside the area a.

The Swing Equation that rules the dynamics of the equivalent system can be written as

$$\frac{H_{COI_a}}{\pi f^0} \frac{d^2 \delta_a(t)}{dt^2} = \Delta P_a(t) = P_{mov_a}(t) - P_{e_a}(t), \qquad p.u. \tag{6}$$

where H_a is the overall inertia of the dynamic equivalent, δ_a is the phase angle of the internal voltage $\mathbf{E}_a$, $\Delta P_a(t)$ is the total power imbalance of Area a and $P_{e_a}(t)$ is the total power injection at BB. The phase angle δ_a may be considered to be an approximation of the mean rotor angle ϕ_a according to

$$\phi_a(t) = \frac{\sum_{i=1}^{n} H_i S_{b_{Gi}} \phi_{G_i}(t)}{\sum_{i=1}^{n} H_i S_{b_{Gi}}} \tag{7}$$

where $\phi_{G_k}(t)$ is the rotor angle of each generator $i = 1 \dots n$ of area a, and the other variables have been previously defined following Equation (3).

3. Methodology

In the technical literature, the definition of equivalent circuit of areas takes place according to coherency criteria, such that generators behaving similarly may be easily grouped together. This proposition has been widely adopted because it facilitates system reduction, diminishing computational effort in dynamic simulations. Here, instead, the aim is not system reduction for simulations, but system reduction for parameter estimation and system dynamic assessment. Therefore, a new way of defining an area is here adopted, based on the location of PMUs in the grid.

To define an area for the study, two requirements need to be fulfilled: the total power injected by the area and the dynamics of its COI must be monitored/estimated. If this is true, it is possible study the dynamics of the area with an equivalent, following the theory in Section 2. Therefore, a methodology is derived in this paper with the aim of modelling the dynamic behavior of an area through the equivalent Equation (6). To do so, $\delta_a(t)$, $P_{mov_a}(t)$ and $P_{e_a}(t)$ have to be monitored or estimated, and H_{COI_a} is obtained as an outcome of the procedure.

Consider an Area a monitored by PMUs installed at its BBs (PMU-1 and PMU-2) and at possible buses inside (PMU-N), as represented in Stage 1 of Figure 2, where dotted lines represent the many connections inside the area. The first step of the methodology consists of reducing the system around the n measurement points, as shown in Stage 2 of Figure 2. Assuming that the interconnections of the system do not change during the period of study, the power exported by Area a through its boundary buses (measured) may be assumed as $P_{e_a}(t)$ in Equation (6).

The dynamics of the hidden machines inside the reduced area are approximated through the estimation of equivalent generators at the points retained (Stage 3 of Figure 2). After the equivalent generators are obtained, it is possible to calculate their mean dynamic behavior (i.e., $\delta_a(t)$) to feed Equation (6). This procedure is described in detail in Section 3.1.

The dynamic behavior of the n equivalent generators above mentioned may be represented together by the COI of the area, as represented in Stage 4. Most methods available in the literature for inertia estimation following perturbations require monitoring the terminal buses where generators are connected, and rely on the assumption that the mechanical power changes slowly in comparison to the electrical power generated by the machine [25,28,29]. However, when monitoring an area through its boundaries, this assumption becomes too strict: due to load behavior and possible perturbations inside the area, the moving power may not behave slow in comparison to the electrical power injected. In this condition, the equivalent moving power of Area a must also be estimated to estimate H_{COI_a} accurately. This procedure is described in Section 3.2.

After $\delta_a(t)$, $P_{mov_a}(t)$ and $P_{e_a}(t)$ are obtained, Equation (6) can be used to estimate H_{COI_a}. This is approached in Section 3.3.

At the end of the section, Section 3.4 presents a flowchart summarizing the full methodology.

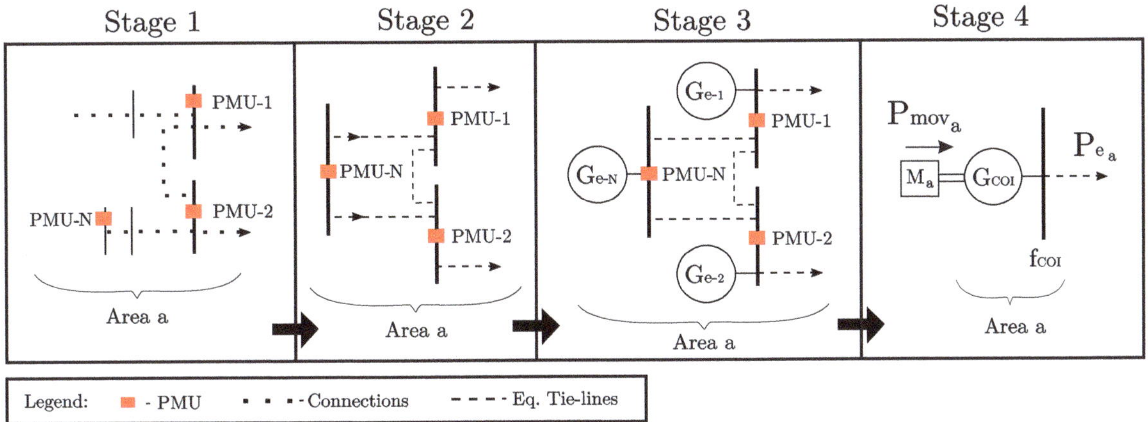

Figure 2. PMU-based system reduction.

3.1. Determination of the COI Dynamics

Starting from Stage 1 of Figure 2, the first step is to reduce the system around the measurement points available. System reduction methods were first developed in the literature for speeding up calculations and simulators [30]. Consequently, some of the criteria to reduce the system may not be reasonable for the goal of this paper, as they were tailored for model simulation, but some techniques may be adopted. The Ward Equivalent method is a generalization of the Thévenin equivalent, widely used with phasorial representation in steady-state studies, that can be extended to represent part of grid in quasi-static conditions.

Assuming the loads modelled as constant impedance, the steady-state operation point and the network topology in this context as known, it is possible to retain the buses where PMUs are available, eliminating the others, as in Stage 2 of Figure 2. At this point, the dynamic of the hidden generators is not yet represented (they will be represented together in the equivalents obtained in Stage 3).

The system reduction starts from the full algebraic equations, which are valid for each time step $t = 1, \ldots, M$ when considering the transient evolving under quasi-static conditions and small frequency variations. By Ohm's law,

$$\mathbf{Y}_{BUS}\mathbf{V}_{Bt} = \mathbf{I}_{Bt} \tag{8}$$

where $\mathbf{Y}_{BUS}$ is the admittance matrix of the system (built not including generator and load admittances), with dimensions $N_T \times N_T$, where N_T is the total number of buses in the system. The vectors $\mathbf{V}_{Bt}$ and $\mathbf{I}_{Bt}$ are respectively the bus voltages and current injections vectors at time step t, with dimensions $N_T \times 1$.

The current injection at a generic bus b $\mathbf{I}_{B_t}^{(b)}$ can be written as:

$$\mathbf{I}_{B_t}^{(b)} = \mathbf{I}_{G_t}^{(b)} - \mathbf{I}_{L_t}^{(b)} \tag{9}$$

where $\mathbf{I}_{G_t}^{(b)}$ is the generator current of bus b, $\mathbf{I}_{L_t}^{(b)}$ is the load current and $\mathbf{I}_{B_t}^{(b)}$ is the injected current in the grid and the superscript denote bus b.

Defining the buses where PMUs are installed as the set of buses to be retained ($R = \{1, 2, \ldots, N\}$), then ($N_T - N$) buses must be eliminated ($E = b \notin R$).

Equation (8) can be reorganized as:

$$\begin{bmatrix} \mathbf{Y}_{RR} & \mathbf{Y}_{RE} \\ \mathbf{Y}_{ER} & \mathbf{Y}_{EE} \end{bmatrix} \begin{bmatrix} \mathbf{V}_R \\ \mathbf{V}_E \end{bmatrix} = \begin{bmatrix} \mathbf{I}_R \\ \mathbf{I}_E \end{bmatrix} \tag{10}$$

where the subscripts R and E are related to the retained and eliminated buses, respectively. For the sake of simplicity, the subscript t is neglected here.

Equation (10) may be manipulated to obtain

$$(\mathbf{Y}_{RR} - \mathbf{Y}_{RE}\mathbf{Y}_{EE}^{-1}\mathbf{Y}_{ER})\mathbf{V}_R = \mathbf{I}_R - \mathbf{Y}_{RE}\mathbf{Y}_{EE}^{-1}\mathbf{I}_E \tag{11}$$

Making

$$\mathbf{Y}_{EQ} = \mathbf{Y}_{RR} - \mathbf{Y}_{RE}\mathbf{Y}_{EE}^{-1}\mathbf{Y}_{ER} \tag{12}$$

and

$$\mathbf{I}_W = -\mathbf{Y}_{RE}\mathbf{Y}_{EE}^{-1}\mathbf{I}_E \tag{13}$$

it is possible to rewrite Equation (11) as

$$\mathbf{Y}_{EQ}\mathbf{V}_R = \mathbf{I}_{R_{EQ}} \tag{14}$$

It is important to observe that $\mathbf{I}_{R_{EQ}}$ in Equation (14) can be obtained with the voltage phasors at the retained buses $\mathbf{V}_R$ and the reduced admittance matrix $\mathbf{Y}_{EQ}$, therefore practical conditions.

With $\mathbf{I}_{EQ}$ and $\mathbf{V}_R$, it is possible to apply a model estimation method at each retained bus to obtain the equivalent generators represented in Stage 3 of Figure 2. The method proposed in [25] requires current and voltage measurements at the terminals of a generator to estimate its second-order model. Here, the method is applied in another context: building dynamic equivalents from the voltages and currents obtained at the retained buses ($\mathbf{I}_{EQ}$ and $\mathbf{V}_R$), which may be generation buses (originally) or not.

At each retained bus, one equivalent generator will be obtained. In general terms, this is done relating the assumptions of a second-order model and the power injected in the grid at the connection point, as shown in Figure 3. In the figure, Bus k is a retained bus, P_{e_k} is the power injected and Geq_k is the equivalent generator, whose parameters are estimated from $\mathbf{V}_k(t)$ and $\mathbf{I}_k(t)$, the terms of the matrices $\mathbf{V}_R$ and $\mathbf{I}_{EQ}$ related to Bus k, respectively. As fast as Geq_k concerned, it is needed to estimate the transient reactance x'_k and the internal voltage E_k; M_k is the moving mass and $P_{mov_k}(t)$ is the moving power.

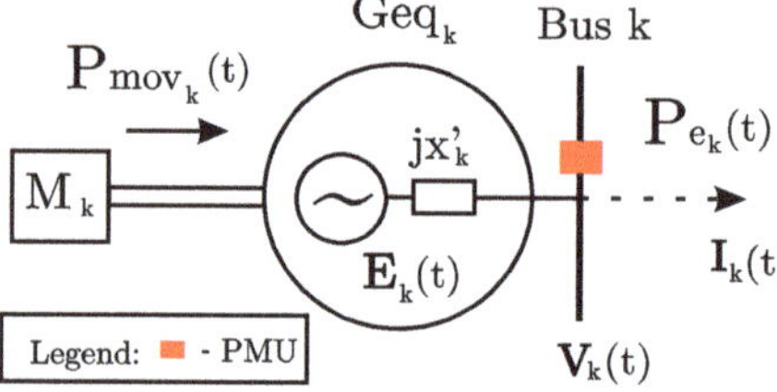

Figure 3. Equivalent generator estimated from the retained bus k.

Assuming that the magnitude of $\mathbf{E}_k$ is constant (which is an intrinsic assumption of the second-order model), the method fits Kirchhoff's equations with input signals to estimate first an equivalent transient reactance x'_k; in fact, the following equation holds,

$$E_k(t)\angle\delta_k(t) = (jx'_k)I_k(t)\angle\alpha_k(t) + V_k(t)\angle\theta_k(t) \tag{15}$$

Since E_k is constant, it is possible to assume the variance *Var* equal to zero

$$Var(E_k) \approx 0 \tag{16}$$

From Equation (15),

$$|\mathbf{E_k}(t)| = \sqrt{x_k'^2|\mathbf{I_k}(t)|^2 + |\mathbf{V_k}(t)|^2 + 2x_k'Q_k(t)} \tag{17}$$

and

$$E_k(t)^2 = x_k'^2 I_k(t)^2 + V_k(t)^2 + 2x_k'Q_k(t) \tag{18}$$

where $Q_k(t)$ is the reactive power injection at Bus k that can be calculated from $\mathbf{V}_k(t)$ and $\mathbf{I}_k(t)$.

Based on Equation (16), it is possible to write

$$Var(E_k^2) \approx 0 \rightarrow \overline{(E_k^2 - \overline{E_k^2})^2} \approx 0 \tag{19}$$

where $\bar{E} = mean(E)$.

Fitting (18) into (19) results in

$$(x_k'^2 I_k^2 + V_k^2 + 2x_k'Q_k) - (x_k'^2\overline{I_k^2} + \overline{V_k^2} + 2x_k'\overline{Q_k}) \approx 0 \tag{20}$$

Rearranging Equation (20) and assuming measurements at every time step t, holds:

$$x_k'^2(\overline{I_{k_t}^2} - I_{k_t}^2) + x_k'(2(\overline{Q_{k_t}} - Q_{k_t})) \approx (V_{k_t}^2 - \overline{V_{k_t}^2}) \tag{21}$$

Expression (21) can be solved for x_k' using a nonlinear LSM. In the present work, the trust-region-reflective method is chosen [31].

After x_k' is calculated, it is possible to come back to Equation (15) and calculate $E_k \angle \delta_k(t)$.

At this point, it is possible to obtain an estimation of $\phi_a(t)$ based on $\delta_k(t)$, adapting Equation (7). As S_{bGi} and H_i are unknown, it is possible to assume weights for the equivalent generators obtained in the previous passages of the methodology. These weights may be adjusted to privilege the equivalents obtained from data collected close to big generation centers or close to the possible physical location of the COI, for example. Therefore, in this context,

$$\delta_a^{est}(t) = \frac{\sum_{k=1}^{n} W_k \delta_{G_k}}{\sum_{k=1}^{n} W_k} \tag{22}$$

where the superscript *est* denotes estimation and W_k is the weight adopted to each equivalent $k = 1\dots n$. Please note that Equation (22) provides an approximation of the mean rotor angle of Area a based on the rotor angles of the N equivalent generators obtained at the buses where PMUs are installed (which might be generation buses or not). Equation (22), instead, is based on the n generators inside Area a.

3.2. Determination of the Equivalent Moving Power

As defined in Equation (5), $P_{mov_i}(t)$ is composed by the contributions of the mechanical power (P_m) of the machines inside the area, the contributions of loads (P_l) and the contributions of losses (P_{losses}). The methodology approaches P_l first, assuming a model for the loads and estimating the total contribution according to measurements available. This step is described in Section 3.2.1. With P_l and the power injected into the area (P_e), it is possible to estimate the total power generated P_g, assuming a typical rate for P_{losses} during the procedure. This step is described in Section 3.2.2. Finally, P_m is estimated through model identification, in the procedure described in Section 3.2.3.

3.2.1. Estimation of the Dynamic Behavior of the Loads

Many different papers discuss the use of PMUs for monitoring, studying or estimating load behavior. In [32] the use of a PMU to monitor and estimate the behavior of induction motors is addressed, detailing practical limitations. Papers [33,34] propose different approaches for dynamic load modelling in distribution grids. In [35], a practical approach based on modelling the load as constant current is presented. The methodology assumes the total load pre-disturbance as known, and by weighting voltage measurements at different buses it can estimate the equivalent load behavior of an area. Papers [36,37], instead, are model-identification-based. They make use of optimization for estimating the parameters of the ZIP model [37] or exponential voltage/frequency dependence model [36].

The fields of load estimation and load characterization is wide and have their own challenges. According to the model chosen to characterize the load, different methods are available. A complex method is both requiring in terms of observability and time consuming. In this work, load estimation is only an auxiliary method necessary for a further goal that is inertia estimation: hence, a simple approach that keeps practical conditions is used. According to [38], the voltage dependence of the loads has a more significant impact on the power imbalance and on inertia estimation studies than frequency dependence. Therefore, a static load model was chosen in this work.

In the constant impedance model, the total power consumed by all loads of an area can be expressed by

$$P_l(t) = P_{ref} \sum_{i=1}^{nl} \frac{V_{l_i}^2(t)}{V_{l_{i0}}^2} \tag{23}$$

where P_{ref} is a reference value of the total load of the area; $V_{l_{i0}}$ is the initial voltage magnitude at bus i; $V_{l_i(t)}$ is the magnitude of voltage at bus i at time t and nl is the number of load buses.

When applying Equation (23), $P_l(t)$ varies according to the variations in $V_{l_i}^2(t)$ if the perturbation is not directly on the loads of the system. If the perturbation is a load step or disconnection, P_{ref} should change to reflect that, but the problem is that in general, that perturbation is neither known nor measured directly. Here, such change in P_{ref} is estimated as follows

$$\begin{aligned} P_{ref}(t) &= P_{L_{total}}^{st1}, \ for \ t < t_e \\ P_{ref}(t) &= P_{L_{total}}^{st2}, \ for \ t \geq t_e \end{aligned} \tag{24}$$

where $P_{L_{total}}$ is the average value of the total load of the area at the steady-state before ($st1$) and after the perturbation ($st2$), and t_e is the time of occurrence of the event. In this paper, $P_{L_{total}}$ is assumed as known and t_e is determined looking at power and voltage measurements. Alternatively, techniques based on the Normalized Wavelet Energy (NWE) [6,39] or score functions [40] may be applied to determine the exact time of occurrence and the duration of the event.

However, it is not practical to monitor all load buses on a system, as in general, $V_{l_i(t)}$ for $i = 1 \dots n_l$ is unknown. Therefore, an approximation of the equivalent behavior of the voltage at the load buses is proposed in terms of the measurements available

$$P_{l_a}(t) \approx P_{ref}(t)\lambda(t) \tag{25}$$

where $\lambda(t)$ is a weighted sum of the voltage measurements at a selected group of buses monitored by PMUs:

$$\lambda(t) = \frac{1}{n_s} \sum_{j=1}^{n_s} \frac{W_j V_j^2(t)}{W_j V_{j0}^2} \tag{26}$$

where V_j denotes the magnitude of voltage and W_j denotes the weight related to the measurement at the $j = 1 \dots n_s$ buses selected. As the accuracy of the approximation depends on which buses are considered in the set, assigning higher weights for load buses and transition buses instead of generation buses enables a better approximation of the equivalent behavior at the area. The impacts of this approximations are discussed in the sections of results. Although in this paper the constant impedance model is described, the same principles can be used for any dependence on the voltage.

3.2.2. Estimation of the Equivalent Power Generated

Considering the power injected at the boundaries ($P_{e_a}(t)$) it is possible to write

$$P_{e_a}(t) = P_{g_a}(t) - P_{l_a}(t) - P_{losses_a}(t) = \sum_{i=1}^{n} P_{g_i}(t) - \sum_{l=1}^{n_l} P_{l_l}(t) - \sum_{m=1}^{n_b} P_{losses_m}(t) \tag{27}$$

where P_{g_a} is the total power generated in the area a and P_{g_i} is the power generated by generator i.

As presented in Section 3.2.1, the total load behavior P_{l_a} can be estimated with Equation (25), and $P_{e_a}(t)$ can be measured, from the PMUs installed at the boundary buses. Therefore, $P_{losses_a}(t)$ and $P_{g_a}(t)$, are missing in Equation (27).

Assuming that $P_{losses_a}(t)$ may be approximated by a share of $P_{g_a}(t)$:

$$P_{losses_a}^{est}(t) = \alpha P_{g_a}(t) \tag{28}$$

where α is a typical rate adopted, and rewriting Equation (27), it is possible to estimate $P_{g_a}(t)$ according to

$$P_{g_a}(t) = \frac{P_{e_a}(t) + P_{l_a}(t)}{1 - \alpha} \tag{29}$$

The estimated $P_{g_a}(t)$ is used in the further estimation of $P_{m_a}(t)$ in the next subsection.

3.2.3. Estimation of the Equivalent Mechanical Power

Substituting Equations (5) and (27) in Equation (6), and assuming quasi-steady-state conditions, it is possible to write

$$P_{m_a}^{st} = P_{g_a}^{st} \tag{30}$$

where the superscript st denotes the steady-state.

Following a perturbation, $P_{g_a}(t)$ changes, and $P_{m_a}(t)$ is adjusted by speed governors, through primary frequency control. The response of the speed governors can typically be modelled by a first order transfer function, such that $P_{m_a}(t)$ follows $P_{g_a}(t)$ slowly according to a time constant.

The block diagram that models a generic primary frequency control scheme of an Area a is represented in Figure 4, where Δf_{COI_a} is the change in the mean frequency of the area, K is the gain, $H(s) = \frac{1}{s}$ is the delay function, R is the frequency regulation of the area and ΔP_{m_a} is the change in the mechanical power.

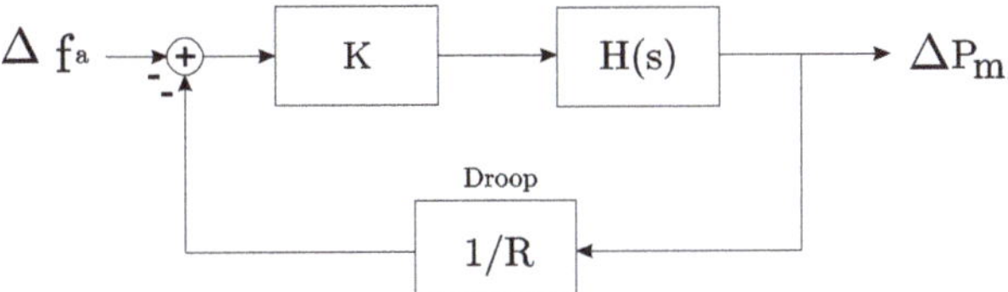

Figure 4. Block diagram of the primary frequency control of Area *a*.

Alternatively, the equivalent transfer function can be written as

$$G(s) = \frac{\Delta f_{COI_a}}{\Delta P_{m_a}} = \frac{T_r}{1 + sT_k} \tag{31}$$

where $T_r = \frac{-1}{R}$ and $T_k = \frac{-T_r}{K}$.

From Figure 4, the main interest is on ΔP_{m_a}, but it is neither possible to measure it directly nor estimate all the parameters in the diagram.

To solve this problem, a three-step strategy is adopted: first, a rough estimation of P_{m_a} (denoted by $P^r_{m_a}(t)$) is obtained through data-fitting, second $T_r = \frac{-1}{R}$ and T_k are obtained through model identification (based on $P^r_{m_a}(t)$), and third the estimation of P_{m_a} is updated through the integration of Equation (31) in the time domain.

In the first step, the data-fitting procedure is applied on $P_{g_a}(t)$, in an attempt to approximate typical mechanical power responses (obtaining $P^r_{m_a}(t)$). However, before applying the data-fitting, there are two typical patterns of $P_{m_a}(t)$ (from the point of view of this methodology) that must be treated differently to determine the procedure to be used. The first pattern is a step change, caused by connection/disconnection of generation. The second pattern includes all the other types of perturbations that do not see a fast change on $P_{m_a}(t)$. These patterns interfere in the number of reference points needed to apply the data-fitting.

In the case of the first pattern, three reference points are needed to guarantee that the data-fitting is applied only after the step, without smoothing it. The reference points are the last point of $P_{m_a}(t)$ in steady-state before the perturbation (Point I), the first point of $P_{m_a}(t)$ in steady-state after the perturbation (Point II) and the first point after the generation step (Point III). The data-fitting is applied between Point III and Point II, in the procedure denominated Strategy 1. In the case of the second pattern, only Point I and Point II are needed, and the data-fitting is applied between them (Strategy 2).

In details, consider the cases simulated with a generic test system and presented in Figure 5. The behavior of the actual P_{g_a} and the actual P_{m_a} of an area that experienced a loss of generation is presented in Figure 5a, while the behavior of an area that suffered a load loss is presented in Figure 5b; both responses are obtained considering the response of a first order generic governor. Please note that the reference points can be fixed also in the P_{g_a} curve.

In this paper, the proposed methodology is tested using computer generated dynamic simulations and, thus, the nature of perturbation and the instant of occurrence of the event are known; such information is used in the selection of the strategy and in the determination of the reference points, according to the procedure previously described. For an automatic real-time procedure, one may make use of event identification methods [6,39,40]. First, a detection of an event triggers the full inertia estimation methodology. Second, the identification of the event as a connection/disconnection of generation triggers the use of Strategy 1, whether the identification of any other event triggers the use of Strategy 2. The automatic determination of the reference points can be done analyzing the second derivative of P_{g_a} (estimated) and adopting a threshold to identify when the signal is in steady state and when it is in transient period. The last sample of the steady-state before and after the perturbation are taken as points I and II, and the first sample of the transient period is taken as Point III (in the case of Strategy 1). Alternatively, one may obtain Point III making use of a technique to estimate the size of a generation loss [41].

After the reference points are selected, the fast frequency transients may be smoothed to allow the electromechanical response to be dominant. To do so, these transients are detected by evaluating the sign of the second derivative of the signal, obtained applying the finite difference method. Then, the fast transients are smoothed with the use of a moving average filter. After a smoother signal is obtained, the responses between points III and II in Strategy 1 and between points I and II in Strategy 2 are approximated with a fifth order polynom

$$P^r_{m_a}(t) = p_1 P^5_{g_a}(t) + p_2 P^4_{g_a}(t) + p_3 P^3_{g_a}(t) + p_4 P^2_{g_a}(t) + p_5 P_{g_a}(t) + p_6; \qquad (32)$$

where p_i, $i = 1, \ldots, 6$ are the coefficients of the polynomial, obtainable using the LSM. In this paper, this procedure is performed with the function *fittype* in *MATLAB*.

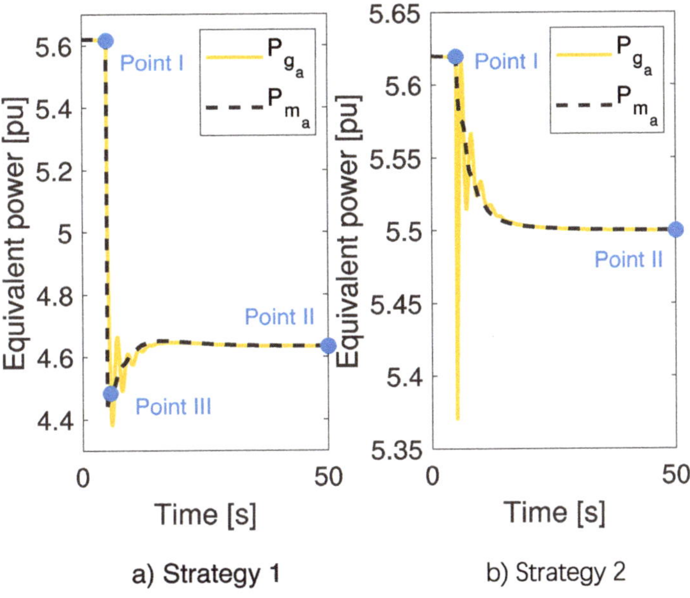

Figure 5. Strategies for the polynomial fitting step.

With the obtained $P_{m_a}^r(t)$, it is possible to calculate $\Delta P_{m_a}(t) = P_{m_a}^r - P_{m_a}^{r-st}$, where the subscript st denotes the steady-state before the perturbation. After that, Equation (31) is used to obtain the parameters of $G(s)$, T_r and T_k, through a transfer function identification method. In this paper, this is done through the function *tfest* in *MATLAB*, which applies a nonlinear LSM.

After T_r and T_k are obtained, the system of differential equations that describes the primary frequency control in time domain can be integrated to obtain the refined estimation of P_{m_a}, denoted here as $P_{m_a}^{est}$. The system can be written as follows:

$$\frac{dP_{m_a}^{est}(t)}{dt} = \frac{1}{T_k}\left(P_{m_a}^{ref} + \frac{1}{T_r}\Delta f_{COI_a} - P_{m_a}^{est}(t)\right) \tag{33}$$

Summarizing, at this point the estimations of the total load power, losses and mechanical power were already obtained. Therefore, it is possible to obtain the equivalent moving power of the Area a:

$$P_{mov_a}^{est}(t) = P_{m_a}^{est}(t) - P_{l_a}^{est}(t) - P_{losses_a}^{est}(t) \tag{34}$$

3.3. Estimation of Inertia

With $P_{e_a}(t)$ measured, $\delta_a^{est}(t)$ and $P_{mov_a}^{est}(t)$ estimated, Equation (6) can be solved for H_{COI_a} through LSM,

$$H_{COI_a} = (\mathbf{A}^T\mathbf{A})^{-1}\mathbf{A}^T\mathbf{B} \tag{35}$$

where $\mathbf{A}$ is the vector composed by each sample of $\frac{d^2\delta_a^{est}(t)}{dt^2}$ and $\mathbf{B}$ is the vector composed by each sample of $\Delta P_a(t) = P_{mov_a}^{est}(t) - P_{e_a}(t)$.

3.4. Summary of the Methodology

A flowchart summarizing the methodology is presented in Figure 6.
The steps of the procedure are described below:

I   –Ward Equivalent method to reduce the system around the n buses where PMUs are installed (described in Section 3.1).

II	–Estimation of the n equivalent generators using the "Variance method" (described in Section 3.1).
III	–Calculation of δ_a^{est} through Equation (22).
IV	–Obtainment of the power injected $P_{e_a}(t)$ from the PMUs at the boundaries.
V	–Estimation of the equivalent load $P_{l_a}^{est}(t)$ through Equation (25).
VI	–Estimation of $P_{g_a}(t)$ through Equation (29).
VII	–Estimation of the equivalent moving power

(a)	–Estimation of $P_{m_a}^r(t)$ through data-fitting (described in Section 3.2.3).
(b)	–Estimation of T_r and T_k through model identification (Section 3.2.3).
(c)	–Estimation of $P_{m_a}^{est}(t)$ through Equation (33).

VIII	–Estimation of $P_{mov_a}^{est}(t)$ through Equation (34).
IX	–Estimation of H_{COI_a} through Equation (35).

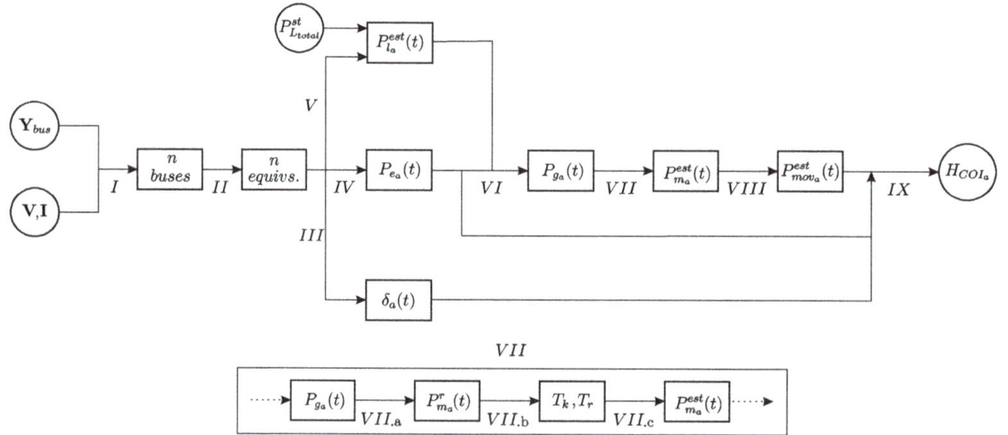

Figure 6. Summary of the methodology.

4. Results

This section is divided in two main subsections. First, tests have been performed to validate the methodology and present details of the procedure; results are presented in Section 4.1. At second, new tests were performed to evaluate the impact of the main steps of the methodology summarized in Figure 6 in the inertia estimation; results are shown in Section 4.2.

4.1. Validation Study

4.1.1. Test System

To validate the methodology, the test system from [26] represented in Figure 7 was simulated in PowerFactory2018, with the two-axis 5th-order IEEE standard detailed Model 2.2 [42] for synchronous machines. Loads were simulated using the constant impedance model, and governors, PSS and AVRs were implemented with generic first order models.

In the simulations, G1 and G2 oscillate against G3 and G4. However, as it can be seen in Figure 7, the areas will be defined around bus 5 (Area 1) and around bus 6 (Area 2), such that Area 2 is composed by three non-coherent generating units (G_2, G_3, G_4). Accordingly, PMUs are assumed as installed at buses 5 and 6, therefore monitoring these two areas. This division aims at testing the methodology proposed with areas containing non-coherent generators. The constants of the primary frequency control simulated are $T_{k_{A1}} = 0.0324$, $T_{k_{A2}} = 0.0108$ and $T_{r_{A1}} = T_{r_{A2}} = 4$, where $A1$ and $A2$ denote Area 1 and Area 2, respectively.

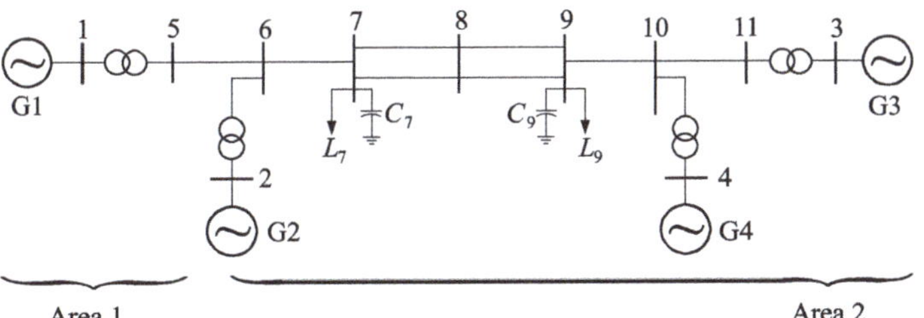

Figure 7. 11-bus test system [26] with areas defined by the proposed methodology.

4.1.2. Estimation of the Mean Frequency

A step increase of 10% in the load of bus 9 (see Figure 7) was simulated at $t = 5$ s, and two tests were performed. In Test 1, PMUs were considered at the boundary buses 5 and 6, measuring voltages and the injected current at the transmission line 5–6, from both ends. In Test 2, a third PMU was considered installed at bus 9.

In Test 1, the "Variance method" was applied with a selected time window of 2 s around the time the perturbation occurred, considering 0.5 s before and 1.5 s after. The initial guesses for the internal reactances were $x_i = 0.1$ p.u. for both areas, and the estimations obtained were $x_{A1} = 0.0312$ p.u. and $x_{A2} = 0.0644$ p.u, for Area 1 and Area 2, respectively. With the equivalent reactances estimated, the internal voltages $E_{A1}(t)\angle\delta_{A1}(t)$ and $E_{A2}(t)\angle\delta_{A2}(t)$ are calculated. The electrical frequencies related to each equivalent generator are determined according to Equation (2), making use of the obtained $\delta_i(t)$ and applying a median filter. In Test 2, Equation (22) is applied to obtain an equivalent behavior for Area 2, taking into account the data acquired at both buses 6 and 9. The weights W_k are assumed equal to one. In other words, the difference between tests 1 and 2 is the consideration in Test 2 of the measurements acquired at bus 9.

To evaluate the dynamic equivalents obtained in both tests, the estimated electrical frequencies of each equivalent machine (denoted by f_{est}) are compared with the true mean frequency of each area (f_{COI}), computed on the simulation results. Both f_{est} and f_{COI} can be seen in Figure 8 for Area 1 and Figure 9 for Area 2, respectively. In Figure 9, f_{G2}, f_{G3} and f_{G4} denote the frequencies simulated of each of the generators of Area 2, $f_{est_{t1}}$ denotes the mean frequency of Area 2 estimated in Test 1 and $f_{est_{t2}}$ denotes the mean frequency of Area 2 estimated in Test 2.

It can be observed in Figure 8 that the estimated frequency of Area 1 is very close to the mean frequency of that area; this is expected, since this area has only one generator. Regarding the simulation, the frequency does not come back to nominal values because secondary frequency control is not implemented.

Regarding the estimations of Area 2 in Figure 9, instead, the behavior of $f_{est_{t1}}$ is significantly different from f_{COI}. This happens because the measurement point is electrically closer to G2 in relation to the geographical COI of the system, such that the influence of the frequency of G2 (f_{G2}) brings the estimated f_{est} closer to its own behavior and far from the COI. Considering a second measurement point (at bus 9) improves the estimation of the mean frequency, as it can be seen with $f_{est_{t2}}$, which is very close to the COI behavior.

This last result show that one may take advantage of PMUs installed inside the area of study. In this case, bus 9 is located closer to the theoretical COI of Area 2; this can be inferred from Figure 7, taking into consideration the radiality of the system and the size of the generators. Therefore, the frequency at bus 9 is a good approximation of the mean frequency of that area, showing that coherent groups does not necessarily have to be individually monitored to provide accurate approximations to the dynamics of the COI.

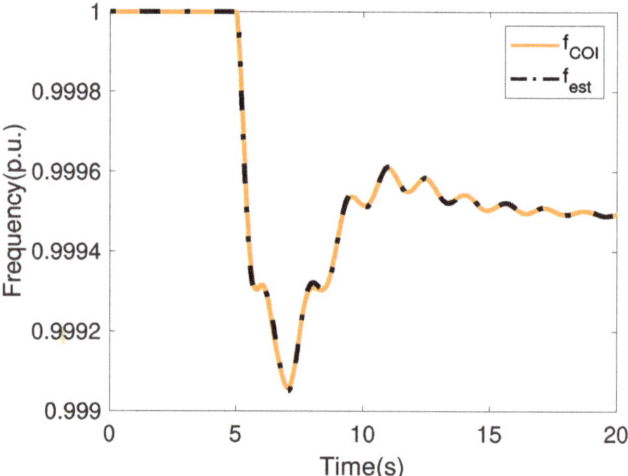

Figure 8. Validation studies—Frequency of Area 1.

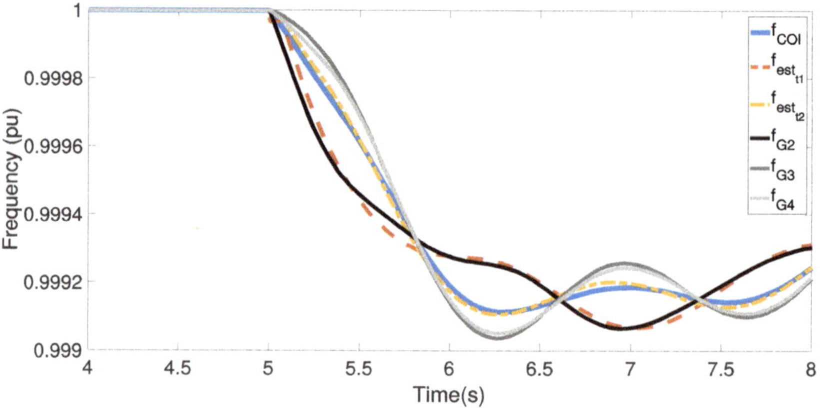

Figure 9. Validation studies—Frequency of Area 2.

4.1.3. Estimation of the Moving Power

According to the procedure proposed and summarized in Figure 6, steps IV to VIII are needed to estimate the moving power of each area. In Step IV, the measured voltages and currents at boundary buses 5 and 6 are used to determine $P_{e_{A1}}(t)$ and $P_{e_{A2}}(t)$, respectively. In Step V, the voltage behavior of Area 2 is approximated by the voltage behavior of bus 9, where a PMU was already assumed to be installed. In this case, bus 9 is a natural choice not because it is located closer to the COI of the area, but because it is a load bus. How available PMUs might provide measurements that are good approximations to the voltage profile of load buses of the area is system-dependent. The interested reader may refer to [32,43,44].

After the load behavior is estimated, $P_{g_a}(t)$ is estimated considering α as 5% in Equation (29) (Step VI). Then, $P_{m_a}^r$ is obtained according to Step VII described in Figure 6. During step VIIa, $P_{m_a}^r(t)$ was obtained with $p_1 = -8.25 \times 10^{-7}$, $p_2 = 3.62 \times 10^{-5}$, $p_3 = -5.80 \times 10^{-4}$, $p_4 = 4.03 \times 10^{-3}$, $p_5 = -9.89 \times 10^{-3}$ and $p_6 = 3.17 \times 10^{-2}$. These parameters provided an index of 0.84 in the R^2 test and 2.9×10^{-3} in the RMSE test. During step VIIb, T_k and T_r were obtained with an error smaller than 5% in comparison to the values simulated.

Results are presented in Figure 10, where $P_{m_{sim}}$ stands for the mechanical power coming from the dynamic simulation, P_m^r stands for the estimated mechanical power after the data-fitting step and P_m^{est} stands for the definitive estimation of the mechanical power after the system identification and the integration processes. As it can be seen, the methodology was able to obtain an accurate approximation of P_m for both areas, with a significant improvement moving from P_m^r to P_m^{est}.

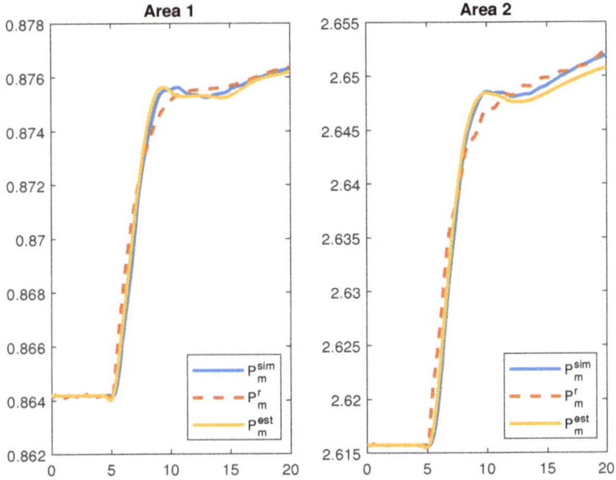

Figure 10. Mechanical Power.

Additionally, results of the estimated P_l and P_{mov} of Area 2 can be seen in Figure 11.

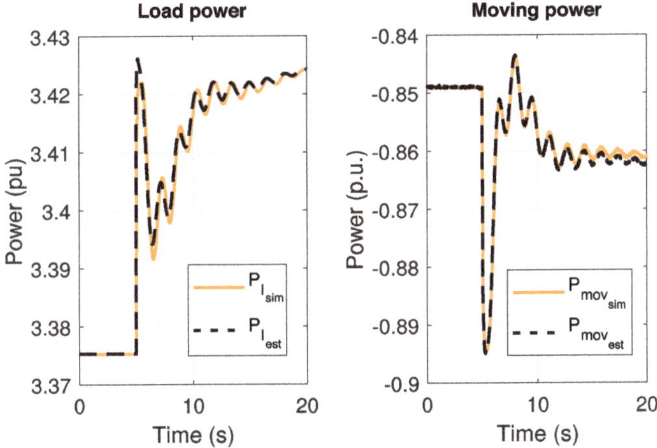

Figure 11. Load power and moving power of Area 2.

4.1.4. Inertia Estimation

A sliding window of 200 samples, with displacement of 20 ms (1 sample per slide) is used to estimate inertia continuously over time. The methodology proposed in this paper is compared with the method proposed in [25] applied at the boundaries of both areas. Results are shown in Figure 12, where 'A1 expected' and 'A2 expected' denote the input inertia of Area 1 and Area 2 in the simulations, 'A1 Prop. Method' and 'A2 Prop. Method' denote the estimations obtained with the proposed method and 'A1 Comp. Method' and 'A2 Comp. Method' denote the results obtained applying the method proposed in [25] directly.

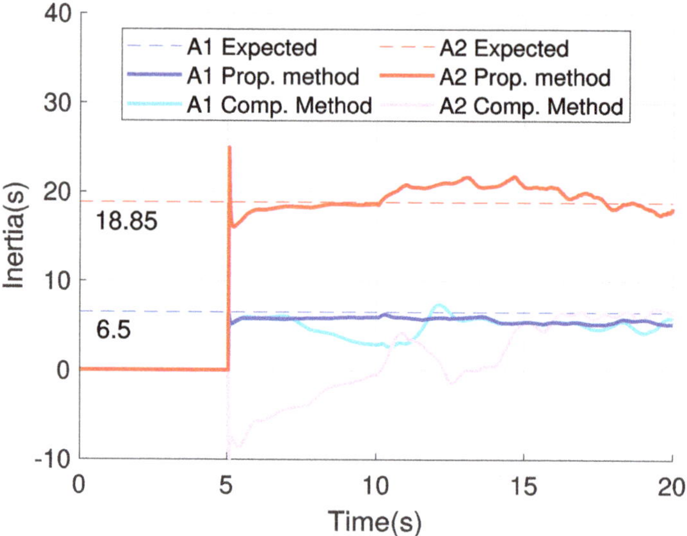

Figure 12. Inertia estimated.

Regarding Area 1, it can be seen that both methods obtain accurate estimations in the first seconds following the perturbation, not surprisingly, as the area is made of a single generator. Regarding Area 2, instead, the proposed method performs well, while the method chosen for comparison obtain wrong results. These results happen because the method proposed in [25] is tailored for estimating machine parameters from measurements obtained at its terminal buses, and it relies on the assumption that the mechanical power has slower behavior in comparison to the electromagnetic power. To apply [25] directly with measurements at boundary buses, it was necessary to assume that the moving power of the area is constant, what does not stand for areas with voltage-dependent loads or perturbations inside, such as Area 2.

Looking at Figure 12, it can also be seen that the estimations obtained with the proposed method lose some accuracy after 5 s. This happens because of two main reasons. First, the moving power estimated also degrades due to the transition of the inertial response to the frequency control (see Figure 10 for Area 2). Moreover, the substantial variations of power and frequency have already faded, and remaining variations are small, impacting on the numerical sensitivity of the method.

4.1.5. Inertia Variation

In this subsection, the simulation has been extended to investigate the behavior of the proposed methodology with the inclusion of RES. At $t = 20$ s, a WP generator is connected to bus 10 (Figure 7) injecting the same amount of power of Generator 4 (approximately 720 MW), that is simultaneously disconnected. The WP generator is not contributing to the inertia of the system, such that a decrease on the equivalent inertia of Area 2 happens. Although the events simulated do not generate a power imbalance at the boundaries of the monitored areas, the reactive power imbalance that follows the substitution of G_4 by G_5 impacts the voltages at the buses of the system and alters the internal flows in Area 2.

The proposed methodology is applied considering PMUs at bus 5, 6 and 9 (as in the previous subsections). In Step VII of Figure 7, the data-fitting method has been applied using the strategy for generation disconnection presented in Section 3.2.3.

The moving power of Area 2 is shown in Figure 13, where $P_{mov_{sim}}$ is the simulated moving power and $P_{mov_{est}}$ is the estimated moving power. As it can be seen, the moving power suffers a perturbation at $t = 20$ s, but of a much smaller magnitude in comparison to the perturbation at $t = 5$ s.

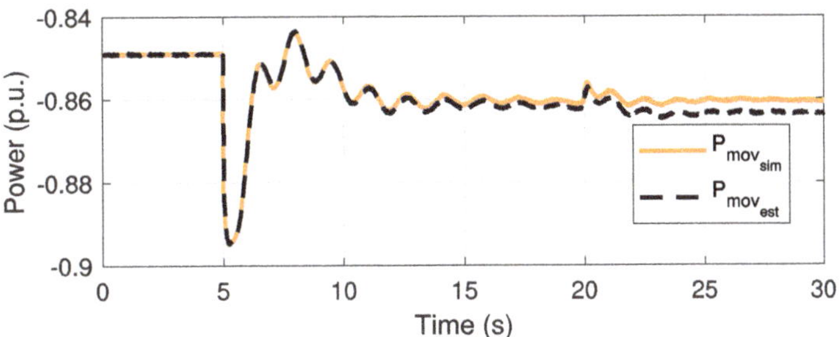

Figure 13. Moving powerestimated (Area 2).

The estimations of inertia can be seen in Figure 14. Please note that the expected value of inertia for Area 2 changes when G4 is disconnected. After the simultaneous events at $t = 20$ s, the estimations obtained with the proposed method for both areas present a higher standard deviation in comparison to the results obtained following the perturbation at $t = 5$ s. Less accurate results (following $t = 20$ s) are natural, since the simultaneous perturbations simulated impose much smaller power imbalance and Rate of Change of Frequency (RoCoF).

At $t = 20$ s, it can be seen that notwithstanding the fact that the measured P_e is slightly affected by the switching of generators, the proposed method was able to react to the event and to identify the change of inertia in Area 2. After 25 s the estimation degrades in consequence of the frequency control, as explained. Please note that the comparison method, assuming slow moving power variation, failed to estimate the inertia of Area 2 through the whole study.

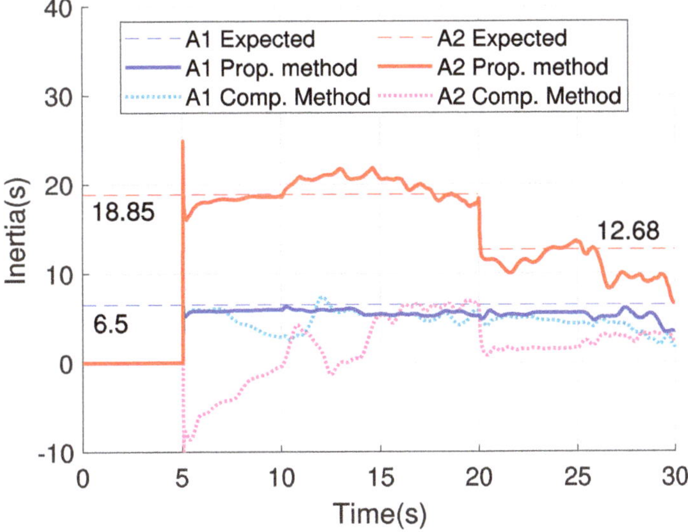

Figure 14. Inertia estimated (inertia variation case).

4.2. Performance Evaluation Studies

4.2.1. Test System and Study Cases

To evaluate the performance of the methodology and its critical aspects, new simulations were performed in PowerFactory2018 with the PST 16 Benchmark Test System [27]. The system is composed of 66-bus, 16-synchronous-generator, with total generation installed of 17.9 GW, and total load demand of 15.6 GW. The two-axis 5th-order model for generators (Model 2.2 [42]) was used in the simulations, loads were modelled as constant impedance and the models for controllers can be found in [27].

The test system is represented in a synthetic way in Figure 15, emphasizing the interconnections between the three Areas (A, B, C) and their boundary buses: A1, A2, B1, B2, C1 and C2, where PMUs are assumed to be installed. Main loads are concentrated in area C and power flows from area A and B through the tie-lines. Regarding coherency, the areas divided are large enough to have internal dynamics, depending on the perturbation. Primary frequency control was implemented considering an equivalent $T_k = 8$ s and $T_r = 0.02$ p.u. for each area.

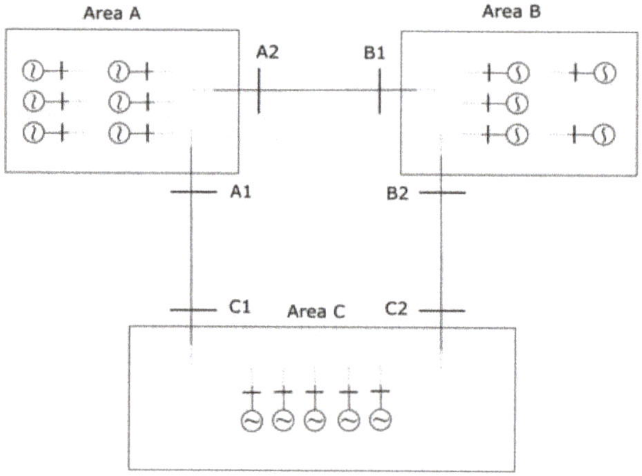

Figure 15. PST 16 Benchmark test system.

Two perturbations have been simulated separately. First, a loss of a generator producing 240 MW and 56Mvar inside Area B (represents about 3% of the demand of this area and around 1.3% of the total demand). At second, a load shedding of 400 MW and 50 Mvar inside Area B. Random noise signals with normal distribution have been added to the simulated data at the terminals were PMUs were assumed to be installed. The papers [12,45] were taken as reference, and the standard deviations adopted were $\sigma_M = 10^{-4}$ for voltage and current magnitudes and $\sigma_A = 10^{-6}$ for phase angles. A median filter is applied considering 5 points for phase angles and 10 points for voltage magnitudes.

Each study has been divided in six different cases to assess the sensitivity of the estimation of inertia in relation to the estimations of mean frequency (f_{COI}), load behavior (P_l) and mechanical power behavior (P_m), according to Table 1.

Table 1. Description of the study cases.

	f_{COI}	P_l	P_m
Case 1	Estimated	Simulated	Simulated
Case 2	Simulated	Estimated	Simulated
Case 3	Simulated	Simulated	Estimated
Case 4	Simulated	Estimated	Estimated
Case 5	Estimated	Estimated	Estimated
Case 6	Estimated	Estimated	Estimated

In Case 1, the mean frequency is estimated while loads and mechanical power behaviors are assumed as monitored. In Case 2, the load behavior is estimated while others are assumed as monitored, and so on in other cases. In Case 5 practical conditions are considered, performing the full methodology with measured data from PMUs at the boundaries and at two internal PMUs in each area (buses A6a, A3a, B3a, B8a, C10a and C2a, referring to [27]). Case 6 assumes full observability.

The selection of the internal buses in Case 5 has been done empirically, aiming at achieving an accurate estimation of the mean frequency with the minimum number of PMUs possible. As one may suppose, the more PMUs the better (Case 6), but the harder to have in practice.

4.2.2. Results

A sliding window of 200 samples (with displacement of 1 sample per slide) is used to estimate inertia continuously over time. To illustrate, the results of Area B in Study 1 are presented in Figure 16. In the legend, H_{B-C1} to H_{B-C6} denote the equivalent inertias of Area B in each of the 6 cases simulated and $H_{B-Expected}$ denote the expected inertia value. As it can be seen, naturally the most practical case (H_{B-C5}) deviated most from the expected value. Still, the absolute error was smaller than 10% during most of the time window.

To evaluate the results and discard deviations, a practical criterion is defined: whenever the inertia estimated does not vary more than 1% for 50 consecutive samples (sampling time of 20 ms), a mean average of the estimations is calculated, and the result is considered to be a representative estimation of the constant of inertia of the area. Results can be seen in Table 2, where $e(H_i)[\%]$ denotes the error on the representative inertia estimated of area $i = A, B, C$ in %, with respect to the constant of inertia at the COI of each area.

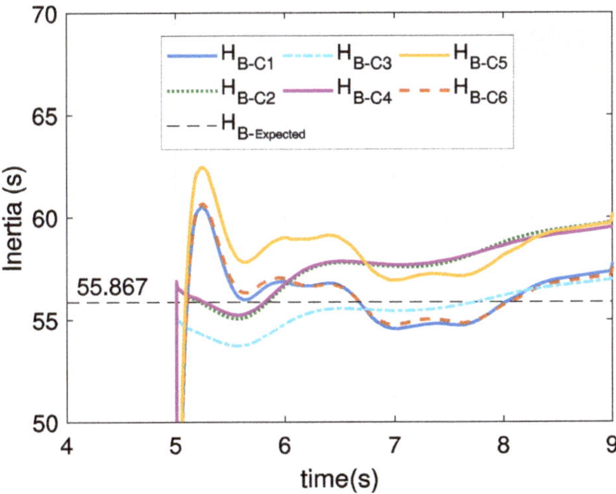

Figure 16. Inertia estimated–Study 1–Area B.

Table 2. Representative inertia estimation results.

	Study 1: Loss of Generation			Study 2: Load Shedding		
	$e(H_A)[\%]$	$e(H_B)[\%]$	$e(H_C)[\%]$	$e(H_A)[\%]$	$e(H_B)[\%]$	$e(H_C)[\%]$
Case 1	−0.54	−1.07	−3.38	−4.81	−3.20	−8.11
Case 2	4.86	2.05	10.79	2.54	4.88	−8.29
Case 3	4.59	4.64	8.51	7.31	4.16	−2.15
Case 4	9.41	1.85	11.14	8.79	4.86	−7.07
Case 5	12.44	−4.47	9.35	6.06	−1.97	−2.90
Case 6	6.87	1.59	0.95	3.71	−3.30	−1.22

Regarding Case 1, it can be seen that it is possible to achieve accurate estimations of the mean frequency with few PMUs, such that the representative inertia obtained presents errors smaller than 10%.

The investigation of the results of Case 2 shows that Area C presented the highest errors. This happens because Area C is the largest area, with larger loads in comparison to the other areas, and therefore estimating the total load behavior based on only 4 PMUs (2 at the boundaries and 2 internal) implies larger approximations. It is worth observing, however, the compensation of this error during the estimation of the moving power in Study 2–Case 4. The same does not happen in Study 1–Case 4.

The results of Case 5 show that the methodology can estimate inertia in practical conditions depending on few PMUs (4 for each area of the test system [27]). The same PMUs have been used to provide data for the mean frequency estimation and to provide data for an approximation of the voltage profile of the area (in the step of load behavior estimation).

The minimum number of PMUs necessary to achieve accurate results depend mainly on the system, the degree of accuracy pretended, and the location where the PMUs will be installed, therefore constituting an optimization problem that may be faced as a topic for future studies. The results of Case 6 show the maximum improvement that can be achieved for this system increasing the number of PMUs.

5. Conclusions

This paper presents a new way of defining areas for dynamic studies, based on synchrophasor measurements. A multi-step methodology is proposed to estimate inertia following perturbations by monitoring an area through its boundaries and possibly a few internal PMUs. First, a measurement-based dynamic equivalent is obtained. At second, an equivalent moving power is estimated, making it possible to study areas that contain perturbations and voltage-dependent loads. At third, equivalent inertia is estimated by the solution of the Swing Equation through LSM.

The determination of the dynamic equivalent brings the first main advantage of the methodology proposed: machines inside the monitored area do not need to be coherent, and in this case the behavior of the COI is estimated. This enables the possibility of delimiting any area by PMUs available on the grid. The methodology also counts with a system-reducing strategy to take into consideration multiple-measurement points. Results show that having spread measurements inside the area provides a better picture of the COI.

Another contribution of the methodology proposed is the estimation of the equivalent moving power of an area. This proposition attends the limitation of methods on the literature that rely on measurements at the terminal of generator groups and on the assumption of slow mechanical power. In the methodology proposed, the estimation of the moving power is done through the estimation of the total load behavior and the estimation of the equivalent mechanical power, which in turn is done through a rough estimation step and an improvement step.

The results obtained with the presented methodology show that accurate estimations of inertia could be obtained considering different types of perturbations with the use of a limited number of PMUs. Future studies include tests with real data, with possible adaptations on Step V of the methodology to estimate the load behavior according to the system.

Author Contributions: Conceptualization: G.R.M., A.B., G.G. and R.Z.; Methodology: G.R.M., V.I. and A.B.; Software: G.R.M.; Validation: G.R.M.; Formal analysis and investigation: G.R.M., V.I. and A.B.; Writing—original draft preparation: G.R.M. Writing—review and editing: V.I., A.B., C.P., G.G. and R.Z. Supervision: A.B., C.P., G.G. and R.Z. All authors have read and agreed to the published version of the manuscript.

Funding: This work was supported in part by Terna SpA, part by the Brazilian National Council for Scientific and Technological Development (CNPq) and part by the INESC P&D Brasil through MedFasee Project.

Conflicts of Interest: The authors declare no conflict of interest.

Abbreviations

The following abbreviations have been used in the paper:

PMU	Phasor Measurement Unit
LSM	Least-Squares Method
RES	Renewable Energy Sources
TSO	Transmission System Operator
COI	Center of Inertia
WP	Wind Power
BB	Boundary Bus
RoCoF	Rate of Change of Frequency

References

1. Arani, M.F.M.; El-Saadany, E.F. Implementing Virtual Inertia in DFIG-Based Wind Power Generation. *IEEE Trans. Power Syst.* **2013**, *28*, 1373–1384. [CrossRef]
2. Bolzoni, A.; Terlizzi, C.; Perini, R. Analytical Design and Modelling of Power Converters Equipped with Synthetic Inertia Control. In Proceedings of the 2018 20th European Conference on Power Electronics and Applications (EPE'18 ECCE Europe), Riga, Latvia, 17–21 September 2018.
3. Rezkalla, M.; Pertl, M.; Marinelli, M. Electric power system inertia: Requirements, challenges and solutions. *Electr. Eng.* **2018**, *100*, 2677–2693. [CrossRef]
4. AEMO. *Inertia Requirements Methodology*; Technical Report; Australian Energy Market Operator (AEMO): Melbourne, Australia, 2018.
5. Malik, O.P. Evolution of Power Systems into Smarter Networks. *J. Control. Autom. Electr. Syst.* **2013**, *24*, 139–147. [CrossRef]
6. Kim, D.I.; Chun, T.Y.; Yoon, S.H.; Lee, G.; Shin, Y.J. Wavelet-Based Event Detection Method Using PMU Data. *IEEE Trans. Smart Grid* **2017**, *8*, 1154–1162. [CrossRef]
7. Bosisio, A.; Berizzi, A.; Moraes, G.R.; Nebuloni, R.; Giannuzzi, G.; Zaottini, R.; Maiolini, C. Combined use of PCA and Prony Analysis for Electromechanical Oscillation Identification. In Proceedings of the 2019 International Conference on Clean Electrical Power (ICCEP), Otranto, Italy, 2–4 July 2019; IEEE: Piscataway, NJ, USA, 2019. [CrossRef]
8. Chow, J.H.; Chakrabortty, A.; Vanfretti, L.; Arcak, M. Estimation of Radial Power System Transfer Path Dynamic Parameters Using Synchronized Phasor Data. *IEEE Trans. Power Syst.* **2008**, *23*, 564–571. [CrossRef]
9. Phadke, A.G.; Bi, T. Phasor measurement units, WAMS, and their applications in protection and control of power systems. *J. Mod. Power Syst. Clean Energy* **2018**, *6*, 619–629. [CrossRef]
10. Chow, J.H. *Power System Coherency and Model Reduction*; Springer: Berlin/Heidelberg, Germany, 2013.
11. Fan, L.; Miao, Z.; Wehbe, Y. Application of Dynamic State and Parameter Estimation Techniques on Real-World Data. *IEEE Trans. Smart Grid* **2013**, *4*, 1133–1141. [CrossRef]
12. Wall, P.; Terzija, V. Simultaneous Estimation of the Time of Disturbance and Inertia in Power Systems. *IEEE Trans. Power Deliv.* **2014**, *29*, 2018–2031. [CrossRef]
13. Ashton, P.M.; Taylor, G.A.; Carter, A.M.; Bradley, M.E.; Hung, W. Application of phasor measurement units to estimate power system inertial frequency response. In Proceedings of the 2013 IEEE Power & Energy Society General Meeting, Vancouver, BC, Canada, 21–25 July 2013; pp. 1–5. [CrossRef]
14. Huang, Z.; Du, P.; Kosterev, D.; Yang, B. Application of extended Kalman filter techniques for dynamic model parameter calibration. In Proceedings of the 2009 IEEE Power & Energy Society General Meeting, Calgary, AB, Canada, 26–30 July 2009; pp. 1–8. [CrossRef]

15. Cao, X.; Stephen, B.; Abdulhadi, I.F.; Booth, C.D.; Burt, G.M. Switching Markov Gaussian Models for Dynamic Power System Inertia Estimation. *IEEE Trans. Power Syst.* **2016**, *31*, 3394–3403. [CrossRef]
16. Petra, N.; Petra, C.G.; Zhang, Z.; Constantinescu, E.M.; Anitescu, M. A Bayesian Approach for Parameter Estimation With Uncertainty for Dynamic Power Systems. *IEEE Trans. Power Syst.* **2017**, *32*, 2735–2743. [CrossRef]
17. Zhao, J.; Tang, Y.; Terzija, V. Robust Online Estimation of Power System Center of Inertia Frequency. *IEEE Trans. Power Syst.* **2019**, *34*, 821–825. [CrossRef]
18. Zografos, D.; Ghandhari, M.; Eriksson, R. Power system inertia estimation: Utilization of frequency and voltage response after a disturbance. *Electr. Power Syst. Res.* **2018**, *161*, 52–60. [CrossRef]
19. Lugnani, L.; Dotta, D.; Lackner, C.; Chow, J. ARMAX-based method for inertial constant estimation of generation units using synchrophasors. *Electr. Power Syst. Res.* **2020**, *180*, 106097. [CrossRef]
20. Garcia-Valle, R.; Yang, G.Y.; Martin, K.E.; Nielsen, A.H.; Ostergaard, J. DTU PMU laboratory development–Testing and validation. In Proceedings of the 2010 IEEE PES Innovative Smart Grid Technologies Conference Europe (ISGT Europe), Gothenburg, Sweden, 11–13 October 2010; IEEE: Piscataway, NJ, USA, 2010. [CrossRef]
21. Angioni, A.; Lipari, G.; Pau, M.; Ponci, F.; Monti, A. A Low Cost PMU to Monitor Distribution Grids. In Proceedings of the 2017 IEEE International Workshop on Applied Measurements for Power Systems (AMPS), Liverpool, UK, 20–22 September 2017; IEEE: Piscataway, NJ, USA, 2017. [CrossRef]
22. Laverty, D.M.; Vanfretti, L.; Khatib, I.A.; Applegreen, V.K.; Best, R.J.; Morrow, D.J. The OpenPMU Project: Challenges and perspectives. In Proceedings of the 2013 IEEE Power & Energy Society General Meeting, Vancouver, BC, Canada, 21–25 July 2013; IEEE: Piscataway, NJ, USA, 2013. [CrossRef]
23. Smart Grid Investment Grant Program. *Factors Affecting PMU Installation Costs;* Technical Report; U.S. Department of Energy–Electricity Delivery & Energy Reliability: Washington, DC, USA, 2014.
24. Schofield, D.; Gonzalez-Longatt, F.; Bogdanov, D. Design and Implementation of a Low-Cost Phasor Measurement Unit: A Comprehensive Review. In Proceedings of the 2018 Seventh Balkan Conference on Lighting (BalkanLight), Varna, Bulgaria, 20–22 September 2018; IEEE: Piscataway, NJ, USA, 2018. [CrossRef]
25. Wehbe, Y.; Fan, L.; Miao, Z. Least squares based estimation of synchronous generator states and parameters with phasor measurement units. In Proceedings of the 2012 North American Power Symposium (NAPS), Champaign, IL, USA, 9–11 September 2012; pp. 1–6. [CrossRef]
26. Kundur, P.; Balu, N.J.; Lauby, M.G. *Power System Stability and Control;* McGraw-Hill: New York, NY, USA, 1994; Volume 7.
27. Gonzalez-Longatt, F.; Rueda, J.L. *Power Factory Applications for Power System Analysis;* Springer Publishing Company: Berlin/Heidelberg, Germany, 2014.
28. Fan, L.; Wehbe, Y. Extended Kalman filtering based real-time dynamic state and parameter estimation using PMU data. *Electr. Power Syst. Res.* **2013**, *103*, 168–177. [CrossRef]
29. Kalsi, K.; Sun, Y.; Huang, Z.; Du, P.; Diao, R.; Anderson, K.K.; Li, Y.; Lee, B. Calibrating multi-machine power system parameters with the extended Kalman filter. In Proceedings of the IEEE Power and Energy Society General Meeting, Detroit, MI, USA, 24–28 July 2011; pp.1–8. [CrossRef]
30. Machowski, J.; Bialek, J.W.; Bumby, J.R. *Power System Dynamics, Stability and Control;* Wiley: Hoboken, NJ, USA, 2012.
31. Byrd, R.H.; Schnabel, R.B.; Shultz, G.A. Approximate solution of the trust region problem by minimization over two-dimensional subspaces. *Math. Program.* **1988**, *40*, 247–263. [CrossRef]
32. Regulski, P.; Wall, P.; Rusidovic, Z.; Terzija, V. Estimation of load model parameters from PMU measurements. In Proceedings of the IEEE PES Innovative Smart Grid Technologies, Istanbul, Turkey, 12–15 October 2014; IEEE: Piscataway, NJ, USA, 2014; pp. 1–6.
33. Alinejad, B.; Akbari, M.; Kazemi, H. PMU-based distribution network load modelling using Harmony Search Algorithm. In Proceedings of the 2012 Proceedings of 17th Conference on Electrical Power Distribution, Tehran, Iran, 2–3 May 2012; IEEE: Piscataway, NJ, USA, 2012; pp. 1–6.
34. Polykarpou, E.; Asprou, M.; Kyriakides, E. Dynamic load modelling using real time estimated states. In Proceedings of the 2017 IEEE Manchester PowerTech, Manchester, UK, 18–22 June 2017; IEEE: Piscataway, NJ, USA, 2017; pp. 1–6.
35. Zografos, D.; Ghandhari, M. Power system inertia estimation by approaching load power change after a disturbance. In Proceedings of the 2017 IEEE Power & Energy Society General Meeting, Chicago, IL, USA, 16–20 July 2017; IEEE: Piscataway, NJ, USA, 2017, pp. 1–5.
36. Bliznyuk, D.; Berdin, A.; Romanov, I. PMU data analysis for load characteristics estimation. In Proceedings of the 2016 57th International Scientific Conference on Power and Electrical Engineering of Riga Technical University (RTUCON), Riga, Latvia, 13–14 October 2016; IEEE: Piscataway, NJ, USA, 2016, pp. 1–5.
37. Vignesh, V.; Chakrabarti, S.; Srivastava, S.C. Power system load modelling under large and small disturbances using phasor measurement units data. *Iet Gener. Transm. Distrib.* **2015**, *9*, 1316–1323.
38. Kuivaniemi, M.; Laasonen, M.; Elkington, K.; Danell, A.; Modig, N.; Bruseth, A.; Jansson, E.A.; Orum, E. Estimation of System Inertia in the Nordic Power System Using Measured Frequency Disturbances. In Proceedings of the Lund Symposium on across Borders–Integrating Systems and Markets–CIGRE, Lund, Zweden, 27–28 May 2015; pp. 27–28.
39. Gao, W.; Ning, J. Wavelet-Based Disturbance Analysis for Power System Wide-Area Monitoring. *IEEE Trans. Smart Grid* **2011**, *2*, 121–130. [CrossRef]

40. Cui, M.; Wang, J.; Tan, J.; Florita, A.R.; Zhang, Y. A Novel Event Detection Method Using PMU Data With High Precision. *IEEE Trans. Power Syst.* **2019**, *34*, 454–466. [CrossRef]
41. Shams, N.; Wall, P.; Terzija, V. Active Power Imbalance Detection, Size and Location Estimation Using Limited PMU Measurements. *IEEE Trans. Power Syst.* **2019**, *34*, 1362–1372. [CrossRef]
42. *Proceedings of the IEEE Guide for Synchronous Generator Modeling Practices and Applications in Power System Stability Analyses*; IEEE Std 1110-2002 (Revision of IEEE Std 1110-1991); IEEE: Piscataway, NJ, USA, 2003; 72p. [CrossRef]
43. Arif, A.; Wang, Z.; Wang, J.; Mather, B.; Bashualdo, H.; Zhao, D. Load Modeling—A Review. *IEEE Trans. Smart Grid* **2018**, *9*, 5986–5999. [CrossRef]
44. Tellez, A.P. Modelling Aggregate Loads in Power Systems. Master's Thesis, KTH Royal Institute of Technology, Stockholm, Sweden, 2017.
45. Wall, P.; Gonzalez-Longatt, F.; Terzija, V. Estimation of generator inertia available during a disturbance. In Proceedings of the IEEE Power and Energy Society General Meeting, San Diego, CA, USA, 22–26 July 2012; pp. 1–8. [CrossRef]

MDPI

St. Alban-Anlage 66

4052 Basel

Switzerland

Tel. +41 61 683 77 34

Fax +41 61 302 89 18

www.mdpi.com

Energies Editorial Office

E-mail: energies@mdpi.com

www.mdpi.com/journal/energies